EMOTION, PLACE AND CULTURE

*This book is dedicated to the memory of Mike Hepworth (1938–2007)
and Val Plumwood (1939–2008), who both contributed generously
to the ideas and events that led to this collection.*

Emotion, Place and Culture

Edited by

MICK SMITH
Queen's University, Canada

JOYCE DAVIDSON
Queen's University, Canada

LAURA CAMERON
Queen's University, Canada

LIZ BONDI
University of Edinburgh, UK

LONDON AND NEW YORK

First published 2009 by Ashgate Publishing

Published 2016 by Routledge
2 Park Square, Milton Park, Abingdon, Oxfordshire OX14 4RN
711 Third Avenue, New York, NY 10017, USA

First issued in paperback 2016

Routledge is an imprint of the Taylor & Francis Group, an informa business

British Library Cataloguing in Publication Data
Emotion, place and culture
   1. Human geography – Philosophy 2. Spatial behavior
   3. Environmental psychology 4. Emotional conditioning
   I. Smith, Mick, 1961–
   304.2'01

Library of Congress Cataloging-in-Publication Data
Smith, Mick, 1961–
  Emotion, place and culture / by Mick Smith ... [et al.].
    p. cm.
  Includes bibliographical references.
  ISBN 978-0-7546-7246-3
  1. Environmental psychology. 2. Spatial behavior. 3. Emotional conditioning.
  4. Emotions. 5. Human geography. I. Title.
  BF353.S64 2009
  155.9'1--dc22

                                                                    2008040008

ISBN 13: 978-1-138-27639-0 (pbk)
ISBN 13: 978-0-7546-7246-3 (hbk)

Contents

List of Figures

Notes on Contributors

Katy Bennett is Lecturer in Human Geography at the University of Leicester, U.K. Her recent research on the former coalfields stems from an earlier project and book *Coalfields Regeneration: Dealing with the Consequences of Industrial Decline* (with Huw Beynon and Ray Hudson, Policy Press 2000). Her current work focuses on women's lives in 'regenerated' landscapes, exploring identity issues and the emotional consequences of change. She has published in a range of journals in sociology and geography, including a paper called 'Emotionally Intelligent Research' in *Area* (2004, 36:4; republished 2008, 40, virtual issue).

Liz Bondi is Professor of Social Geography and Head of the School of Health in Social Science at the University of Edinburgh, Scotland, U.K. She has published extensively in feminist geography and is founding editor of *Gender, Place and Culture*. Her most recent publications include the co-authored volume *Subjectivities, Knowledges and Feminist Geographies* (Rowman and Littlefield 2002), *Emotional Geographies* (co-edited with Joyce Davidson and Mick Smith, Ashgate 2005) and *Working the Spaces of Neo-Liberalism* (co-edited with Nina Laurie, Blackwell 2005). Her current research focuses on counselling/psychotherapy, psychoanalysis, voluntary sector activism and emotional geographies.

Emilie Cameron is a Doctoral Candidate in the Department of Geography at Queen's University, Kingston, Canada. Her research focuses on imaginative geographies of the Canadian Arctic and their intersection with political and economic power. She is particularly interested in story and in articulating a 'critical narrative geography' of the Arctic.

Laura Cameron is a Canada Research Chair in Historical Geographies of Nature at Queen's University in Kingston, Canada and her recent work addresses cultures of nature, art and psychoanalysis. She is the author of *Openings: A Meditation on History, Method and Sumas Lake* (McGill-Queens University Press 1997), and her essays appear in a number of books and journals including *History Workshop Journal*, *Radical History Review* and *Society and Space*. She co-organized the 2006 international conference on Emotional Geographies and is currently completing a book with John Forrester entitled *Freud in Cambridge* (Cambridge University Press).

Joyce Davidson is an Assistant Professor of Geography, cross-appointed to Women's Studies, at Queen's University, Kingston, Canada. Her research and teaching focus on geographies of emotion and embodiment, and her publications include *Phobic Geographies: The Phenomenology and Spatiality of Identity* (Ashgate 2003) and *Emotional Geographies* (co-edited with Liz Bondi and Mick Smith, Ashgate 2005). She has published her research in books and journals aimed at philosophy, women's studies and sociology, as well as geography audiences. She has co-organized two conferences on Emotional Geographies (Lancaster, U.K. 2002; Kingston, Canada 2006).

Frances Dyson is an Associate Professor in Technocultural Studies, University of California, Davis, and serves on the editorial board of C-theory. Dyson's current research and writing focuses on sound, digital media and posthumanism, and can be found in *The Critical Digital Studies Reader* (Kroker and Kroker, eds.,University of Toronto Press forthcoming), *Convergence* (Sage, Winter 2005), and Catherine Richards *Excitable Tissues* (Ottawa Art Gallery 2004). Her book *Sounding New Media: Rhetorics of Immersion in the Arts and Culture* is forthcoming with the University of California Press, and her web project *And Then it Was Now*, published by the Daniel Langlois Foundation for Art, Science and Technology, Montreal can be found at (http://www.fondation-langlois.org). Dyson has also exhibited installation/performance works in the U.S., Canada, Japan and Australia, and for over a decade has been a regular contributor to Australia's premier audio arts program, *The Listening Room* (Australian Broadcasting Corporation).

Jennifer Foster is an Assistant Professor of Environmental Studies at York University and is Coordinator of the Urban Ecologies certificate program at York. Her research focuses on urban planning, ecological restoration, environmental aesthetics and social justice. She has published in urban design, geography and environmental philosophy journals. Her current research examines the ecological politics of post-industrial urban greenspace.

Alexandre Gillet is a research assistant in the Department of Geography at the University of Geneva. His research is mainly directed towards a better understanding of the notion of 'open world' as it has been developed by Kenneth White for more than forty years. Following this idea, he has conducted doctoral research on the erratic spatiality of the cairn at University of Geneva (Switzerland). He has authored articles published in the fields of geography, literature and geopoetics.

R. Darren Gobert is an Associate Professor of English and Theatre Studies, York University, Canada. He specializes in modern and contemporary Western drama, and in dramatic and performance theory. As a critic, he has published on French, English, and German drama; as a practitioner, he has directed plays by Albee, Beckett, Chekhov, and others. He is a member of the Editorial Advisory Board and the Book Review Editor of *Modern Drama*, for which he has edited a recent special issue on contemporary playwriting from the U.K. In 2007, the Government of Ontario awarded him the John Charles Polanyi Prize for Literature.

Anh Hua is an Assistant Professor of Women's Studies at San Diego State University. She has taught as a visiting scholar in the Department of Gender Studies, Indiana University, Bloomington. She teaches in the areas of sexual politics, gender and the body, critical race feminism, and women and violence. Her research focuses on memory, cultural trauma, diaspora studies, critical race feminism, cultural studies, and literary and film studies. She has published in the collection *Diaspora, Memory and Identity* (University of Toronto Press 2005) and the journals *Canadian Woman Studies*, *J-spot*, and *Politics and Culture*.

Alphonso Lingis is Professor Emeritus of Philosophy at Pennsylvania State University. He is the author of *Excesses: Eros and Culture*, *The Community of Those Who Have Nothing in Common*, *Abuses*, *Foreign Bodies*, *Dangerous Emotions, Trust, Body Modifications: Evolutions and Atavisms in Culture*, and *The First Person Singular*.

Cheryl Lousley is a post-doctoral fellow in the School of English at the University of Leeds, UK, and a lecturer at Wilfrid Laurier University, Canada. Her research appears in *Journal of Environmental Philosophy, Canadian Literature, Essays in Canadian Writing, Interdisciplinary Studies in Literature and Environment*, and *Environmental Education Research*. She is the series editor for the *Environmental Humanities* book series published by Wilfrid Laurier University Press.

Avril Maddrell is Senior Lecturer in Geography at the University of the West of England, Bristol. Her research is centred on social, cultural and historical geographies and geographical thought. She is author of *Complex Locations. Women's geographical work in the UK 1850–1970* (Royal Geographical Society-Institute of British Geographers/Blackwell, forthcoming) and co-author of *Charity Shops. Retailing, consumption and society* (Routledge 2002), as well as numerous publications on geographical thought, gender and bereavement. Her current work on 'Mapping Grief' explores ways of conceptualising spatial experiences and expressions of bereavement, and relating these to specific cultural, place and belief contexts.

Jolene McCann writes and teaches in Seattle, Washington. She has published on women's curatorial work in the field of science-fiction in *Foundation* (forthcoming). Her current research interests focus on the history of education.

Dianne Newell is a Professor of History and Director of the Peter Wall Institute for Advanced Studies at the University of British Columbia, Vancouver, Canada. Among her various research interests, she has published on women's auto/biography in *Biography* and *Life Writing*, and on women's postwar science fiction in *The European Journal of American Culture*; *Journal of International Women's Studies, Science Fiction Studies*; and *Foundation*, with chapters in *Gender and Knowledge* and *On Joanna Russ* (forthcoming).

Mary O' Neill is Senior Lecturer in Cultural Context and Higher Education Academy Teaching Fellow at the University of Lincoln, U.K. Her doctoral research was on the relationship between ephemeral practices in contemporary art and the behaviours associated with bereavement (see 'Ephemeral Art: Mourning and Loss' in *Impermanence: Cultures in/out of Time*, Penn State University Press 2008). Her current research interests span a variety of disciplines and fields and include loss, failure, boredom, rejection and sorrow (see 'Art and Money: Experience Destruction Exposure' in *Money and Culture*, Peter Lang 2007). She is interested in the methodology of communicating these subjects and the intersection between academic writing and creative narrative. She was organizer of the symposium Telling Stories: Theory and Narrative, Loughborough University, 2007. Her interest in challenging and difficult material has also led to an examination of ethics and contemporary art.

Richard C. Powell is a Lecturer in Human Geography at the University of Liverpool. He has written on the geographies of scientific practice, the cultural economy of the Circumpolar Arctic and the epistemic politics of fieldwork. He is completing an ethnographic monograph on post-1945 environmental science in the Canadian Arctic. He currently holds an ESRC Research Fellowship (2007–2010) to investigate the politics of changing Arctic environments. This work takes forward themes from geographies of science into discussions about the political economies of hydrocarbons. He was awarded the *Area* Prize 2002 and the *Environment and Planning A* Ashby Prize 2007.

Mick Smith is an Associate Professor of Philosophy and Environmental Studies at Queen's University, Kingston, Canada. His current research is focused on questions of environmental ethics, responsibility and citizenship, and he has published widely in journals such as *Environmental Ethics, Environmental Values, Ethics, Place and Environment*, and *Environmental Politics*. He is author of *An Ethics of Place* (SUNY 2001), co-author of *The Ethics of Tourism Development* (with Rosaleen Duffy, Routledge 2003) and co-editor of *Emotional Geographies* (with Joyce Davidson and Liz Bondi, Ashgate 2005).

Deborah Thien is an Assistant Professor of Geography at California State University, Long Beach. Her research interests include geographies of gender, emotion, health and well-being, particularly in rural and northern communities; feminist geographic theories and practices; feminist, cultural and psychoanalytic theorizations of emotion; and recent engagements in place-making politics in Long Beach, via her involvement with the Spaces of Democracy, Democracy of Space network. She has published in *Area*, and has forthcoming work in *Gender, Place and Culture, The Canadian Journal of Public Health* and *The Journal of Geography in Higher Education*. She contributed a chapter to *Emotional Geographies* (Joyce Davidson, Liz Bondi and Mick Smith, eds., Ashgate 2005) and she is a Book Review Editor for *Emotion, Space and Society*.

Nigel Thrift is a Professor and Vice-Chancellor at the University of Warwick. His main research interests are in international finance, cities, non-representational theory and the history of time. His most recent books include *Knowing Capitalism* (Sage 2003), *Non-Representational Theory: Space, Politics, Affect* (Routledge 2007) and *Shaping The Day: A History of Timekeeping in England and Wales 1300–1800* (Oxford University Press 2009).

Acknowledgements

We would like to thank the Social Sciences and Humanities Research Council of Canada, and also Queen's University Faculty of Arts and Sciences and Office of Research Services for providing financial support for this project. Thanks to the team at Ashgate, and especially Valerie Rose, for vital encouragement (and patience!). The invaluable research and editorial assistance of Victoria Henderson is also gratefully acknowledged.

Introduction

Geography and Emotion – Emerging Constellations

Mick Smith, Joyce Davidson, Laura Cameron and Liz Bondi

Astronomy tells us that the light from the Pleiades, the Seven Sisters star cluster, takes approximately 400 light years to reach the Earth. And yet those few photons which, having travelled this unimaginable distance in time and space, impinge upon our eyes can excite the most profound and varied of feelings, can change, dispel, or deepen our moods, incite and admix awe, wonder, and utter loneliness, and even, if we should happen to be so fortunate, entangle themselves in those universally expansive feelings of inseparable closeness impressed upon us as the very moment of love's realization in a starlit kiss. Suddenly the whole world appears differently. What could possibly be more romantic?

Well, clearly such responses are not simply matters of cause and effect, automatic reactions to the impact of light-waves on the retina. Some might, for example, be altogether immune to romantic feeling, regarding such astral associations as expressions of self-indulgent affectation rather than heartfelt affection. Perhaps romance is a fiction best left between the covers of True Love stories – 'for women'. And after all, doesn't romantic love (and its association with the experience of nature as sublime) have a recent and fairly well documented history? Isn't it, at least in part, socially and culturally constructed, and usually along gendered lines? And then there are other circumstances to be considered. What if it happens to be -40 degrees out there or the stars are obscured by smog and the glare of sodium lamps? What if those concerned are too hungry, angry, tired, or distracted, to pay attention to heavenly bodies? What if they regard stargazing as merely a metaphor for wasting valuable time given their busy schedules and the accumulating business in their in-tray?

Such differences and difficulties raise the question of what it might mean to think about emotional geographies when emotions also have a culture, history, seasonality, psychology, biology, economy, and so on: When to detail the intricate entanglements of even a single emotion (if one could ever be analyzed apart from others) might take longer than those photons' earth-bound journeys and still leave more undiscovered – just as the Pleiades contains many hundreds more than the *six* stars usually visible to the naked eye. (According to mythology Merope, the seventh sister, married Sisyphus, a mere mortal and so shines less brightly than her sisters – so much for romantic love!) How might emotions be described 'geographically' when such feelings can connect us to places so far beyond the confines of the *earth* and yet still evade capture by the *writings* of any discipline?

Of course, we have to start somewhere. Even if the shifting grounds of emotions are not easily identified, they can't be studied in a vacuum – unless by this we mean the experiments conducted on those beings we occasionally fling into circular orbits far above the planet's atmosphere. Emotions might need to be understood as events that take-place in, and reverberate through, the real world and real beings, and so far as we know the existence of living, breathing, creatures is a pre-requisite for emotions to exist at all. (Perhaps Merope knows differently.) We can, at least, be sure that emotions are amongst the most important ways in which such creatures, including humans, are both connected with and disconnected from their world and their mortality. Indeed, emotions are a vital ingredient in the very composition of the world as a world, as something more than a concatenation of causes and affects, as those places, people and incidents, that become meaningful to us, that we care about, fear, disdain, miss, hate, and sometimes, inexplicably, love. Without emotions we might indeed survive in (but hardly experience) a world that resembled the empty space envisaged by Newtonian mechanics and measured by Cartesian coordinates (see below).

Such a space is not, of course live-able, indeed it doesn't even *exist*, in the same way that the 'empty' inter-stellar medium of outer-space exists. Yet its origins and effects are, unfortunately, much closer to home. It is, as Lefebvre reminds us, a secondary abstraction (abstract space), a historically and culturally specific rationalization, the purpose of which is the 'reduction of the "real" on the one hand to a "plan" existing in a void and endowed with no other qualities, and, on the other hand, to the flatness of a mirror, of an image, of pure spectacle under an absolutely cold gaze' (Lefebvre 1994, 287). This modern, Western, 'formal and quantitative' understanding of space 'erases distinctions, as much as those which derive from nature and (historical) time as those which originate in the body (age, sex, ethnicity)' (Lefebvre 1994, 49). It claims to be neutral, universal, apolitical, value and emotion free, and this claim is precisely why this understanding of space facilitates political, bureaucratic, and technological interventions of a kind that regard emotional involvements as either faults to be corrected, as vestigial and worthless remnants made defunct by the linear march of 'progress', or as resources to be manipulated in order to ensure more efficient forms of (human) resource management (see, for example, Hochschild 1983).

This abstract understanding of space, which, according to Lefebvre, increasingly dominates the modern world, facilitates the emotionless reduction of a diverse and beautiful planet to that 'raw material' (Lefebvre 1994, 31) necessary to reproduce a society – which itself is now re-envisaged in terms of a social 'system'. It thereby also serves to conflate the reproduction of complex, creative (excessively emotional) social relations with the crude 'biological' reproduction necessary for the system's survival (Lefebvre 1994, 50). Lefebvre's analysis can thus be connected with Foucault's claims about modernity's increasingly 'bio-political' management of populations and, via this, with Agamben's (1998) chilling account of the increasingly pervasive reduction of real human beings to the category of 'bare life', that is, to human beings stripped of their political possibilities, subject only to a 'cold gaze' that, for example, reduces people to abstract information stored in governmental computer systems, to faces recognized technologically.

As this imaginary abstract space is *realized* around us it wields, says Lefebvre, (1994, 52) an 'awesome reductionistic force vis-à-vis "lived" experience', a violence recently exemplified by the 'Shock and Awe' campaign of the U.S military machine in Iraq. But, since 'shock' and 'awe' are also emotions even this example of reductive force remains open to critique in terms of its emotional *duplicity*, the fact that even the 'empty' abstract space supposed by the 'military precision' of such plans still requires and utilizes its own emotional 'proxemics' (288). These are suited to treating the world and its inhabitants as abstractions (as *collateral* damage rather than as diverse, living, individuals or beings) to cutting emotional 'ties', and celebrating emotional distance (a very modern, masculine, and in Lefebvre's (and Lacan's, 1992) terminology 'phallocratic' ideal) now implicated in leaving hundreds of thousands in Iraq dead, angry, mourning, despairing, furious, hopeless, vengeful and bereft – hardly the results originally predicted by abstract planning, but hardly surprising to anyone with any understanding of the *affects* of violence.

Emotions are then intimately and inescapably caught up in the current re-writing of the earth, the production of new, transformed, geographies, and New World Orders, that affect us all, albeit in very different ways. And yet the question of how we might *feel* as well as think about these transformations is seldom addressed and this is (only one reason) why the world needs emotional geographies and why geography needs to take emotions seriously. Unfortunately, emotions have often been deliberately excluded from, or habitually suppressed within, many geographical discourses, especially those which understand geography as an attempt to provide objective, narrowly scientific, and quantitative models of the world – that is those that exemplify and employ Lefebvre's 'abstract space' more or less uncritically.

Geography and Emotions: Feeling our Way

The fact that emotions are not themselves easily located, defined, or measured should not be allowed to detract from their crucial importance to human (and more-than-human) geographies and lives. They are, paradoxically, both inordinately diffuse and all pervasive and yet also heart- and gut-wrenchingly present and personal: Sometimes moods seem to envelop us from without, sometimes emotions seem to surge up from within and comprise the very core of our being-here in the world, they inform every moment of our existence.

And if emotional responses de-limit places as much as people, if they *insist* that we regard the world about us in more than abstract terms, then their absence from most geography texts seems very strange indeed. Where they are present at all they usually appear only as part of geography's background scenery rather than as intimate, indeed indispensable, components of almost every worldly understanding. Political geographers, for example, will regularly employ theoretical terms like 'alienation' to describe the woeful conditions wrought by contemporary capitalism. Although definable in terms of modes of economic production, class, the division of labour, and even a metaphysics of 'human nature' (species-being), alienation still relies for both its descriptive potential and its political bite on the emotional resonances associated with the misery of feeling exploited, the loneliness of being detached from the companionship of other humans, despair at the seemingly senseless routines of labour, and the desire and hope to feel more at home and live life's possibilities more fully in the face of a heartless economic system.

The apparent absence of geographical attention to emotion therefore needs explanation. Perhaps, after all, it is more apparent than real. Perhaps we just need to reappraise past works in the light of our newly discovered interests to find geography's hidden treasures sparkling behind the indifferent surface of its texts – like constellations masked by city lights. To some extent this may be so. After all, if emotions are as all-pervasive as just suggested then traces of them should be revealed almost anywhere we care to look. But there again, perhaps like stars, their distribution will be uneven and recognizing them as emotional constellations might require more intellectual effort (to identify, relate, interpret, and configure them) in some instances than others. In some texts emotions are very deeply hidden indeed under tables of data and figures, maps, detached observations, third person or impersonal narratives, and so on. And thus an emotional geography, one that takes emotions as its subject matter, would still need to reflect upon and interpret this unevenness, suggesting reasons why certain authors, geographic fields, historic periods, methods, and so on, express, or alternatively suppress, emotions in the many various ways they write (about) the world.

This suggests then that emotional geographies might, as it becomes an increasingly familiar term, serve to denote an emergent and critical field of studies where a new focus also, quite rightly, throws the history and current state of the discipline into relief, illuminating what it excludes and overlooks. Just as feminist geographies continue to critique and recompose geographical accounts of a world where women had been noticeable largely by their absence, so emotional geographies too might be thought of as a recognition and response to a lack, that is, geography's failure to represent our emotional lives. Perhaps we could even understand this term 'lack' in ways reminiscent of Lacan's psychoanalytic usage. If geography has defined its self-identity as a discipline partly by excluding the emotional, as something it regards as 'Other' than geography (for example, as belonging to psychology), then the recent emotional turn could be understood as the exercising of a geographical imaginary, the acting out of a desire to make geography more complete by giving graphic (written and symbolic) form to what now, upon reflection, seems lacking in its relations to the world.

This, no doubt, would be a partial and controversial analysis but one that might at least spark further thought about what emotional geographies can and can't do. For example, in Lacanian terms the human subject's experience of lack and subsequent desire for completion is ultimately unachievable precisely because this subject is originally defined over and against what it is not, what is Other. To change the relation to the Other, to try to own it in some way, inevitably changes and redefines, rather than completes the self. It thereby also reveals to the human subject the dizzying absence of any essential fixed self-identity, any final answer to the question of who they really are. In terms of our analogy, recognizing geography's emotional lack makes apparent the fluidity of what might previously have been thought to be hard and fast disciplinary boundaries and the absence of an essential centre or core that identifies what geography, as an academic subject, is. To appropriate Livingstone's (1994, 28) words the 'idea that there is some eternal metaphysical core to geography independent of historical circumstances will simply have to go'.

There is an irony here. Even if we accept that the motivation for the relatively recent emergence of emotional geographies (Anderson and Smith 2001; Davidson et al 2005; Parr 2005) arises from the recognition that there is something crucial missing in geography as it is currently instantiated, their development will not make geography whole, restore its integrity, or fill a gap. Rather they make this lack available for reflection, make it seem problematic, and on this basis facilitate a form of self-critique – a negative dialectics. This threatens to decompose as well as recompose its object, breaking down current disciplinary boundaries and challenging fixed ideas of geography. Some geographers, already feeling their discipline threatened, might not be happy with such a situation, they might try to deny that emotional geographies are real (proper) geography but this, after all, is partly the point. This too was why feminists often received a frosty reception when they attempted to reconstitute geography by recognizing the importance of spaces traditionally associated with women's existence, for example, through geographies of the home (Blunt and Dowling 2006). Emotional geographies will, if this analysis is plausible, not just extend the remit of current geographical research, they will reveal something lacking at the centre of geography, fundamentally challenging geography's self-identity; they will recompose it in terms of emotionally, as well as socio-historically, situated theories.

So, if re-composing geography is something emotional geographies can do, something that aligns them with other critical geographies, then we need also bear in mind what, on this same analysis, they can't do. They cannot actually succeed in showing us exactly what geography really is because, as we have argued, there is no fixed essence of geography qua an academic subject, no-thing to make whole, just more or less traditionally accepted discourses and practices in constant flux with Others. Nor can they provide a complete representation of emotions. Why? Because our Lacanian analogy would also suggest that trying to put geography's emotional lack into words can only ever be only a partial (incomplete) solution because of another, parallel, kind of lack, a failure of representation that, Lacan argues, operates at the level of the symbolic order itself. Words (indeed all symbols) lack a fixed signifier, something solid and permanent that we might point to as anchoring their meaning or, reciprocally, that might be fully represented by them. Words' meanings, like subjects' identities, are only constituted over and against Other words, discourses, and practices. Words too are in flux; they allow us to reflect on the world, to situate our knowledge within various contexts but they can't capture the world's essence – if such a thing exists. Something of the world we try to describe always eludes those descriptions. Indeed, this seems paradigmatically the case with (words for) emotions which overlap and melt one into another, that allude to each other and their contexts in complex but often vague and mysterious ways, that operate in close association with other emotional states or exist through attempts to dissociate oneself from them – think of shouting loudly at someone 'I'm not angry!'.

Emotions are conceptually linked together only by our perception of what Wittgenstein (1981) originally referred to as family resemblances, overlapping patterns of similarities and differences, between them and the contexts in which they are used, the situations that give them meaning. Emotions have no common defining feature, a point emphasized by Thomas Dixon's (2003, 1) recent claim that 'emotions did not even exist until just under two hundred years ago'. That is to say that emotion only comes into use as an over-arching category after about 1820, earlier texts tending to refer to the affections, passions, interests, appetites, and so on, each with their own associations.

This certainly doesn't mean that prior to 1820 people didn't feel desolate or happy but that the way particular emotions were understood (and perhaps even something of the ways they were experienced) varied accordingly. For example, Adam Smith, who was by no means just concerned with *The Wealth of Nations*, and like many writing in the eighteenth century, focused attention on what were termed 'moral sentiments', feelings – like remorse – regarded as intimately associated with ethical responsiveness to Others and with judgements about right and wrong (Smith 1861). How remorse is defined, occasioned, and experienced, perhaps even whether it is experienced at all, are all dependent upon historically, culturally, and geographically, varying norms. And here again it might be helpful to think in terms of constellations, since they too are differently composed across times and cultures with different storied associations that include and exclude specific stars – Great Bear, Plough, Big Dipper or Revolving Male – the Navajo constellation that includes *Ursa Major* plus the pole star – the Greek Pleiades or the Navajo's Dilyehe. So, although geographies can, quite literally, add new dimensions to the understanding of these patterns of emotional differences – the spaces betwixt and between boredom and irritation, anger, rage and jealousy – they re-construe rather than complete the definition of them.

Emotional geographies are not then, on this understanding, indicative of a new science that will focus on how to map emotional landscapes, to provide a demographics of despair or a cartography of embarrassment. Indeed emotional geographies work against such attempts to pin emotions down, to define them in such ways that they might become mere objects of quantification, comparison and manipulation. After all, were we to accept such a view we would merely have provided geography with a new academic resource and have accepted the very understanding of geography as an entirely rational (in the sense of being emotion-free) enterprise that we seek to disrupt. As Wittgenstein (1981, 174) himself pointed out, emotions can't properly be understood by such mechanistic reductions: '"Grief" describes a pattern which recurs, with different variations, in the weave of our life. If a man's bodily expression of sorrow and of joy alternated, say with the ticking of a clock, here we should not have the characteristic formation of the pattern of sorrow or of the pattern of joy'.

This is not to say that emotional geographies are not concerned with empirical matters, far from it. The phenomenal experiences of emotions are the starting points for most researchers, as are the ways emotions are embodied and serve to colour our experiential world such that we interpret and value aspects of it in particular ways – we see red watching Canada's pointless mass-slaughter of seal pups, we feel dark despair at the emotional disconnection that those involved in such killing exhibit, their absence of com-passion for Other's lives. But attending to the phenomenology of emotion also leads us away from attempts to reduce the importance of such feelings to simplistic lines of causation or mere numbers, the thousands killed, the paltry few thousands of dollars they represent to the Canadian economy. This experiential aspect is precisely why an emotional geography of seal slaughter would be so very different from an economic geography of the same issue, or a geography framing the question in terms of resource distribution and allocation.

This is why it is important for any emotional geography to understand how the suppression and repression of emotions in contemporary geography originated and how emotional geographies can respond to this lack. It is, hopefully, already more obvious why certain forms of critical geography, namely, feminist, non-representational, psycho-analytic and phenomenological, have been most closely associated with the emergence of emotional geographies. This is precisely because these approaches have been concerned to critique a world-view that accepts the 'masculine' ideal of an essentially rational, unchanging, autonomous and emotion-free or emotionally controlled human subject, who has the ability to fully represent the external world within a universally applicable, objective, and rationally determined symbolic order. It is hopefully also obvious that this ideal is itself dependent upon a radical separation between 'emotional life' and the logic of a supposedly dispassionate 'reason', a binary distinction that emerges most explicitly (along with an ideal of abstract space) in the work Descartes, often referred to as the first *modern* philosopher.

Emotional Knowing

While Descartes' philosophy certainly plays a key role in the development of a modernist world-view that radically separates reason from emotion, the tendency to lay the entire blame for this binary division at his doorstep massively oversimplifies matters. After all, we could trace the conceptual separation of reason and emotion (and the attempted exclusion of the latter) much further back than the seventeenth century. The meditative practices of the Stoic philosophers of ancient Greece and Rome were specifically designed to encourage the eradication of emotions and subjective value-judgements while developing a detached and uncompromisingly materialistic view of the world, a *phantasia kataleptike* which Pierre Hadot (1995) translates as 'objective representation'. For the Stoics emotions clouded our rational abilities, they were the cause of our being distracted by and caught up in worldly cares and concerns about which we could do nothing. Epictetus, amongst others, provided meditative exercises that people might follow in order to foster this rationalistic and emotion-free approach to life – 'So-and-so's son is dead. What happened? His son is dead. Nothing else? Not a thing' (Epictetus in Hadot 1995, 188).

Such callousness finds echoes in Descartes too because of similar, although not identical, philosophical purposes, namely separating the confusions associated with our involvement in the physical world from the objective mindful clarity supposedly obtained by employment of pure unadulterated reason. Descartes posited an absolute divide between two very different substances – mind or soul (*res cogitans*) and matter (*res extensa*), a distinction so influential that it is now referred to as Cartesian dualism. For him, the human self is essentially a thinking and reasoning being, one who is only contingently connected to their body and the wider world (and hence capable of surviving bodily/worldly death). Indeed, it is thinking, the ability to rationally reflect on oneself, that ultimately provides the surety of our individual existence, hence Descartes' famous *cogito ergo sum* which proclaims that it is because 'I am thinking, therefore I am' (I can be sure that I, at least, must exist).

While Stoicism attempted to foster a largely passive acceptance of life's inevitable hardships (hence the present meaning of the term) Descartes' philosophy led in another direction, one that was, in the very different social climate of seventeenth century Europe, used to justify active intervention in a world now re-envisaged mechanistically. The absolute separation of human minds from nature, and the supposed superiority of the former over the latter, divided the universe into (independent and self-contained) human subjects and non-human (mindless) objects. The detached human observer was thereby free to manipulate the world experimentally and dispassionately in order to obtain useful knowledge. Hence Descartes' infamous views on animal vivisection, justified on the basis that animals were merely machines, their howls of pain just automatic responses to stimuli. They had, Descartes assumed, no mind that might feel pain.

So Descartes' philosophy has a number of important geographical and emotional implications. First, in terms of coming to see the world materialistically, as composed of mere objects at the disposal of human subjects. Second, in terms of how subjects come to understand this world through abstract rational representations of its mechanical properties. Objects come to be defined in terms of their physical extension in a supposedly empty space, a space that can be mapped and measured using mathematical coordinates thereby providing objective mental representations of how material things stand in relation to each other. Third, Cartesian dualism thus gives rise to a representational epistemology, an understanding of knowledge as a mapping or reflection of the physical world within an entirely separate world of mental constructs, concepts, symbols, and so on. A paradigmatically masculine subject mentally surveys the map which represents the outside world from which he averts his eyes. These intellectually coordinated representations are all he needs to know to master that world. For Descartes, unlike Lacan, there is no underlying reason why all that the world contains cannot be fully (completely) represented in every material aspect in the transparency of thought. Indeed with the benefit of hindsight Descartes might be regarded as a key source of the rather different view, often associated with scientific positivism, that all that really matters, all that there actually is, can, in theory, be so represented. From this perspective what can't be measured, manipulated, predicted, or accurately defined by a concept doesn't exist in any meaningful sense. And this, of course, is the fundamental principle underlying what Lefebvre refers to as abstract space.

So where does this leave the emotions? Interestingly, they seem to be suspended in a kind of no-man's land, they inhabit an overlooked territory in between mind and matter, subject and object, resisting attempts to entirely appropriate them by one side or the other. They are not alone here since this interstitial space is also where the senses reside, both having about them a necessarily ambiguous status because, given the absolutism of Descartes' dualism, there should be nothing that is not either entirely mind or entirely matter, no in-between. And yet without such a ground all connection between these two distinct realms is simply impossible, there can be no representation, no worldly knowledge. If this was so then what would it mean for me to touch something or be (emotionally) touched by what someone does? Where exactly do these effects and affects take place? Such questions expose what is lacking in Descartes' conceptions of both his human subject and his theoretical edifice, a space for senses and emotions to operate. The rather bizarre location he does propose, the brain's pineal gland, is hardly convincing!

Although Descartes' substance dualism is largely jettisoned in later philosophical and scientific developments, the basic structure of his thought remains. Today's geographer sitting before their computer screen is, if anything, even more dependent on the visualization and manipulation of mathematical representations and coordinates. The emotions consequently find themselves further side-lined, as even the senses, required to play the role of a trustworthy intermediary running back and forth between thought (concept) and object, become cleansed of any remaining emotional encumbrances.

Sight, for example, is reduced to scientific observation, a mode of visual communication with the sole function of information retrieval. But to this extent geography becomes both detached from the real world, and dehumanized. There is no space here for, say, the emotional and ethical insight a loving gaze might offer into Others' lives. Nor, for that matter, can an emotionally cleansed notion of the senses explain the feeling of anxiety or the uncanny [unheimlich] that an-Other's gaze can provoke in us, a look that according to Sartre makes us aware both of our own individual existence and the fragile, groundless, nature of that existence, the nothingness that 'lies coiled in the heart of being – like a worm' (Sartre 1993, 56). Thus, even sight, the most paradigmatically detached, distanced and objective of the senses, affects us intimately, magically expressing something of our mode of being-in-the-world and providing insight into Others'; how else could a stare be aggressive, a look malicious, a gaze lustful or leery, or a glance wistful, capricious, or yearning? How else would starlight ever affect our hearts?

It is just these emotionally mediated senses and relations that compose the fabric of our existence, that make our lives meaningful, or, in their absence, hopeless. Emotions are vital (living) aspects of who we are and of our situational engagement within the world; they compose, decompose, and recompose the geographies of our lives. And even Descartes recognized this to some extent. After all, in his last book, fittingly entitled *The Passions of the Soul*, he reflected upon what was lacking at the centre of his divisive philosophy, the emotional no-man's land in and between the subjective and objective worlds he had broken apart. Here he recognizes that even his rationalistic philosophy is, after all, motivated by desire, a desire for knowledge and that this passion is, in its turn, sparked into being by another, by wonder. 'Wonder', says Descartes (1952, 308), 'is a sudden surprise of the soul causing it to consider with attention those objects which seem to it novel and unexpected'. And while Descartes (1952) argues an excess of wonder is harmful and must be suppressed having no remedy 'other than acquiring the requisite amount of knowledge' this is, nonetheless, an explicit admission that the 'sweetest pleasures of this life' we feel only because we are not pure mind but hybrid beings, profoundly passionate creatures also composed of flesh and blood. Contra Cartesian dualism we are always already emotionally engaged beings-in-the-world.

Approaching Emotional Geographies

We have argued that emotions are not just a new topic for geographical study. They trouble yet inspire its very existence just as they trouble and inspire individuals. Without passions, sentiments, and their affects there would be no geography worth the name. Phenomenologist, feminist, psycho-analytic and non-representational geographers have all recognized this and have consciously tried to situate themselves in this interstitial no-man's land, dissolving the absolutism required by inter-related dualisms between thought and world, male and female, reason and emotion, culture and nature.

Phenomenological approaches achieve this by focusing, as the term suggests, on what is experienced rather than on what those experiences might be supposed to represent. The phenomenological world is not the abstract empty space of Cartesian coordinates but the lived world perceived and produced through our emotionally laden activities. Distances, for example, would not be thought of as fixed measurements between objects but as differently experienced movements affected by myriad influences – the frightful depths of the night through which we walk, the wearisome weather we struggle against, the lightening of our mood and steps in good company. This is not a form of subjectivism, but a recognition of the self's entanglement with the world since all such emotions are intentional, that is, they are inherently about something – the dark, the weather, the stars, and so on – even when their root causes are far from clear to us. Being alive we will feel, but never fully own, these emotions which come and go without our having called them into being. It is not surprising that geographic phenomenologists, like Yi Fu Tuan (1974) and Edward Casey (1993), have often focused on emotional responses and attachments to particular places. Places like people can thus be understood as being constituted within an emotionally charged middle-ground, one neither entirely subjective nor objective (Jones, 2005; Smith 2005). As Merleau-Ponty (1942, v) wrote: 'Truth does not only inhabit "the inner man", or more accurately there is no inner man, man is in the world, and only in the world does he know himself'.

This quotation, in its turn, emphasizes the need for, and achievements of, feminist geographies, since it is not only man that is in the world! Feminist geographers' interests often overlap with phenomenology but bring a critical awareness of the gendering of emotions (indeed of emotionality in general, so often dismissed as a feminine trait), such as the particular fears felt by women in night-time urban environments (Koskela and Pain 2000; Valentine 1989) or the gendered incidence of phobias (Davidson 2003). Feminist theory has been particularly critical of absolute binary distinctions (Bondi 1990; Rose 1993) and, though rarely explicitly focused on emotion, feminist explorations of embodiment have proved fruitful for emotional geographies. They uncover feelings about experiences of, for example, sexuality (Browne 2004), pregnancy (Longhurst 2001), disability (Dyck 1999), chronic illness (Moss 1999), and the consumption of foods (Heenan 2005; Mathee 2004) and fashions (Colls 2004). In investigating these taken-for-granted emotional aspects of embodied experience feminists illustrate the intimate connections between physical (material) and mental health. Rather than treating individuals in isolation feminists also make explicit theoretical connections with broader social systems of politics and power. Feminist methodologies reflect these theoretical concerns emphasizing relational, reflexive, and inter-subjective approaches that are typically qualitative, attending closely to the emotional relations that permeate geographical practices (Moss 2005; Rose 1997).

Psychoanalytic and psychotherapeutic perspectives offer geographers another field of emotionally relevant theories and practices that are capable of more than simply providing Lacanian analogies (though this hopefully illustrates something of their wider theoretical potential) (Bondi 2005; Rose 1996). Some recent work exhibits an increasing historical and geographical sensitivity to the ways these various theories and practices have intersected with bodies and spaces in different places and moments in time (Gagen and Linehan 2006; Philo and Parr

2003). Psychoanalytic geographers like David Sibley (1995) have paid particular attention to the factors mitigating the creation, maintenance or dissolution of the more or less stable and permeable psycho-social boundaries that are constitutive of both individual subjectivity and of social identities. Emotions clearly play a pivotal role in these formative processes which impact both individual mental health and socio-political cohesion, shaping social relations through the mediation of the self's relations to those deemed Other, drawing and redrawing boundaries between you and me, them and us, via feelings of attraction, repulsion, and so on. Psychotherapeutic approaches too, although apparently focused on reaching and treating the contents of individual minds through discursive processes, concern themselves intimately with subjects' relations to other people and places, not least the therapist and therapeutic setting (Bondi with Fewell 2003). A key aspect of these, as with psychoanalytic, approaches is the need to recognize the difficulties inherent in composing a necessarily incomplete emotional self-understanding and of expressing this (representing it) in language. Composing and expressing oneself emotionally are both necessary and yet, ultimately endless (lifelong and incomplete) tasks.

If psychoanalytic and therapeutic approaches focus on the emotional lack at the heart of subjectivity then, crudely speaking, Non-Representational Theory (NRT) focuses on the necessary incompleteness (the lack) within the act of representation itself, that is, on that which eludes expression in language. NRT's preference for the term affect over emotion is, as Nigel Thrift (2004) suggests, linked to this concern for the ineffable; affect is supposedly indicative of those more immediate embodied engagements with the world beyond language. Hence NRT tends to focus on performance, on what people do rather than what they say they do, challenging the Cartesian priority accorded to thinking in and through words (McCormack 2003; Paterson 2005). This is clearly important and NRT offers some radically different insights into emotional flows and conjunctions although there may, ironically, be something of a performative contradiction in NRT's approach, insofar as some of its practitioners employ some of the densest and most abstract philosophical language (such as that associated with Deleuze) to theorize and interpret such performances (Thien 2005). There is also an important question about how far even non-verbalized bodily movements, in say a ballet performance, are both implicitly and explicitly representational (Nash 2000). The dance of the dying swan is actually overburdened with layers of cultural (for example, gendered) meanings as well being indicative of presentationally constrained and rehearsed emotional content – something true even of less overtly stylized manifestations. NRT is surely right that not everything about an emotional performance can be captured in language; actions, like bursting into tears, often speak louder than words. However the fact that words cannot completely represent emotions is not necessarily a problem for representation; it is, as Lacan might argue, of the nature of representation. And representation in some form or other seems to be an inescapable aspect both of a symbolic order that encompasses far more than words, and of our self-understandings. It is not immediately clear that the anger expressed in giving another driver the finger is any less representational (or indebted to a masculine symbolic order) than giving vent to irritability through a sarcastic remark. Words too are always performances and all performances, however non-verbally immediate, are meaningless unless they can be related to certain interpretative pre-figurations (Bondi 2005).

There are then important debates still emerging within and between these different approaches to emotional geographies despite their overlapping interests. These debates are an encouraging sign of the creative fluidity of an emergent field that will hopefully continue to resist the desire to present a single unifying theoretical totality, one that would belie the intricate complexities of its subject matter. Indeed, what makes emotional geographies so fascinating is their potential to critique and re-construe almost everything that geography has so far taken for granted, a point made explicitly and implicitly in the essays included in this volume.

Advancing Emotional Geographies

Emotion, Place and Culture is organized in five sections, each of which explores a theme that absorbs contributors to the field of emotional geographies and that expresses their efforts to explore the emotional textures of our lives and worlds. These themes – remembering, understanding, mourning, belonging, and enchanting – necessarily overlap and together represent how we are collectively feeling our way forward in the production of new emotional geographies. In so doing, this volume both extends areas explored in its fore-runner *Emotional Geographies* (Davidson et al 2005) and advances in some new directions. For example, this collection highlights the impact of geography's emotional turn on historical geographical enquiry as well as contributing to the work of historicizing emotion. It also engages extensively with creative writing and artwork, including novels, science fiction, poetry and installations. Compared to its predecessor, this volume moves from a concern to articulate emotional dimensions of social and cultural geography to consider how doing emotional geographies might transform the assumptions and actions embodied in such ordinary practices as walking, driving, reading and living in the wake of traumatic losses and displacements. Essays variously seek to convey, caution, convince, commemorate as well as 'simply' communicate from heart and mind, together constellating terrains recognizable at least temporarily as emotional geographies.

Part 1, 'Remembering', begins with a meditation on a part of everyday automotive life which modernity might rather forget, our fatal encounters with Other animals on the road. Mick Smith explores the emotional phenomenology of these (un)expected events and raises questions about their affective meanings as they relate to claiming or avoiding ethical responsibility for our environmental actions behind the wheel. Focusing on ways the living remember the dead, Avril Maddrell investigates the emotionally redolent memorial landscapes of St. Patrick's

Isle, Isle of Man, both past and present. She highlights the recent practice of creating memorial benches and the ways these artefacts act as liminal spaces in the spatialities of bereavement. In a different approach to the theme of remembering, R. Darren Gobert underlines the importance of the specific historical contexts in which concepts of emotion emerge. In his reading of different historically influential theories of 'katharsis', he alerts us to an appreciation of emotions as inescapably historical and cultural constructs and reminds us how we, as historical and cultural agents, are always implicated in their emergence.

How are people primed to act? In the second part, 'Understanding', Nigel Thrift thinks through concepts of affect and 'affective contagion', challenging any easy separation of the 'social' and the 'biological', to shed new light on contemporary political action. He emphasizes the crucial importance of addressing the transmission and reception of affect if we are to understand and not merely be unknowingly swept along by the tides of change. Jennifer Foster explores affective dimensions of environmental aesthetics and, using different theoretical resources, develops a related argument for deepening our understanding of the feelings that lie at the core of environmental decisions. Such endeavours, she argues, are crucial to fostering social justice and environmental sustainability. For Foster, as for Thrift, understanding affect is key to new political formations and a source of hope. In Richard Powell's account of some of the practices of environmental science in the Canadian Arctic, he offers an ethnographic approach to the challenges and contradictoriness of playfulness and fun. Feelings about 'free' time in this context are revealed never to be care-free, and Powell advocates a concerned and careful approach in his reflections on seeking to understand the emotional lives of others.

Part 3 focuses on 'Mourning'. Anh Hua discusses Dionne Brand's *What We All Long For*, a novel that explores how the diasporic Vietnamese community has contended with the trauma of war. She examines remembrance practices of loss and mourning arguing that art has the potential to translate trauma. Hua's account also shows how a work of fiction may function as an emotional geography that configures new democratic forms of community through its creative transformation of voices and associated places repressed or suppressed in official history. Mary O'Neill also explores the transformative power of creative arts, visual and written. She explores how ephemeral artworks may act as vehicles for understanding feelings of grief and mourning. Her engagement, which refuses aesthetic detachment and the withholding of deeply felt response, highlights her personal knowledge as both artist and mourner. Mourning is approached from a different angle in Emilie Cameron's discussion of questions arising from the naming of an Arctic flower, *Senecio lugens*, derived from the Latin word '*lugeo*', meaning 'to mourn'. As well as demonstrating how scientific and imperial naming practice may be inflected with European explorers' emotional struggles, Cameron argues that emotional geographies need to register the occluding effects of colonial power and the fact that in many local knowledges, this flower may have no association with mourning at all.

Opening the fourth part, 'Belonging', Katy Bennett offers a study of an ex-mining village in County Durham, U.K., focusing on the role of nostalgic feelings and performances in the (re)creation of collective identity. Bennett argues that as people cope with change, their nostalgia for particular versions of the past allows for a sense of continuity. But as Bennett elaborates in relation to gendered forms of nostalgia, the means through which communities cultivate feelings of belonging always also serve to exclude. This theme is explored further in Deborah Thien's discussion of the emotional geographies of the Royal Canadian Legion, a largely masculine institution where the 'boundaries of belonging' are carefully guarded. In her consideration of how gender, geography, and emotional well-being intersect to affect life and health in British Columbian rural and northern communities, Thien finds that Legion branches have made spaces for the expression of 'at least some feelings' and that exclusionary practices are nonetheless subject to ongoing revision. Cheryl Lousley also discusses cultures of masculinity and emotion, focusing on the ways in which men have been represented and constructed in the novels of David Adams Richards and Matt Cohen. Lousley highlights powerful associations between masculinity and nature, and explores how masculine subjectivity, envisaged as inarticulate but emotionally embodied in compassionate acts towards animals, might inform rather different understandings of environmental ethics.

For many of those involved in the development of emotional geographies a key task is to evoke the vitality of the lives and worlds we study. This ambition might be described as 'Enchanting', hence the title of the fifth and final section of the volume. Frances Dyson discusses research into Human Computer Interaction (HCI), criticising the 'en*chant*ing data' of affective computing and the potential for empathetic relationships between humans and embodied, emotionally 'intelligent' machines. In an analysis that intersects with Thrift's discussion of the biotechnological political sphere, Dyson draws on contemporary art engagements and the discourse of posthumanism to argue that meaning and affect inevitably evade clear transmission between bodies and technologies. At the moment of being captured by technology, 'the "motion" in emotion enters another mode of circulation, an economy of simulation where affect is, in a sense, always at the same time dis-affected'. Although 'enchanting' means to 'to cast a spell on' or to make half-awake, it also has connotations of fascination and delight. The remaining three essays that constitute this section express such ideas in different ways, exploring the *motion* in e-*motion* and the fascinating places that move us. Following Dyson into the realm of science fiction, Dianne Newell and Jolene McCann trace the bio-geography of author Judith Merril (1923–1997) and her frequent movements and dislocations in her pursuit of new sites of enchantment and 'portals for discovery'. Two chapters offer geopoetical journeys into landscape, spatiality and motion. Alexandre Gillet explores his fascination with stone cairns and the case for *erratic* encounters. Drawing on a range of poetic sources, Gillet evokes the emotional texture of relationships between people, paths, stones and symbols enabled and invited by cairns. Responding to the summons 'to go on',

Alphonso Lingis narrates a trek through Mongolia and reflects on trust, travel and impassioned experience. He reminds us of, to borrow Arendt's phrase, the 'human condition', experienced and communicated across far-flung times and places, and thereby, echoing a theme running through several chapters, insists on the sociality of emotion.

We hope that that this volume helps to spur and inspire research and scholarship in and of emotional geographies. Through the constellation of contributions offered in this volume, with its particular blend of theoretical, methodological and empirical concerns, we seek to offer some redolent and heart-felt examples that highlight absences, silences, limits and impossibilities as much as defining and addressing new agendas. While we might wish to influence the sense, feel and shape of things to come, we do so from specific locations and perspectives in which our limited capacity to know and understand is vital to our openness to what is uncertain and unknown.

Note

Some of the material used in this introduction has been reproduced from Davidson and Smith's entry, 'Emotional Geographies', in the *Encyclopedia of Human Geography*, with kind permission from Elsevier.

References

Agamben, G. (1998), *Homo sacer: Sovereign Power and Bare Life* (Stanford CA: Stanford University Press).

Anderson, B. (2006), 'Becoming and being hopeful: towards a theory of affect', *Environment and Planning D: Society and Space* 24, 733–752.

Anderson, K. and Smith, S. (2001), 'Emotional geographies', *Transactions of the Institute of British Geographers* 26, 7–10.

Bendelow, G. and Williams, S.J. (eds.) (1998), *Emotions in Social Life: Critical Themes and Contemporary Issues* (London: Routledge).

Blunt, A. and Dowling, R. (2006), *Home* (New York: Routledge).

Bondi, L. (2005), 'Making connections and thinking through emotions: between geography and psychotherapy', *Transactions of the Institute of British Geographers* 30, 433–448.

Bondi, L. (1990), 'Feminism, postmodernism and geography: space for women?' *Antipode* 22, 156–167.

Casey, E. S. (1993), *Getting Back into Place: Towards a Renewed Understanding of the Place-World* (Bloomington and Indianapolis: Indiana University Press).

Davidson, J. (2003), *Phobic Geographies* (Aldershot: Ashgate).

Davidson, J. and Bondi, L. (2004), 'Spatialising affect; affecting space: an introduction', *Gender, Place and Culture* 11, 373–374.

Davidson, J. and Milligan, C. (2004), 'Embodying emotion, sensing space: introducing emotional geographies', *Social and Cultural* Geography 5, 523–532.

Davidson, J., Bondi, L. and Smith M. (eds.) (2005), *Emotional Geographies* (Aldershot, UK: Ashgate).

Descartes, R. (1952), 'The passions of the soul', in *Descartes' Philosophical Writings* (London: Macmillan).

Dixon, T. (2003), *From Passions to Emotions: The Creation of a Secular Psychological Category* (Cambridge: Cambridge University Press).

Gagen, E. and Linehan, D. (2006), 'From soul to psyche: historical geographies of psychology and space', *Environment and Planning D: Society and Space* 24:6, 791–797.

Hadot, P. (1995), *Philosophy as a Way of Life: Spiritual Exercises from Socrates to Foucault* (Oxford: Blackwell).

Harré, R. and Parrott, W.G. (1996), *The Emotions: Social, Cultural and Biological Dimensions* (London: Sage).

Heenan, C. (2005), '"Looking in the fridge for feelings": the gendered psychodynamics of consumer culture', in Joyce Davidson, Liz Bondi and Mick Smith (eds), *Emotional Geographies* (Aldershot: Ashgate), 147–160.

Hochschild, A. (1983), *The Managed Heart* (Berkeley: University of California Press).

Jones, O. (2005), 'An ecology of emotion, memory, self and landscape', in Joyce Davidson, Liz Bondi and Mick Smith (eds), *Emotional Geographies* (Aldershot: Ashgate), 205–218.

Koskela, H. and Pain, R. (2000), 'Revisiting fear and place: women's fear of attack in the built environment', *Geoforum* 31, 269–280.

Lacan, J. (1992), *Écrits: A Selection* (London: Routledge).

Lefebvre, H. (1994), *The Production of Space* (Oxford: Blackwell).

Livingstone, D. N. (1994), *The Geographical Tradition: Episodes in the History of a Contested Enterprise* (Oxford, U.K. and Cambridge, U.S.A: Blackwell Publishers).

Lupton, D. (1998), *The Emotional Self* (London: Sage).

McCormack, D. (2003), 'An event of geographical ethics in spaces of affect', *Transactions of the Institute of British Geographers* 28, 488–507.

Merleau Ponty, M. (1942), *La structure du comportement* (Paris: Presses Universitaires de France).

Moss, P. (2005), 'A bodily notion of research: power, difference and specificity in feminist methodology', in Lise Nelson and Jonie Seager (eds) *A Companion to Feminist Geography* (Malden MA: Blackwell), 41–59.

Nash, C. (2000), 'Performativity in practice: some recent work in cultural geography', *Progress in Human Geography* 24, 653–664.

Parr, H. (2005), 'Emotional Geographies', in Cloke, P. Crang, P. and Goodwin, M. (eds.) *Introducing Human Geography*. pp. 472–485 (London: Arnold).

Paterson, M. (2005), 'Affecting touch: Towards a "felt" phenomenology of therapeutic touch', in Joyce Davidson, Liz Bondi and Mick Smith (eds), *Emotional Geographies* (Aldershot: Ashgate), 161–173.

Philo, C. and Parr, H. (2003), 'Introducing psychoanalytic geographies', *Social and Cultural Geography* 4:3, 283–293.

Rose, G. (1997), 'Situating knowledges; positionality, reflexivity and other tactics', *Progress in Human Geography* 21, 305–320.

Rose, G. (1993), *Feminism and Geography* (Cambridge: Polity).

Rose, G. (1996), 'As if the mirrors had bled. Masculine dwelling, masculinist theory and feminist masquerade', in Nancy Duncan (ed.), *BodySpace: Destabilizing Geographies of Gender and Sexuality* (London and New York: Routledge), 56–74.

Sartre, J.P. (1993), *Being and Nothingness* (London: Routledge).

Sibley, D. (1995), *Geographies of Exclusion* (London: Routledge).

Smith, A. (1861), *The Theory of the Moral Sentiments* (London: Henry G. Bohn).

Smith, M. (2005) 'On "being" moved by nature: geography, emotion and environmental ethics', in Joyce Davidson, Liz Bondi and Mick Smith (eds), *Emotional Geographies* (Aldershot: Ashgate), 219–230.

Thien, D. (2005), 'After or beyond feeling?: A consideration of affect and emotion in geography', *Area* 37, 450–54.

Thrift, N. (2004), 'Intensities of feeling: towards a spatial politics of affect', *Geografiska Annaler* 86, 57–78.

Tuan, Y-F (1974), *Topophilia: A study of Environmental Perception, Attitudes, and Values* (New York: Columbia University Press).

Valentine, G. (1989), 'The geography of women's fear', *Area* 21, 385–390.

Williams, S. (2001), *Emotion and Social Theory* (London: Sage).

Wittgenstein, L. (1981) *Philosophical Investigations* (Oxford: Basil Blackwell).

PART 1
Remembering

Chapter 1

Road Kill: Remembering What is Left in our Encounters with Other Animals

Mick Smith

The automobile might, in so many ways, be regarded as an avatar of modernity, a metallic manifestation and incarnation of this globally dominant social form (Smith 1998). Intimately involved in driving processes of urban change, mass production, commodification, and pollution it is also inseparable from ideals of private property, social distinction, mobility, and representations and feelings of personal freedom. And, as with modernity, the image of humanly controlled and managed power hides the constant emergence of unintended side-effects and risks. These are exemplified by the human deaths and injuries caused by vehicle 'accidents' but also by the daily toll exacted on wildlife and companion animals – the road kill literally flattened by passing automobiles.

The emotional aspects of these fatal contacts between driver/vehicle hybrids and the non-humans unfortunate enough to cross their paths are, however, rarely expressed or addressed directly in public, despite their devastating effects/affects. The ways in which these animal deaths touch, or apparently fail to touch, so many of our lives, and our evasive silences about our responses to, and responsibilities for, such impacts actually says much about our strained relations to those animal Others now permitted to share so little of 'our' world. It seems, in so many ways, that their suffering surpasses the limits of our sufferance: we simply cannot afford the time to care. It also speaks to the constant social re-creation and mediation of emotional (and ethical) distances, the management and control of which, however total, occasionally falters, ruptures, and fails in the face of these supposedly unavoidable collisions between inexplicably different beings and worlds. The ethical task of an emotional phenomenology of such events is to find words that might express and convey something of these incommensurable experiences, the fleeting contact, the slight swerve, marking the end of animal experience and, perhaps, also the permanent loss of a familiar(s) face. Such words are necessary not in order to rationalize these events, or to explain them away, but to release otherwise suppressed concerns about the consequences of that project of world rationalization, that indefensible denial of emotional involvement in non-human nature, and that 'forgetfulness of Being' under the aegis of technology, that Heidegger deems so characteristic of modernity (Smith 2001, 23–53). Indeed Heidegger's phenomenology, so attentive to moods, mortality, language, the question concerning technology, and so problematic in terms of the differences he presumes between human and animal existence in the 'world' offers a way to approach the complex inter-weavings of these issues.

Ground Zero

Rounding the last corner before home two vehicles had stopped, parked awkwardly on the opposite side of the road, one a few yards behind the other. On the asphalt to the left of the first I saw the dark shape of the cat lying on her side, stretched full length. After turning on the hazard lights I walked towards her; only a young cat, breathing quick short adrenalin breaths but unable to move. She looked almost untouched; hardly any blood. It must have been a glancing blow, just a split second too soon, or too late.

The driver 'responsible' hadn't even stopped but these otherwise unconnected people had. One had found the tabby's owner who now ran from her neighbour's front door. As she ran she called the cat's name in a panicked howl, tearing the surrounding air in disbelief and terrible realization, a quivering expression of a crumbling world of lost attachments. It can't be, don't let it be, let it be alright; the hopeful/hopeless plea to recompose the world as it had been only moments before. She was followed by her distraught son. I got up and stepped back. 'I can drive her to the vets.' She didn't seem to hear me, oblivious to all but immediate concerns, but the son did. Almost accusingly he blurted out 'She's a he.' What could a stranger know of his cat, of anything that might matter now?

The cat was part lifted and slid onto an old board furnished from the second driver's truck. He was placed in the back of the owner's car, still breathing, I think. Owner, son, cat, and car drove off at speed towards the veterinary surgery 5 kilometres away, leaving us behind. We spoke only briefly about what had happened. We had cats too. 'I don't understand why people let their cats outdoors', she said, 'its bound to happen'.

And so it seems. I don't know, but hardly dare hope, that this cat, whose name I heard but have now forgotten, still lives. I could never bring myself to stop again at the house and ask, and now they have sold-up and moved. But I know for certain that the small white cat we saw and spoke to just months later was to die a hundred metres further down this same road; that the farmer three doors further down had lost almost every one of his many cats in the same way; that the porcupine we saw cross the grass a few feet from our window in March, or one identical in size, died the very next day; that the fox we watched at play in the neighbours garden too would end up mangled in the ditch. And then there are the chipmunks, the snakes, the deer, the squirrels, the snipe that make such eerie, endearing, whirring sounds in their descending flight, and on, and on. Each of these deaths may connect with many others as abandoned young die in nests, forms, and burrows, slowly starving. Then there are the cadavers we never see, picked over and removed by coyote and crow. And this is just one stretch of road in rural Eastern Ontario less than a kilometre long, and not even a busy stretch.

I have seen deer, beaver, muskrats, hare, gophers and coyote too; turtles their shells squashed and split almost beyond recognition: Once a massacre on a quiet country back-road where someone had driven straight through a flock of birds leaving many dead. Another time a mother and young racoon lying just a few feet apart. I can only think she had ventured out to the body of her young and suffered the same fate, perhaps from the very next car. They lay there for days gradually being ground formless into the pavement. A snake writhing frantically in the air, its tail squashed into the road surface. On yet another occasion a pigeon hit at a busy junction and with broken wing flapping was flattened by a second vehicle that could easily have avoided it. Of course *these* animals aren't companion animals and our encounters with other animals – even whether we recognize them as individual animals worthy of concern – depends so much on classifying them according to an effective history/geography of previous encounters (Jones 2000; Smith 2005). The estimated 41 million squirrels killed per annum in the United States and 15 million racoons may not have the emotional resonance accorded to at least some of those 26 million cats and 6 million dogs that share their 'fate'.[1] For many they are all 'pests'. But then it seems to me that this definition is applied to any and all creatures that appear without our bidding, that are not at our beck and call, that somehow intrude into 'our' world and just happen to get in the way.

I know that I am not the only one who cares about this, that feels the same apprehension at every approaching shape lying on the morning's road-side, the same 'relief' if it turns out to be yet another sack of litter dumped from passing truck windows, the same tugging anguish on spotting any recognizably animal features. A friend tells me about the time she stopped two men putting a dazed otter 'out of its misery' with a spade. Thanks to her intervention the otter survived relatively unscathed and was re-released. But two weeks later another (or perhaps the same) otter was run over and killed in almost the same place. There was the rare occasion on highway 38 when a driver pulled over to move a crossing turtle to the verge. But such activities are accompanied with a palpable feeling of embarrassment at being seen on a public highway ushering a water snake across the road. Is it a sign of emotional weakness to be caught caring about such 'things', to be drawn outside the private confines of your steel box, to be seen in the flesh on the other side of the windshield?

How might we venture into the phenomenology and ethics of these mechanistically mediated human/animal encounters, usually fatal for Other animals but all too often just a matter of callous disregard or awkward 'humour' for humans; where the final frame of the cartoon or the surprise ending of the story's punch line imitates the unexpectedly violent demise of the animal. The

---

1 These figures were generated using the Roadkill (1993) survey data. Along with their numerous other deleterious effects on ecological communities, Trombulak and Frissell (1999, 18) note that '[v]ehicle collisions affect the demography of many species, both vertebrates and invertebrates, mitigation measures to reduce roadkill have been only partly successful'.

title sequence of the sci-fi comedy *Men in Black* begins by following the acrobatic flights of a seemingly care-free animated and anthropomorphized insect that is suddenly cut-short by its impact with the wind-shield of the car involved in the 'real' story. Resting overlooking a beaver dam a fellow hiker subjected us to the apparently amusing story of her first encounter with a beaver. 'That beaver was dithering, edging into the road then retreating, edging and retreating, wondering about when to cross the road. Then he finally went for it but – Wham – splat he didn't know what hit him. That 18 wheeler just came out of nowhere! There was blood and guts everywhere.'

Is it a failure, or alternatively a denial, of our empathic abilities that such imaginative narratives of unexpected death often represent the limit of both hermeneutic understanding and of permissible public expressions concerning such events? Is this as far as anthropomorphism can take us? Or is there, even here in the forced laughter, sometimes a tinge of sympathy? Does getting such a 'joke' actually imply some subliminal recognition of the ultimate absurdity of situations shared by both humans and animals, where the habitually banal activities of our everyday lives can suddenly be so drastically and unwittingly transformed without reason? Is there more than a sense of dramatic irony at play here? Perhaps, even, as Freud might have suggested, a form of cathexis of pent up emotions concerning our own, all too frequent, mortality at the hands of the machines that *we* are supposed to control?

Such questions lack simple answers and there are innumerable, various, context dependent, forms of emotional displacement occurring in all such narratives (including, no doubt, this one). But this anthropocentric analysis seems ethically lacking since it subjugates the reality of the animal death itself by reducing it primarily to an allegorical substitute for, and illustration of, the futility of our human modes of existence. It also over-emphasizes the role of chance or fate. If the animal just happened to be in the wrong place at the wrong time then responsibility for their death becomes detached from any individual's actions. It is an unpredictable 'accident', an uninsurable act of God.

Clearly the focus on instant and unexpected death can also operate in other ways as a form of emotional and ethical suspension, absolving us from any requirement to imagine terror, pain, shock, or sense of loss. We become both witnesses and accomplices in an 'accident' now redefined as unlucky, inevitable, and, in terms of its swiftness almost a form of euthanasia, a 'good' death. At best one might say that this is a form of ethical coping strategy, a way of diminishing responsibility for such acts. It lessens the wrongs we do or are complicit in. At worst it can exemplify the kind of macho andro- and anthropo-centric displacement of emotional involvement in the world, the re-assertion of the centrality of the individual's autonomous and automotive interests above all else. In such cases the point seems precisely to brag to a presumably compliant audience about his lack of concern, his ability to commit or witness any atrocity as a mark of his independence from emotional or ethical concerns.[2]

This seems evident in the numerous recipes for roadkill to be found in print and on the web, which celebrate such emotional distantiation with what passes for wry humour, for example, 'Ted's original roadkill chilli'.[3] These exploratory menus might, alternatively, be thought of as the final, rather pathetic, cannibalization of Western frontier mythology. The modern Grizzly Adams drives a 4x4 with double rear axles, he's Daniel Boone in a drop-tail wagon living wild and free off what his pick-up runs over. What he is strong enough to stomach, hot chilli and potentially rancid meat, is implicitly and explicitly a mark of his masculinity and his triumphal eradication of any emotional sensitivities towards other creatures. The 'good death' is no longer defined from, what the coping strategy at least, attributes to the animal's perspective, but wholly in terms of its utility in satisfying the driver's voracious appetites.

Of course there may just be something of a waste not, want not, mentality in eating roadkill. Certainly many other carrion feeding species benefit from the year-round pile-up of carcasses. But the emotional resonances here, the combination of black humour and the literal consumption of bodies, the parading of deliberate *distastefulness* in both senses, is surely intended to denote emotional self-mastery and isolation. This same flaunting of emotional insensitivity recently appeared on a large scale over the issue of the Canadian seal hunt. Here too numerous web pages were filled with juvenile macho posturings and jokes about clubbing 'baby' seals. Here too the same refusal to recognize the possibility of ethical concerns for non-humans, the same designation of the creatures concerned as pests, even the same visceral emphasis on the rough and tough character building edibility of, in this case, seal-flipper pie. Overcoming any potential feelings of disgust is again utilized as an alimentary form of social 'distinction' (Bourdieu 1998). This was not confined to student blogs but 'contagiously' infected the mass-media, exemplified in the sardonic *witlessness* characterizing the Canadian Broadcasting Company's radio reporting.

In this sense, points made about roadkill have a much wider resonance, standing as an expression of modern Western societies devastating environmental relations as a whole. Here too emotional distantiation is a mechanism whereby responsibility is divorced from individual action, displaced systemically amongst constantly mobile hybrid constituents that are only ever part human and can therefore, whenever politic, shift guilt as easily as they automatically shift gears. As we drive we 'pass the buck' – all equally guilty or guiltless as participants.

---

2 Though of course I'm not arguing such expressions are confined to men, rather that they epitomize a certain dominant understanding of masculinity. In fact the story of the beaver was, as noted, recounted by a woman.

3 Or you might want to peruse Roadkill rugs (2006) site which markets flattened toy animals with 'nothing gory about them. They are simply fun and absolutely hilarious'. Or the supposedly 'funny' roadkill raps alongside the roadkill pictures (2006).

'I just couldn't stop in time. I needed to think of the traffic behind.' In such circumstances we come to occupy an uneasy, unethical, space where it is never our fault – for as Hannah Arendt (1993, 21) remarked 'where all are [deemed] guilty no one is'. The myth of the 'accident', which is *always* waiting to happen, to leap out on the innocent, combines seamlessly with the inescapable historical 'reality' of current social and economic systems. 'What else can one do?' Chance and historical necessity, two sides of the same corporate coin in a world where all associations are more than human in everything but their ability to feel for Others and the possibility of their taking responsibility for their actions.

As Neil Evernden suggests (1999, 14), people actually have to be acculturated and trained to be so emotionally insensitive to the world's other inhabitants, a process he refers to (invoking the practices of some nineteenth century vivisectionists) as 'cutting the vocal chords'. Our 'transformation [he says] from beings with an interest in mysteries and animate nature to beings with an interest in a mechanical order did not come easily and quickly and still does not' (14). The influence of scientific 'objectivity' in particular, he argues, trains us to regard the world as 'made of parts, just like a car' (14). As a child in Britain the public safety programmes instructing us how to cross roads featured, without any apparent sense of irony, 'Tufty' the talking red squirrel – 'look right, then left, then right again'. Compare this to the rather different ethical ambiguities in using roadkill as educational tools in schools (see also Knutson 1987). The Roadkill (2006) website, originally facilitated by a grant from the U.S. National Science Foundation suggests various class exercises including using simple statistics, graphing of weekly roadkill data, and producing roadkill games and story books. It also offers a roadkill mystery page concerning 'the adventures of Dr. Splatt and Ms. Roadrash [Ms. Roadrash is a geography teacher]. They are travelling round the world to encounter roadkill or a hapless animal that has been involved in a terrible situation [...] It is up to your students to figure out where Dr. Splatt and Ms. Roadrash are located and what animal they ran over or found dead in a certain ecosystem' (1) 'Enjoy!' (2).

Worlds in Collision

The animal, says Heidegger (1995, 185), is 'poor in the world' while the human is 'world-forming'. By this he means, amongst other things, that the animal is captivated by the world in which it finds itself, under its ecological spell, affected only by certain disinhibiting features of the world which its sensory and emotional capacities are attuned to recognize as cues to engage in its characteristic activities and behaviour. The snake responds to the warmth of the spring asphalt lying prone as though basking on a sunlit rock, the startled hedgehog curls rapidly into a ball. A 'being such as the animal, when it comes into relation with something else, can only come upon the sort of entity that '*affects*' or initiates its capability in some way. Nothing else can penetrate the [disinhibiting] ring [*Enthemmungsring*] around the animal' (Heidegger 1995, 254 emphasis in original). Thus, the rabbit is held fast by the headlamps that suddenly eradicate the sensory environment to which it is attuned. Of course, these 'poverty struck' animals can still be deprived of the 'world', can be road-killed, but even this deprivation, Heidegger would say, is one to which the animal has limited accessibility in the sense that it is in-capable of any understanding of the inevitability of its eventual death, of its mortality, its finitude. The moment of 'death' cannot even come as a surprise to the animal since it has never been anticipated. Only mortals, those who can recognize and comport themselves towards their own erasure, can die, animals simply perish.

Of course, the frequency of road deaths might also suggest that part of the problem is actually a lack of human attunement even to our very own, world-formed, road conditions, a situation where our supposed ability to suspend ourselves from nature's disinhibiting ring, to attend to the things around us as they appear, has, all too often, only resulted in its replacement by mechanical means – the anti-lock brakes, air-bags, seat belts that offer a spurious feeling of invulnerability belying the fact of our own mortality. This false feeling of safe enclosure in the automobile's replacement ring of steel is, of course, a primary cause of this cybernetic organism's potential for extra-vehicular destructiveness.

Here too the car encapsulates modernity in miniature, parodying and exemplifying what Heidegger (1993) refers to as the technological enframing (*Gestell*) of the world. Here everything in the world is forced to appear one-dimensionally, *only* as a standing reserve, a *store*-house, a potential resource for its human use, a calculable ordering. In this sense the automobile is a manifestation of the 'forgetfulness of Being', the enframing that 'drives out every other possibility of revealing' (Heidegger 1993, 332), every other understanding of how the world might make its presence felt. Instead, the world is set upon and used up. Mountains, forests, rivers, are dissected, flattened, drained to 'save' travelling time thereby reducing even time itself to a resource to be spent. Travelling too is no longer an encounter with others in the world but becomes a matter of eating up the miles, of speed, efficiency, and conditioned air, all of which mark a radical dislocation from the places travelled through. The possibilities of our worldly existence, of our being-there (*Dasein*), of our sensory and emotional involvements are *thrown away*, discarded like litter from the car window, reduced to an endless self-involved getting-there on these roads to nowhere. Animal intrusions into this world are inevitably regarded only as obstacles to be over-run.

Nevertheless, if Heidegger is right, doesn't the worldly poverty of the animal mean that any concern we might have for their 'deaths' is only misplaced anthropomorphism, only the ridiculous result of transposing our own feelings and understandings onto the animals being? Is the racoon's parental concern for its struck offspring, its risking and losing its own life to attend to a being already passed saving, only 'apparent' not 'real' – only the result of a programmatically followed instinct long evolved but now ill-suited to modern conditions? Isn't the passing motorist's complete indifference to this scene the only sensible policy, the refusal of an inappropriate (sym)pathetic fallacy, the denial of a tragedy not only of our making but actually only made present in some of our overly-sensitive minds? Should the hollow sigh I express, the drawn out 'oh no' that cannot, for all its repetition, now deny the reality of the feline shape deposited along the roadside, be focused not on the cat's now motionless corpse, but entirely on the subsequent impoverishment of the human world – the child who, searching for the missing object of their affections, will now, at best, find only empty space, suddenly, heart-wrenchingly, evacuated of a life-enhancing presence?[4]

But then, if this is so, even the child's affection might be misplaced, its tears and the tear in his/her world, the consequence of a failure to reach the right kind of adult emotional detachment, the result of an, as yet incomplete, ability to resist needless, pointless, sentimentality, a lack of understanding concerning the unbridgeable gap between human and animal existence. If this were so shouldn't it be sensible to aspire to the kind of detachment displayed by those unswerving drivers who, in their own ways, epitomize the instrumentality of the vivisecting physiologist Claude Bernard, the man of science, who, 'no longer hears the cry of animals, …no longer sees the blood that flows…'? (Bernard in Evernden 1999, 16).

Not at all: Heidegger's philosophy in no way encourages such instrumental detachment, indeed, it is a whole-hearted critique of it, especially as it appears in the technological enframing of modernity. Certainly Heidegger, a humanist despite himself, cautions against 'the uncanny humanization of the "creature" i.e. the animal, and a corresponding animalization of man' (Heidegger 1998, 152) but he regards this reduction of both to merely a shared biology, to scientifically established similarities in, say, their nervous circuitry, as itself a consequence of the 'forgetting of Being' characteristic of the technological *Gestell*.[5] Heidegger, by contrast, understands the relation and gap between animality and humanity phenomenologically, in terms of the kind of opening in/on the world that experience grants us. For him the key difference is that only human *Dasein* has the possibility

---

4   The child, of course, may not yet be reflectively aware of death or of her/his own mortality. Ironically, they may often begin to come to this (supposedly uniquely human) self-awareness through experiencing the loss of an animal companion. They may also be open to such experiences precisely to the extent that they have not yet learned how to cut the world's vocal cords.

5   This term, the 'forgetting of Being' denotes Heidegger's critique of the metaphysical tradition of Western philosophy which, he argues, increasingly takes the world as something given (present), and therefore 'ours' to be manipulated, avoiding, rather than thinking, the more originary question concerning existence (Being) itself. As Gadamer (1994: 133) notes 'the oft quoted "forgetfulness of Being", with which Heidegger had originally characterized metaphysics, proved to be the fate of an entire age. Under the sign of positive science and its translation into technology the "forgetfulness of Being" is carried towards its radical completion. For technology allows nothing else beyond itself to be noticed …' This is why Heidegger's notion of a technological *Gestell* refers not to technology (for example, automobiles) *per se*, but to an all-encompassing enframing of the world, an understanding that can only envisage it as a 'standing reserve', that is, instrumentally, as an actual or potential resource. The automobile only exemplifies this enframing.

of seeing the world *as a world*, of re-collecting and re-membering the question of Being, rather than being wholly captivated within a sensory environment, (*Umwelt*).

But this means that while the animal may be 'poor in the world', in this specific sense, in not being aware of it *as a world* or of its own finite existence within this world, in other ways its life is incomprehensibly rich. The life of animals does not represent 'something inferior or some kind of lower level in comparison with human Dasein. On the contrary, life is a domain which possesses a wealth of openness with which the human world may have nothing to compare' (Heidegger 1995, 255). The animal's relation to its environment is not a mechanical relation (214) but one of stimulation, disinhibition, involvement, of the kind of close immersion in nature of which we can have only an inkling. This difference and this richness is precisely why the animal is 'that which is most difficult to think' (Heidegger in Agamben 2004, 50) When we think of the way the racoon holds what excites it in her paws, or softly touches her perished offspring, it is difficult not to think in our terms, to envisage the racoon conceptualizing her behaviour, giving it meaning in terms of a world she strives to tightly-grasp hold of in thought, consciously mourning her loss in terms of a ravaged future. But while, from Heidegger's perspective, this kind of anthropomorphism is a mistake this does not mean that her involvement is, in any sense, emotionless, any less heart-felt, and in what sense can we actually consider her any less caring in her involvement than the attitude of the drive-by killers who rend her 'world'?

Heidegger exaggerates, makes absolute, the gap between human world and animal environment, but these 'worlds' do, nonetheless collide in all manner of ways all of the time. We do possess a limited potential to transpose ourselves into the animal's situation, indeed the human 'already finds himself transposed into the animal in a certain manner' (Heidegger 1995, 211). We have a *'peculiar transposedness into the encompassing contextual ring of living beings'* (Heidegger 1995, 278, original emphasis) in that we exist *'in the midst* of beings' (278, original emphasis). Yet Heidegger is more concerned with what the animal lacks, namely the wor(l)d, that which, because of its absence, refuses any complete transposition, any 'going along with' the animal – that which creates an apparently insuperable difference between us. Still, while it may be difficult to put into words what the animal feels, or to enter the animal's 'world', emotions are precisely what offer the possibility of bridging this gap. Perhaps this is why Heidegger is so reticent to discuss the emotionality of animals or the evolutionary heritage, the common origins, of our own emotions, speaking only of our 'barely conceivable, unfathomable physical kinship with animals' (Heidegger in Gadamer 1994, 187). Perhaps this would bring the phenomenal 'worlds' he has so carefully separated crashing together?

For Heidegger moods are the constantly changing modes of our attunement to the world, '*Dasein* always has some mood' (Heidegger 1988, 173) they offer a *primordial* disclosure of our Being-there *'prior to* all cognition and volition and *beyond* their range of disclosure' (175, original emphasis). Moods ground our being-open to the world. The open attunement they provide is pre-logical, before *logos*, the word (Heidegger 1995, 341) but also primordial in the sense that moods provide the 'fore-having' (Heidegger 1988, 275) that precedes and underlies the possibility of any interpretative understanding of existence, that is, the possibility of Being becoming a question for us. They are thus the necessary phenomenological preconditions of our becoming aware of the prospective unity of the world, of '*Being-in-the-world as a whole*' (176) and of the very possibility of our awareness of our own mortality – the nothing that is the limit of our Being-there surfaces through the experience of anxiety. The difference here between animal and human modes of attunement, their and our 'moods', seems only that the latter involves 'letting oneself be bound' (Heidegger 1995, 341) rather than remaining captivated of necessity. But knowledge concerning this opportunity to 'let oneself be' appears, as an opportunity, only in certain very specific moods (like profound boredom) and only for *authentic Dasein*, the kind of Being-there that is not entirely absorbed in everyday activities, that is not 'completely fascinated by the "world"' (Heidegger 1988, 220). It also depends upon questionable assumptions about the nature and ascription of 'instinctual' behaviour to animals. As Evernden (1999, 153) remarks, instinct is 'a term that signifies nothing but permits us to dismiss subjectivity', that is used whenever and wherever there is a need to cover over the fact that we do not understand others' mode of Being but want to distinguish ourselves from them.

So even if we accept that the peculiarity of the word, of human language, is its ability to express some of these emotions differently, to engender new emotional possibilities, to allow the possibility of a self-reflexive transformation of our emotionality, to 'through knowledge and will, become master of its moods' (175) primordial forms of emotional involvement nonetheless stay with us as modes of attunement to the world that are akin to, even if not identical with, the disinhibiting ring of the animal's sensitivities. And even if animals lack some or all of specifically human emotional possibilities it is nonetheless true that these attunements, which are never fully mastered or explicable, remain necessary for any worldly engagement They are what makes the world a concern for us, what calls us to care about that world and, even occasionally, what offer the possibilities of transposing ourselves, in however limited a way, into the emotional experiences of the animals that inhabit the world. Even if we chose to align ourselves with Heidegger's anthropocentrism – and while accepting that we may not be able to *entirely* transpose our Being-there into the animal's place, to *feel along with* the animal – we are still granted an emotional ability to *feel for*, to concern ourselves with, animal others, this possibility of feeling for and attending to, is, after all, what, for Heidegger, indicative of our Being-there.

How so? Well, we might say that ethically and ecologically Heidegger's work shows us that, as human beings, we have the possibility of comporting ourselves toward the world in a way that, (for him) unlike animality, can choose to *let other beings be*. That is to say, our ability to escape the disinhibiting ring of environmental stimuli, and to understand other beings *as beings*, opens up the possibilities that our worldly concerns, our involvement in the world, might also become concerns for other beings as such. '[O]nly where there is the manifestness of beings as beings, do we find that the relation to those beings necessarily possesses the character of *attending to…* whatever is encountered in the sense of letting it be or not letting it be' (Heidegger 1995, 274, original emphasis). We are not, then, like the cat who, quite innocently, tortures a captured mouse – precisely because, insofar as we become human, we have lost our innocence but also gained the possibility of being ethical, of concerning ourselves with other beings as other beings rather than simply as disinhibiting stimuli surrounding us. We have the possibility of taking on or shirking the responsibilities that arise from our worldly concerns and actions, that is, from our capacity to let things be or not be.

What then happens emotionally in a society dependent on auto-mobility? Certainly our enclosure in the automobile's metal sheath, the speed of travel, the narrow beams of the headlights, the wall of sound from the stereo-system, the touch-button control of air flow and *temper*-ature, disrupt our possible attunements to the non-human world, they conspire to reduce the potential for any emotional openness that might regard the animals encountered on the pavement as ethical Others, as beings in themselves. This is not to say that the phenomenology of the automobile is emotion free, it creates its own modes of concern, but these too are often oriented within the technological *Gestall*, the 'rush' of acceleration, or, more frequently, the frustration at the red traffic light or stop-go of road-works. Even the fury of road rage is directed at those who 'cut' us (the cybernetic organism we have become) up, that push in front of us in line, that make us deviate from prescribed routes, that interfere in our everyday projects of getting-there – to our place of *employment*, or the department *store*. And even the most macho posturings of 'Wild Bill' Roadkill do not in any sense denote a return to a primitive or primordial emotional state. Their celebratory public expressions are not a case of a civilizationally repressed individuality and a now resurfacing animality as they would have us believe. Rather it indicates, like flashing hazard lights, only a thoughtless intentionality, an instrumental going-along-with how things 'have to be' framed, that is nonetheless deliberate in its being devoid of any com-passion.

And since we are not, in Heidegger's convoluted sense, animals, since we have the wor(l)d this also means that we cannot claim innocence concerning the everyday slaughter under our wheels. Yet, for the most part, those of us who retain, at least in other less constrained circumstances, some connection to other beings will still try to satisfy ourselves with an ethical *avoidance* of the 'issue' of our responsibilities directly proportional to the extent to which we can rationalize the claim that we were unable to *avoid* hitting the animal. I could have done nothing else. We all have to get to work. I couldn't brake because of the car behind.

It came out of nowhere.

And, of course, the claim that the automobile is, in a sense, like the technological *Gestell* of modernity itself, 'de-humanizing' has a ring of truth about it. But the sense that we give to the notion of de-humanizing here is precisely one that seeks to alleviate our feelings of responsibility. It marks a dis-placing of feelings, of compassion, concern, even guilt, a coming to regard oneself only as a cog in the machine, enframed and encircled by technological necessities, by instrumentality. This dis-placing places responsibility outside of our Being-there (*Dasein*) – but then this is the whole problem we find ourselves framed within, and not at all the solution.

It came out of nowhere.

And this 'it' too is part of the whole problem for it marks a forgetting of the Being of other beings, their being turned after a sudden appearance, into that which is present at hand, as an object in our line of vision, one more dead skunk. Only some kind of emotional involvement can ever broach this mode of presentation, this objectification of other beings so typical of everyday life within the technological *Gestell*. At 'first, and for the most part in the *everydayness* of our Dasein [there-being] we let beings come towards us and present themselves in remarkable undifferentiatedness ... a levelled out uniformity of the present at hand' (Heidegger 1995, 275 original emphasis). It's just an insect we think to ourselves at the noise of its impact, the smudge on the windscreen. The sudden start the animal's appearance gave us, the rush of concern even if only for our own continued involvement in the world, falls just as quickly back into a remarkable undifferentiatedness – just one more, hardly noticed, stain on the road, a persistent odour.

It came out of nowhere ...

But this 'coming out of nowhere' is precisely what we need to think in our relations with animals and what thinking Being means. While what the 'it' occludes is an(O)ther being, what the animal's sudden appearance in our everyday life offers is a possibility of seeing 'things' differently. It offers a possibility of re-membering the animal in terms of its Being, its existence and extinction, which also opens the possibility of ethics *and* offers an opportunity for us to momentarily break the hold of the technological *Gestell*.[6] The animal's seeming to 'come out of nowhere' is a revealing appearance of the truth of our and 'its' situation, of the danger to thinking and to life of this enframing of the world.[7] Though bringing this to mind we again open the possibility of dwelling in the midst of other beings, our re-collecting our kindred relations to others, our concern with and for the world. This is why, for Heidegger remembrance (*Andenken*) is a way of thinking that brushes against the unthinkable (Being) and, why we might say, a way of thinking

---

6   As Gadamer (1994: 194) describes it: 'The forgetfulness of Being is always accompanied by the presence of Being, sporadically illuminating in the instance of loss and constantly superjacent to Mnemosyne, the muse of thinking ... The recollection of Being is the contemplative accompaniment of the forgetfulness of Being'.

7   Heidegger understands truth as *aletheia*, appearance, or unconcealedness (not accurate re-presentation) which always also involves a withdrawal from appearance, a concealedness. This movement of revealing/concealing is referred to by Heidegger as the *clearing* of Being, or the 'event'.

that might find its source in 'that which is most difficult to think' – the animal's being.

The mysterious unlooked for appearance of living animals is one of the Earth's greatest gifts. Such encounters make every journey different, a surprise that, if we don't foreclose our emotional responses, if we attend to how subtly things may manifest themselves, and if we refuse to be artificially isolated from the world, leads to wonder. Life is wonder-full – though our current social form, including its automotive avatar, works incessantly to contain the realization of such experiences, to cut ecological/emotional ties and the vocal chords that might express such attunements, such feelings. As Luce Irigaray suggests, even the architect of modernity's rationalistic disinterest, the vivisector Descartes, realized that philosophy and ethics, that understanding and appreciating of that which is 'different' from us, originates in that first passion 'wonder'. '[I]t appears to me that wonder is the first of all passions; and it has no opposite, because if the object that presents itself has nothing that surprises us, we are in nowise moved regarding it, and we consider it without passion' (Descartes in Irigaray 1993, 13).

And so it is that so many consider those creatures that appear before us on the highway, without passion. The surprise they offer is nothing more than the possible danger that they might divert us from going about our daily business, that they might dent the fenders placed between us and world, scratch the superficial surface of our incessantly mobile lives. We pass what remains of them by without concern for their demise, without wondering about their existence and appearance. In such a society as ours 'awe' is now politically reduced to something distant Others are to be shocked into feeling by the militaristic adventurism and wanton destructiveness 'necessary' to keep oil flowing and traffic moving. But it doesn't require a philosopher to understand that all pain and death is awful (terrible) not awe inspiring. How can any but those afflicted with maladies of the soul be inspired by death, by the cessation of breathing, the loss of animus, by a beings' falling away into nothing. How can a failure to express feeling for the pain and loss of Others, however different, constitute anything but a recipe for disaster?

References

Agamben, Giorgio (2004), *The Open: Man and Animal* (Stanford, CA: Stanford University Press).

Arendt, Hannah (2003), *Responsibility and Judgement* (New York: Schocken Books).

Bourdieu, Pierre (1998), *Distinction: A Social Critique of the Judgement of Taste* (London: Routledge).

Evernden, Niel (1999), *The Natural Alien: Humankind and Environment* (Toronto: University of Toronto Press).

Gadamer, Hans Georg (1994), *Heidegger's Ways* (Albany, NY: SUNY).

Heidegger, Martin (1993), 'The Question Concerning Technology', in *Basic Writings* (London: Routledge).

Heidegger, Martin (1995), *The Fundamental Concepts of Metaphysics: World, Finitude, Solitude* (Bloomington IN: Indiana University Press).

Heidegger, Martin (1998), *Parmenides* (Bloomington and Indianapolis: Indiana University Press).

Irigaray, Luce (1993), *An Ethics of Sexual Difference* (London: Athlone Press).

Jones, Owain. (2000), '(Un)ethical Geographies of Human-non-human Relations: Encounters, Collectivities and Spaces', in C. Phil and C. Wilbert (eds.), *Animal Spaces, Beastly Places* (London: Routledge).

Knutson, Roger M. (1987), *Flattened Fauna: A Field Guide to Common Animals of Roads, Streets, and Highways* (Berkeley: Ten Speed Press).

Roadkill (2006) <http://roadkill.edutel.com/rkmysteries.html> Last accessed 15 May 2006.

Roadkill pictures (2006) <http://www.livingpictures.org/roadkill.html> Last accessed 15 May 2006.

Roadkill rugs (2006) <http://roadkillrugs.com/> Last accessed 15 May 2006.

Roadkill survey (1993) <http://roadkill.edutel.com/rkdataarchive.html> Last accessed 15 May 2006.

Smith, Mick (1998), 'The Ethical Architecture of the Open Road', *Worldviews, Environment, Culture, Religion* 2:2, 185–99.

Smith, Mick (2001), *An Ethics of Place: Radical Ecology, Postmodernity, and Social Theory* (Albany, NY: SUNY).

Smith, Mick (2005), 'On "Being" Moved by Nature: Geography, Emotion and Environmental Ethics', in J. Davidson, L. Bondi and M. Smith (eds.), *Emotional Geographies* (Aldershot: Ashgate).

Trombulak, Stephen C. and Frissell, Christopher A. (1999), 'Review of Ecological Effects of Roads on Terrestrial and Aquatic Communities', *Conservation Biology* 14:1, 18–30.

Chapter 2

Mapping Changing Shades of Grief and Consolation in the Historic Landscape of St. Patrick's Isle, Isle of Man

Avril Maddrell

Introduction

This chapter explores the ways in which burial and mourning practices have manifested themselves in a particular landscape, historically and in the present, and the relationship between these in producing an emotional landscape reflecting personal and collective geographical imaginations. The case study is St. Patrick's Isle and the adjacent Peel Hill, in the Isle of Man, historically significant as the reputed site of the foundation of Christianity in the Isle of Man, as well as being the location of heritage-status castle ruins, and a working fishing harbour. A brief outline is provided of the conceptual and theoretical placing of the study, the methodology and a note on my own positionality in relation to the subject matter.

Just as there are particular geographies of dying and deathscapes (see Hallam and Hockey 2001, Morris and Thomas 2005), 'place' is central to giving the bereaved a focus for locating grief (Hartig and Dunn 1998; Kong 1999); 'embodied emotions are intricately connected to specific sites and contexts' (Davidson et al 2005: 5), which can be expressed at various scales, including the physical home (Hockey, Penhale and Sibley 2005) and wider notions of 'home' such as transnational belonging (Blunt and Dowling 2006). Grief can be both triggered and ameliorated in relation to particular places that take on meaning in relation to the dead. This chapter combines forms of representational analysis such as reading landscape as 'text', with reference to what Lorimer (2005) refers to as 'more-than-representational' geographies, including those of emotion and affect, sense of place and sacred space. Ways in which we can 'map' meaning or the 'invisible landscape' (Jackson 1989; Ryden 1993) of grief and thereby understand more of the spatialities of bereavement are explored, with particular reference to continuities and changes in practices in the locale.

Physical spaces of memorialization are part of the fabric and meaning of place (see Gough 2000, Johnson 1994, Howard 2003); some of these forms of memorialization and memory inscription are very familiar cultural markers within the landscape of the British Isles, notably cemeteries, war memorials and park benches. Others such as spontaneous roadside or mountainside memorials are less familiar in areas where the cultural context is predominantly Protestant, and are relatively new additions to the lexicon of remembrance in these areas. Spontaneous memorials reflect a trend to highly individualized expressions of memorial-making and location, whether associated with a formal belief system or not, and reflect a growing awareness of and desire to mark one of the most powerful examples of what Anderson and Smith (2001, 8) describe as 'emotionally heightened spaces'. It is through specific contextually located studies of such places (Kong 2001) that insight can be gained to 'the complex arrangement between the living and the dead in changing modern societies' (Worpole 2003, 12).

The notion of 'sacred' is central to understanding the emotionally heightened spaces associated with death and memorialization. The idea of the 'sacred' is contested within theology, anthropology, between different cultures, and within contemporary Western society. Milton (2002, 104) argues that the sacred takes many varied forms: 'what is sacred to someone is simply what they value most highly, be it their mother's memory, their religious traditions, the mountain scenery near their home or the football team they support'. In the discussion here the 'sacred' encompasses both spaces defined by formal religion and more individualized locations found outside places and practices of those formalized beliefs, places which become endowed with meaning and are seen as 'sacred' to the bereaved. For many, this sacred quality is associated with a sense of 'continuing bonds' with the deceased. The idea of continuing bonds emerged in bereavement literature in the 1990s, principally through the collection of essays edited by Klass, Silverman and Nickman (1996) which explored new approaches to understanding grief, challenging the received wisdom in Western bereavement counselling that grief resolution was achieved through 'closure' and the severing of bonds with the deceased (Silverman and Klass 1996, 3). Subsequent work has suggested that experience of continuing bonds is not new to the bereaved but only to the dominant Western masculine-rational model of grief therapy (Walter 1999). While further studies have unpacked many complexities in the 'continuing bonds' model (see Bonanno 2006), numerous studies suggest that a sense of continuing emotional bonds with the deceased can help the bereaved to adapt to the loss of a loved one (e.g. Riches and Dawson 1998, Hockey, Penhale and Sibley 2005, Francis et al 2005). While the sense of 'continuing bonds' and/or notions of the sacred are often difficult to convey representationally, they can be read or inferred to a degree from material forms of memorial and the spaces chosen for those memorials and associated rituals, and this idea is used to analyze the St Patrick's Isle area as a more-than-representational space of memorialization. Both belief and grief are expressed through representational means in material form (e.g. place of worship, memorial), but these stand in for/ point to that which is difficult or impossible to articulate (see Laurier and Philo (2006) on communicative aporia and Harrison (2007) on the inarticulable for different positions on what can/cannot be said).

Memorials represent a fixed and concrete record of someone's life: a spatial fix which 'facilitates relationships between the living and the dead' (Hallam and Hockey 2001, 85, 90; also see Francis, Kellaher and Neophytou 2005). They are places of representation, sites of identity markers and as such are socially, culturally, economically and politically embedded. They can be read as 'text', part of the 'performative utterances' (Davies 2002) of death rituals, but they need to be read critically, aware that they potentially represent more than any textual analysis can reveal. They are symbolic spaces invested with meaning: respect for the remains of kith and kin, and/or symbolic remembrance of them. Whilst in the UK cemeteries and crematoria are the most obvious material spaces associated with death and remembrance, this is changing. These changes, occurring over the last two decades are credited to the 'informalisation of emotional expression …[which] has contributed to a more permeable public/private boundary' (Hallam and Hockey 2001, 99); but the growing rate of cremation in the British Isles also plays a part in facilitating personal rites and ash disposal: 70 per cent of the deceased are now cremated in the UK and 67 per cent in the Isle of Man.[1] Parallel to the changing understanding of 'sacred', this blurring of 'public/private' space and the flexibility of ash dispersal has facilitated increasingly individualized vernacular and informal memorial practices and spaces which are used to achieve a personalized and meaningful farewell to the bodies of loved ones and/or the creation of sacred sites of remembrance where absence/presence can be mediated. This sense of 'continuing bonds', continuities, and changes in expressions of remembrance and belief will be considered in the case of St Patrick's Isle.

Methodology and Positionality

St Patrick's Isle was chosen as one of two study sites in a small research project looking at continuities and discontinuities of grief-belief practices in areas recorded historically as foundational sites for Christianity in the British Isles, as part of a wider study of space/place and bereavement.[2] Kong (2001) has argued that research on 'mainstream' religious spaces and practices have been neglected in the UK and USA, and this study aimed to explore particular intersections of historical Christian sacred spaces and contemporary bereavement practices. The methodology used in this research has been based on field research including mapping historic and contemporary memorials, and observation of landscape use. This study of the visible representational marks of mourning was complemented by in-depth interviews with local government representatives, archaeologists,

---

1   2005 figures, UK figures from *The Guardian*, Isle of Man figures based on raw data on body disposal 2006 from the Isle of Man Government Civil Registries.

2   The other place was the Isle of Whithorn which is not dealt with here for reasons of space. St Patrick's Isle and the Isle of Whithorn were chosen as study sites for continuities and changes in bereavement practices in historic Christian sites because of a number of common factors: their status as foundational sites for Christianity in their localities, their common loss of status in the post-Reformation era, their similar topography, and contemporary links through shared fishing grounds and use of harbours.

clergy, other stakeholders and community members, in order to access contextual, qualitative and implicit meanings associated with the landscape. Given the sensitive nature of contemporary memorial practices, especially for the recently bereaved, it seemed appropriate to feel a way forward (to borrow a phrase) in the fieldwork, rather than to track down all those who erected memorials, but I was fortunate to gain access through 'snowball' contacts to one memorial maker, and accounts of some of the other memorials' histories were gleaned through interviews.[3] The historical context was of particular importance to understanding contemporary practices in the light of continuities and discontinuities of grief-belief practices and will be discussed further below. In order to place the recently collected field data in the context of historical spaces and practices in the locality, archive materials and secondary sources were studied and a brief account of the history of St Patrick's Isle and surrounds follows below.

As for my own history with this area, my childhood home was in Port Erin in the Isle of Man and my relationship to Peel, the chosen site for the study, was that of day visitor, a local 'tourist'. I grew up in the Manx Methodist Church, a tradition with a strong sense of place but little attention to the biographies of saints, and a theological emphasis on resurrection of the dead which downplayed the significance of bodily remains. Similarly, my knowledge of local history proved patchy when it came to this research. In terms of my own situatedness, I might be considered as an insider in relation to place in the Isle of Man (and having a Manx name certainly eased several initial 'cold-call' enquiries), but not having been resident there for over twenty years and never having been part of the Peel community, made me something of an insider-outsider at best. Personal experience of bereavement has been the starting point for my wider interest in the field of death studies, not least in the ways in which particular spaces become emotion-laden places, both those we choose to identify and those affective spaces which can unexpectedly interpellate us.

The next section outlines the historical context of St Patrick's Isle and hinterland, with particular reference to the related histories of belief and burial practices.

---

3   Two periods of field study were undertaken over three months in 2007. St Patrick's Isle, including Peel castle and cathedral, and Peel Hill were visited, visible memorials were mapped alongside secondary sources (maps, archaeological reports etc.) indicating earlier burial grounds or memorials. Memorials were photographed and context recorded for subsequent analysis. Comparisons were made with memorials in Peel town and promenade, as well as wider Manx cultures of memorialization. Initial interview contacts were made with people who had formal roles in the town pertinent to St Patrick's Isle (e.g. the lifeboat coxswain, the curator of the Leece [Peel] Museum, Peel town clerk, the Manx National Heritage curator of St Patrick's Isle, and the Bishop of Sodor and Man); these interviewees provided snowball contacts to representatives of local churches, local historians and in one case to someone who erected a memorial bench, as well as an account of a community initiative behind another memorial. Some follow up interviews/ queries were conducted by phone after leaving the island.

St Patrick's Isle: A Brief Historical Context

St Patrick's Isle is adjacent to the small town of Peel on the West coast of the Isle of Man[4] and is the place where St Patrick is reputed to have landed in 441 to evangelize the Isle of Man. Archaeological evidence has shown permanent occupation of St Patrick's Isle from the late Bronze Age settlement until the nineteenth century, the land overlaid by successive communities including the Vikings, ecclesiastical authorities and military garrisons. The present day ruins include the curtain wall castle, keep and St. German's Cathedral, named for the local Celtic saint. St Patrick's Isle includes several *keeils* or small chapels, one of which is named for St Patrick, as well as St Patrick's church, and may have been the site of a seventh century monastery made up of several cells, similar to the one at Maughold in the North West of the Isle of Man (Freke 2002). The prolific number of keeils on the Isle of Man reflects both the eremitic tradition of the early Irish church and the practice of treen (smallholding) chapels whereby a cleric would be attached to each extended family, the sacred and secular being interwoven in everyday life in early Celtic Christianity (Joyce 1998; Silf 2005). Other local place names on the Isle of Man testify to the influence of St Patrick, e.g. Ballakil*pheric* and nearby Keeil *Pharic* (Pheric/ Pharic being Manx versions of Patrick) and St. Patrick's Chair located in Magher-y-Chiarn (the Field of the Lord), traditionally held as the site where Patrick preached (Dugdale 1998). There are also several holy wells (chibbyr) named for Patrick, including Chibbyr Pharick below Peel Hill (Kinvig 1975). Irish-style Ogham stones and other markers on the island are dated to the earliest period of Celtic Christianity and Patrick's lifetime, and 'In a hundred ways his [St Patrick's] name has been woven into the texture of Manx life, in customs, in superstitions, in ballads and folklore, in the names of wells, islets, stafflands, etc. No other saint, not even the island's patron St Maughold, has been so honoured by the Manx People' (Dempsey 1957, 21). However, there is no textual evidence that Patrick visited the Isle of Man, and there is debate whether the name given to St Patrick's Isle predates records from the thirteenth century (Freke 1995, 2002).

Excavations on St Patrick's Isle during the 1980s revealed an unanticipated extensive and successive use of the islet for burials (Interview, Andrew Johnson, 2007). The Vikings ruled the Isle of Man during the ninth to eleventh centuries and graves from this period included remnants of mixed Norse and Celtic clothing and jewellery styles as well as a hybrid of Viking and Christian burial rites, thought to represent a transitional group. 'Some of the [tenth century] group are accompanied by grave goods which might indicate a pagan ritual. The presence of these graves in an earlier Christian cemetery however, aligned with the earlier Christian graves, with an inhumation rite and body posture consistent with Christian practice,

---

4   The Isle of Man is located in the Irish Sea; it has an independent government and is a Crown Dependency rather than part of the UK. The name 'Man' is derived from the Norse 'Mannin' ('Isle of Mountains') and is sometimes shortened to 'Mann'.

Figure 2.1   St Patrick's Isle taken from the highest memorial bench on Peel Hill 2007

Source: Photo by author.

together with their likely mid- to late 10ᵗʰ-century date, might equally indicate the influence of Christian ideas … possibly first generation converts, or Christians buried by pagan partners, or pagans buried by Christian partners' (Freke 2002, 73). Graves included various forms of interment, including stone lintel graves, evidence of wooden coffins, clinker built coffins and shroud burials. Some graves included small rounded white quartz pebbles, another traditional practice common in the Celtic West before and during the Christian period (Freke 1995), and examples of vernacular Christian grave markers were found, engraved with rough crosses (the likelihood being that more elaborate carvings may have been removed and headstones recycled in new buildings) (Trench Jellicoe 2002).

The earliest burials in the locality are Bronze Age tombs found on Peel Hill overlooking St Patrick's Isle; burial areas on St Patrick's Isle itself have been dated from the seventh century and are located around early Christian chapels (Freke 1995).[5] The thirteenth century *Chronicles of Man* (thought to have been written by the Cistercian monks at Rushen Abbey) record that two Kings of Mann died and were buried on St Patrick's Isle, namely Godred II and Olaf II (in 1187 and 1237 respectively), and that the last Norse King of Mann gave the islet to the Church about 1257, at a time when defensive focus had shifted to the new political centre of Castletown on the East coast (Freke 1995). Under the diocesan system St Patrick's Isle was always divided between the parishes of Patrick and German, and while St Patrick's Isle became the seat of the Bishop of Sodor and Man, a centre of ecclesiastical ritual, taxation and incarceration, this was relatively short-lived, and by the fourteenth century it had moved to Bishopscourt a few miles up the West coast of the Isle of Man. After the Bishop's removal, the church struggled to have control over the cathedral, as the islet became alternately a centre for defence, for local administration and a home to the Stanley dynasty who ruled the Isle of Man for over three hundred years from 1405, signalling a shift of power from the church to the state (Kinvig 1975). Although Bishop Crigan was installed at the cathedral in 1784, it was without roof, windows or doors by 1791 (Moore 1900; Freke 1995). The first attempt to restore the cathedral and castle came about in the mid nineteenth-century with the foundation of the Peel Castle Preservation Committee in 1858, which was followed by the intervention of the Island's Lieutenant Governor Loch who commissioned repairs to improve the site principally as a fee paying tourist attraction. While the road around the castle to the harbour completed in 1882–4 improved access from Peel to the previously tide-bound St Patrick's Isle, the restoration of the castle's curtain wall and the introduction of an entrance fee reduced day-to-day access to the castle and cathedral for locals.

The cathedral on St Patrick's Isle, was replaced by St German's Church built in the adjacent Peel town in 1879–84 for reasons of accessibility, but the cathedral itself

---

5   Although Christianised, there is evidence of hybridity of Christian and pre-Christian beliefs (e.g. see Garrard (1989) on house charms), with aspects of the latter persisting in contemporary folklore practices, such as greeting the 'Little People' at the Fairy Bridge.

was never deconsecrated, and the shell has continued to be used as an occasional place of worship since.[6] However, the islet tends to be represented discursively as the 'Peel castle' heritage site rather than as 'St Patrick's Isle, an historic sacred space. This may be attributed in part to the visual and metaphoric dominance of the castle on the islet, with its historic defensive and custodial functions, as well as the pre 1980s spatial, economic and social prominence of the fishing fleet in the harbour and quayside. The dominance of the Protestant church on the Isle of Man in the Post-Reformation era must also be a factor in the cathedral's status: Moore (1900) estimated that of the Isle of Man's population of 54,000 in 1865, only about 2,000 were Roman Catholics (but this was a significant increase from 25 in 1781). The Catholic community in Peel grew in the nineteenth century as a result of Irish famine emigrants, largely fishermen, who settled and married local women; and it was this denominational community which continued the use of St Patrick's Isle as a burial ground in preference to the Protestant parish churchyard in the mid-nineteenth century. Coffins were carried on what was known as the Funeral Path along the side of Peel Hill and, prior to the construction of the Harbour access road, mourners often had to wait for the tide to turn to allow crossing to the islet. This same community built the new St Patrick's Church in Peel in 1866 near to the 'Irish cottages' (Interview, Mr. Ledley Senior 2007).

Manx National Heritage impose strong controls on landscape use and change at St. Patrick's Isle and Peel castle. The site has a high aesthetic value locally and is accessible by car and well used for leisure purposes, principally walking, dog walking, fishing, refreshments, boat excursions and bathing at Fenella Beach, as well as visits to the castle. The Isle of Man's Christian heritage has been developed as an educational and tourism theme within wider cultural and historical offerings, but discursive orientation to other sites of Christian heritage on the island downplays St Patrick's Isle as spiritual site. Perhaps as a matter of strategic policy emphasizing distinct tourist and leisure offerings, Manx National Heritage promotes the recently acquired remains of Rushen Abbey (founded in 1134) as the 'highlight' of Christian heritage on the Isle of Man (Manx National Heritage, no date). In 2005 an ecumenical movement entitled 'Praying the Keeils' was initiated by Bishop Graeme Knowles to draw on the rich spiritual heritage of the Island through pilgrimage walks and meditations to early sites of Christian worship; this has resulted in more contemporary recognition of and engagement with the keeils, their history and meaning, including those on St Patrick's Isle. The following sections explore the emotional landscape of current memorial practices, and how such 'sacred' activities are negotiated in relation to heritage, historic practice and planning regulations.

---

6   After several failed attempts to restore the cathedral on St Patrick's Isle, St German's was designated cathedral church in 1980 <http://www.isle-of-man.com/manxnotebook/parishes/gn/german.htm>.

Memorialization on and near St Patrick's Isle

While there is free access to the footpath which circles the castle on St Patrick's Isle, the castle itself, including the space of the original cathedral and earlier chapels, is largely an enclosed space under the surveillance of custodians, where access must be paid for and actions are monitored and ultimately regulated by Manx National Heritage. Burials on the islet continued into the mid-nineteenth century after the cathedral was disused, and named monuments can still be read within and around the cathedral precincts from this period. Interview evidence revealed that a significant number of these nineteenth century graves were Roman Catholic (Interview Mr Ledley Senior 2007). Memorial stones also mark the graves of those who died on passing ships and were brought in to Peel Harbour, for example the Rev. Edmund Violet, minister of the Independent Church of St John's Newfoundland, who died en route to Liverpool in 1810. Archives also record the unmarked grave of a stillborn baby born onboard a passing ship (Freke 2002). No interments have taken place since the nineteenth century, but anecdotal evidence suggests there has been recent ash-scattering outside the castle perimeter on St Patrick's Isle, reflecting the increasingly individualized departure rites facilitated by cremation, including at historic and prehistoric sites, causing concern to archaeologists tasked to preserve the integrity of historic and especially archaeological remains (Interview, Andrew Johnson, 2007). Recent requests for memorials on St Patrick's Isle have been turned down (except one, see below), but both non-Christian hand-fasting and Christina marriage blessings have been permitted (Personal communication, Yvonne Cresswell 2008; Roy Baker, Interview 2007).

Peel Hill overlooking St Patrick's Isle, noted above for Bronze Age burials, is also significant for Corrin's Tower, its nineteenth century summit landmark, and benches which punctuate the paths on the hill. The tower was built by local land owner Thomas Corrin c. 1806, as a family memorial and retreat; inside it bears plaques and inscriptions and has a small family graveyard enclosure outside. Corrin, a Dissenter and founder of the Isle of Man's Congregational Church, buried his wife and two of their children there, commissioning an 1839 inscription: 'This mound and within the enclosure upon its top rest the remains of Alice Corrin and her two beloved children. This pillar, tower and mound were erected by Thomas Corrin to perpetuate her memory'. Corrin wished to be buried there himself; but when he died in 1845 his son preferred to have him buried in the consecrated ground of the parish churchyard. In the event Corrin elder's body was moved at least twice, and his son was only reconciled to his burial at the tower when the ground was consecrated by the Bishop.[7] Thomas Corrin donated the tower and surrounding area to the Manx government before his death, and the Tower is now commonly referred to as 'Corrin's Folly' and the surrounding area is known as Contrary Head. Thus both St Patrick's Isle and the Peel Hill area were used in the

---

7   There are various accounts of this story, see <www.IsleofManguide.com/corrinstower.php> and <www.peelheritagetrust.net/wordocs/corrin-tower>.

nineteenth century for individual expression of belief and burial rites, in the form of minority anti-established church practices. There is no visual or oral history evidence for late nineteenth or early twentieth century memorials in the locale, a hiatus reflecting the enclosure of the castle, the cessation of Roman Catholic burials on the isle, a cultural trend to collective war memorials after the 1914–18 and 1939–45 wars, and the containment of memorials in churchyards, except for financial or social notables.

Contemporary memorials are prohibited on St Patrick's Isle proper because of its heritage status, and consequently memorials are limited to the periphery of the locale (several applications for memorials on the isle have been refused in recent years (Personal Communication, Frank Cowin 2008)).[8] Three memorial benches are located on the harbour side and eight on the publicly owned Peel Hill adjacent to St Patrick's Isle. The paths up the steep incline of Peel Hill from Fenella Beach to Corrin's Tower are now peppered with eight contemporary memorial benches, all in different styles. This contrasts with only three memorial benches on Peel promenade and two on the opposite side of the bay. Four of the Peel Hill benches are clustered near the bottom of the hill where there is easy access from the car park and where there is a tradition of benches as part of the tourist amenity of visual encounter with landscape dating back to the Victorian heyday of tourism (Manx National Heritage, archive photographs of St Patrick's Isle and Peel Castle). Most of the benches look to the castle and cathedral, with a few facing the River Neb and Peel town on the direction of local planners attempting to disperse the benches (Interview, Peel Town Clerk 2007). Their inscriptions memorialize four men, one couple, two women and one child. The only other memorials on the Isle proper are found on the harbour in and outside the Lifeboat House, which sits in the shadow of the castle. More recently, contemporary memorial plaques have been erected inside and three memorial benches outside the lifeboat house. Several of these memorials were donated by Edwin Waterworth, a Lancashire undertaker who visited the Isle of Man every year for the Tourist Trophy (TT) motorcycle races. He began by improving facilities for the lifeboat crew, building benches and other amenities in the rest room and then proceeded to memorialize his nearest and dearest in the vicinity over subsequent years (Interview, Paul Jones (Peel Lifeboat Coxswain) 2007). Memorial items include a brass clock inside the lifeboat house given in memory of his wife and two benches on the harbour side, one in memory

---

8 After this research was completed, one bench was permitted on St Patrick's Isle proper in 2008, the bench situated outside the castle wall, looking to the sea in the West. The Peel Heritage Trust (PHT) recorded in their 2006 annual general meeting: 'Thanks to the co-operation of Manx National Heritage, we are erecting a bench, behind the castle, in memory of Robert Forster. His work for the P.H.T., local history, charities, the castle and first Head of QE2 [a local secondary school] are well known and fondly remembered' (Peel Heritage Trust Reports 2006–7 at www.pht.iomwebs.net). These social and cultural roles are reiterated on the memorial plaque and are indicative of the rationale based on contribution to local heritage for the privileged location of this memorial bench.

of a female friend and finally one which records himself as 'Long time supporter and friend of Peel lifeboat'. The density of these memorials represents a particular form of naming and claiming of emotional identity with this micro-space, even if it was only visited once or twice a year, as well as the reciprocal relationship which can be established through the process of gifting (see Miller 1998). As Jones noted, 'Even my five year old son commented that his [Waterworth's] name is on every wall' (Ibid.).

All of the memorial benches were located outside the designated heritage area, but planning permission was treated loosely outside this heritage zone, and a blind eye has been turned to 'spontaneous' benches in the past, however, near saturation point has been recognized and the local authority has intervened in both the spacing and number of the benches (Interview, Peel Town Clerk 2007). In the UK some local authorities have banned memorial benches outside cemeteries because of the contested use of public space, perceptions of 'mawkish memorials', private ceremonies and attached non-ephemeral commemorative goods such as football t-shirts and wind chimes causing offence to members of the wider public (see the case of South Lanarkshire, *The Telegraph* 24 August 2007). Others have regulated and homogenized benches in order to maintain quality control, eradicating traces of local culture and individuality (e.g. Ardur Council in West Sussex which asserts discursive control through a limited choice of four types of bench and three types of plaques), supporting Petersson's (2005) argument that informal memorial practices or 'tactics' tend to be formalized over time. However, disputes over grave types and decoration *within* churchyards and cemeteries show that the growth of vernacular practices is increasingly making these contested spaces too (e.g. 'Row over photos on family grave' (expressandstar.com, posted 4/1/08; 'Cemetery bans 'socialist' headstone' *The Times*, 7 January 2006).

As benches represent by far the majority of contemporary memorials in the study site, their location and discursive meaning will be analyzed in more detail below.

The Memorial Bench as Vernacular Expression of Emotion, Identity and 'Continuing Bonds'

I argue here that these benches represent a loose form of vernacular memorial. The vernacular refers to the local, the indigenous, the homely, the domestic, the everyday, the informal, in contrast to high art, literature and formal liturgy (Oxford English Dictionary) and I am using the term in this wider sense rather than the more specific meaning of vernacular architecture. The focus here is on benches, but other contemporary vernacular memorials include online memorial pages, spontaneous roadside or mountain memorials, domestic shrines, tattoos, graffiti, t-shirts and bumper stickers. Memorials need to be placed in socio-economic and political context because they reflect the social status of the dead (Williamson and Bellamy 1987; Hallam and Hockey 2001; Kong 2001); benches are both financially affordable for most families or groups of friends, as well as being culturally accessible and meaningful. While none of the memorial benches studied are homemade, hand crafted or distinctly local in character (although one was sourced from the local forestry board), their vernacular character comes from the fact that they reflect popular culture and the discourses of everyday life: they are familiar and homely items, intended to be sat upon, to be a resting place, to be a meeting or picnic venue. It has been suggested that vernacular inscriptions diverge from professional memorials as a result of creativity and/ or disrespect for established practice and can have a 'naïve visual liveliness lacking in more formal memorials' (Thomson 2006, 1). The memorial benches will be analyzed in relation to the said and unsaid (see Harrison 2007), representational 'liveliness', alternative discourses and local character, and placed in the context of the locale's longer term historical status as sacred place/spiritual site and landscape of consolation.

As outlined above, memorials represent an attempt to mark a life and its impact on the lives of others. If, as Jones has argued, memory is 'clearly bound up with processes of place and emotional attachments to place' (Jones 2005, 213), landscapes of memorialization distil these processes and attachments. Francis et al suggest that 'Funerary and domestic landscapes generate complementary resources for emotional processes' (2005, 105); they also maintain that funerary and domestic spaces represent two spaces where there is no time limit to grief, which intimates a lack of social tolerance for grief beyond the immediate period of bereavement. The growth of memorials such as benches in public spaces suggests they constitute a mediating space between those funerary and domestic spaces, a Third Emotional Space, at a distance from both, but where ongoing experiences of bereavement can be located and negotiated in the medium term. The growing number of memorial benches can be seen as a reflection of the wider practice of using benches as memorials in the British Isles; and/or the specific relation of benches as places to physically stop and sit in landscapes of beauty and landscapes of reflection and re-creation/recreation. However, as Hockey et al (2005, 135) have argued, significant 'objects and spaces have their own agency' and are capable of animating the presence of the deceased: 'Past presence and present absence are condensed into the spatially located object … its materiality feeds memory' (Hallam and Hockey 2001, 85). Indeed the very form of the bench, as seat, implies not only resting, but also visiting and the maintenance of continuing bonds.

Textual inscription on benches are typically inscribed on a small plaque and is usually brief; messages have to be written in a form of shorthand, leaving hiatus to be bridged and depth and nuance of emotions to be guessed. Despite these limitations, several readings of the benches as 'text' are possible. Discursively benches show personal identity in terms of relationships to others (father, child etc.) e.g. 'In loving memory of Elsie Owen of Peel. A loving mother, grandmother and great-grandmother 1912–2000'. While the chosen identity markers for the deceased are explicit, these also implicitly refer to the (often changed) identity of the bereaved; for example, the adult child who no longer has a parent, the memorial representing at least in part a version of what Jones describes as 'the loss of past geographical selves' (2005, 217). As Walter (1999) has noted, storying the deceased tells us about who they were, but also who the *storymaker* is. The inscription for Elsie Owen also points to the significance of place identity. Attachment to place and associated implications of authenticity, belonging and entitlement to memorialization in this place are commonly expressed implicitly and explicitly on memorials. Whilst benches have been erected in memory of visitors whose entitlement to memorialization there was justified through testimony to strong place attachment (e.g. 'friend of Peel lifeboat'), these need to be set alongside the memorial texts which emphasize the 'authenticity' of local credentials. One bench memorializes a young boy as 'A child of the Island' and another states 'To the memory of my Father Captain John Edward Bell (Jack). Born in Peel 1899–1972'. The latter maximizes the identity characteristics of the deceased: 'my Father' (parental), Captain (occupation and social status), 'John Edward Bell' (full official name), 'Jack' (informal moniker), 'Born in Peel 1899–1972' (spatial and temporal fixing in the locality).

There is a greater presence of women than is typically found in public memorials, which historically have tended to memorialize the unexpected or untimely death in the public world, as in accident or war, giving public memorials a masculine bias (Kong 2001). Benches could be considered a domestic item, making them seem an appropriate form of memorial for someone such as Elsie Owen principally remembered for her motherly roles; her particular bench is also situated in a walled enclosure giving it a greater sense of sheltered domesticity. On the other hand, the two benches located overlooking the harbour are dedicated to fishermen (Bell and Reid), the positioning of their memorials apparently reflecting their working lives. The memorial bench dedicated to Reuben Reid, fisherman and local footballer who died in his thirties, is situated at the base of Peel Hill looking towards the harbour and reads: 'Donated by the fishing community and friends in fond memory of Reuben Reid', identifying the local collective impetus for this memorial. Despite tragic loss of life at sea in the past, there are no earlier visible memorials to fishermen, and only one recently erected civic collective memorial on the harbour side dedicated to all those from the Isle of Man lost at sea during peace and war (one plaque in Manx, one in English). In part this may reflect the fact that St Peter's church in Peel is known as the 'fisherman's church' and functions as the centre for rituals and memorialization; but also reflects a cultural antipathy to memorials, as the lifeboat coxswain noted 'they're not into that sort of thing' (Interview, Paul Jones 2007). Contemporary memorials and acknowledgement of the emotions of grief, could reflect the shrinking cohort of fishermen, which makes the loss of an individual more marked; it could reflect the position of an individual within the community; it could also reflect a wider construction of masculinity within that fishing community which allows for greater acknowledgement of loss and emotion than was culturally acceptable in the past. As one respondent noted, 'These benches seem to be a recent thing' (Interview, Mr Ledley Senior 2008).

Benches are liminal spaces, sites of mediation within the spatialities of bereavement in five key ways. Firstly, they mediate between the pain of loss and the pleasure of happy memories, often situating those memories as in the case of the bench plaque 'X loved this place'. Secondly, benches mediate between absence/ presence for the bereaved, with plaques sometimes inviting others to enjoy the place as they did. Others address the deceased directly, asserting continuing emotional bonds, as, for example, the bench on Peel Hill which reads 'We all love you and miss you so much. Family and Friends. *Traa da liooar*' (a popular Manx saying meaning 'time enough', 'in good time'). Still others are an invitation to remember the deceased, as in the case of the four year old boy memorialized on a bench on Peel Hill: 'A child of the Island beloved by family and friends. "Rest awhile and think of me"'. Thirdly, as the last plaque also testifies, benches mediate between private 'domestic' memory and public place, a public 'mapping' of private emotion, by offering the amenity of a seat, but in this case also simultaneously petitioning for an emotional contract on the part of the sitter to remember the deceased. Fourthly, within local cultures of memorial practice benches mediate between the ephemerality of flowers and the permanence of a stone monument. While funerary architecture creates 'libraries in stone' in which the beliefs and identities of past individuals and cultures are inscribed for future generations' (Worpole 2003, 11), the 'Third Emotional Space' of the vernacular memorial, in this case the bench, is rooted in and speaks to present emotions and needs, including what Walter (1996) describes as storying the deceased. Finally, they can mediate between sacred and secular space and practice, through inscription and/ or location. While some of the memorial benches studied express 'continuing bonds', none express explicit religious faith. However, the benches are located in and around an historic Christian site which is still used occasionally as a place of worship and which some local discourses refer to as a spiritually 'thin space … where that gap between heaven and earth is almost non-existent' (Interview Bishop Graeme Knowles 2007), places where the 'presence of the invisible and the spiritual in those places is almost palpable' (Silf 2005, 8). This lack of faith-based text on memorial benches may be taken as a sign of secularization, but needs to be set alongside other faith-context memorials. For example, none of the seven memorial benches in Hereford Cathedral's Chapter House garden have explicit statements of faith, rather the context is part of the wider 'text' of the memorial; its meaning is greater than the few words inscribed upon it. This seems pertinent to this case study where the nineteenth century working class emigrant Roman Catholic community have a 'hidden history' in the memorial practices on St Patrick's Isle and where anecdotal evidence indicates the use of the 'thin' spiritual space of keeils for contemporary informal rites, both Christian and non-Christian.

The variety of benches represent a degree of 'visual liveliness' characteristic of vernacular memorials, and there is a degree of textual liveliness expressed on the bench plaques, exemplified by Manx sayings and nicknames; however, despite greater textual freedom than allowed in many churchyards and cemeteries, the wording on the benches was still set within clear discursive parameters. By comparison the three memorial plaques on local authority benches on Peel promenade, all for men, seem to represent a greater concentration of 'vernacular liveliness'. All had strong local references ('The Peel Viking', 'A true Peel man who loved this town', and Harry Quirk … who used this bench as his Garden Seat'). One used Manx colloquial dialect greeting 'Alright yessir, rest awhile' and another, the Manx language: '*Sie Sheeshe*'[9] to represent the character or values of the deceased and reinforce their national identity.

The proliferation of memorial benches may be taken as a marker of any or all of several changes in memorialization practices: increased entitlement to and acceptance of public expressions of private emotions (see Clark 2006 on roadside memorials); a need for a 'spatial fix' for the bereaved, given the increased numbers of cremations and ash scattering (Hallam and Hockey 2001; Harting and Dunn 1998) which means 'we've lost the bit of standing by the grave and talking to the dead person' (Interview, Bishop Graeme Knowles 2007); or increased secularization whereby mourners seek an alternative landscape of consolation and 'performative utterances' outside religious spaces. As I have argued elsewhere (Maddrell, forthcoming), green cemeteries in the UK are in part presented as idealized landscapes, where in death the deceased attains through their final resting place the sort of environment which few could aspire to during their lifetime. In many ways benches and other memorials in beauty spots echo this discursive location of a loved one in an ideal setting that they previously accessed periodically and temporarily, but was nonetheless highly significant. One of the highest benches on Peel Hill reads 'Treasured memories of John Sydney Kelly 23-2-39–23-8-05. To know him was to love him'. John's widow, Patricia Kelly, explained the significance of the location of the bench in terms of their grounding in the local community, and the hill's panoptican-like view of a shared history:

'When I sit there I can see everything I want to see … Memories … I lived on the harbour side in Peel for twenty six years, in the hotel there. That's the area my husband and I spent our time – it's also the area of my family and forbears. I can see my mother's house, where she was born, my son's house, my brother's house ... the cathedral, the town, its beautiful up there … I can see the bench from the town too, from places I frequent … We used to take the dog for a walk up there on nice nights, we'd often just sit and look around, places look different from a different angle. We just used to like being up there … it was our choice of place … I think he would have approved …' Mrs Kelly commented further on the bench as a place for reflection and resolution: 'if I have problems when I go up [the hill to the bench], I don't have them when I come down' (Interview, Mrs Patricia Kelly, 2008).

While there does not appear to be evidence of an explicit link between the contemporary memorials studied and historic Christian practices such as those of

---

9   Roughly translates as 'down with badness', the deceased was renowned as a teetotal sailor (Interview, Roy Baker 2007).

the nineteenth century Roman Catholic community, the clustering of these mostly informal memorials around or near a place of historic faith practice is hard to ignore. There is a resonance between the many successive generations who have chosen this place as site of burial and memorialization and the group of contemporary memorials. The question Bishop Graeme Knowles posed: 'Why did the builders of the keeils very carefully go out of their way to build keeils ... in what have clearly always been stunningly beautiful sites' might equally be addressed to those erecting contemporary memorials: are the memorial benches there simply because there is a 'good view'? Knowles added 'There is a definite link between the natural environment and being willing to worship there, and wanting to worship there' (Interview 2007), indicating an interplay between aesthetics and spiritual experience of landscape (note nature is commonly defined as sacred (Milton 2005, 139)). However, the view is equally good from the footpath on the opposite side of the bay, and yet there is a definite clustering of memorial benches on Peel Hill. Although this could be seen as an example of a performative (memorial) 'themed space' (see Lukas 2007), I suggest there is a deeper significance to the locale associated with both its pre-Christian and Christian status, but given fresh meaning through contemporary expression. Dewsbury (2003) describes witnessing space as 'knowledge without contemplation' and while there is evidence that for some there is explicit deep contemplation of place and its meaning within both expressions of faith and/or memorialization on and around St Patrick's Isle, there is also a sense in which people can have a deep-seated implicit understanding of the locale's historically grounded meaning. This may be seen in the continuities of burial and memorial in this place since the Bronze Age, and its deep-seated (if often downplayed) significance as a place of Christian worship and 'thin' Celtic spiritual space. Even the folklore of the ghostly black dog (the *Moddey Dhoo*) haunting Peel castle is indicative of the liminal status of the whole locale, underscored by its topography, set apart form the town, with access in the past controlled by tides. While the recent phase of memorialization coincides with a cultural trend to greater public emotional expression, it has also followed the 1980s archaeological digs which revealed the extent to which the isle had been successively used for burial: as one local visitor to the dig attested 'You could be digging up my ancestors' (Interview Andrew Johnson, 2007). It is this implicit historical, emotional and/or spiritual understanding of the nature of this place, intertwined with aesthetics and personal identity which attracts memorial-makers. This reciprocal relationship between individual identity and material place has been described as 'constitutive coingredience' (Casey 2001, cited by Anderson 2004, 255), whereby 'Time alongside practice sediments meaning onto places, with personal memories meshing with cultural meanings on an individual and (potentially) societal scale ... As a consequence of the reciprocal relations between place, human identity and time, individuals engender meanings and significance for particular places' (Anderson 2004, 256). Dewsbury further suggests we experience 'the folded mix of our emotions, desires, and intuitions within the aura of places, the communication of things and spaces, and the spirit of events.

Such folds leave traces of presence that map out a world that we come to know without thinking' (2003, 1907). Whether as 'thin' spiritual place or site of sacred historical depth, favourite walk or view, St Patrick's Isle and Peel Hill represent an emotional landscape deeply rooted in local place meaning and belonging, associated aesthetics of landscape and links beyond the present material world.

Conclusion

It has been suggested that the dead have been under-integrated into Anglo-American society, and that societies, as well as bereaved individuals, have to find place/s for bereavement, although those places may change over time (Walter 1999). While Worpole (2003) described cemeteries and burial places as 'last landscapes', a growing trend to individualized memorials in other public spaces suggests firstly, a search for mediating and alternative landscapes of consolation that go some way to reintegrating the dead into contemporary society; and secondly a democratization of public memorialization which in the past has reflected social standing and benefaction (e.g. the memorial hospital or local dignitary's municipal statue) but has become more widespread in recent years through the creation of informal and medium term memorials such as benches. The clustering of memorial benches around St Patrick's Isle reflect the locale's historical, spiritual and emotional landscape which is rooted in its 'Politics and poetics of sacred place, identity and community' (Kong 2001, 212) in all its multifaceted hybrid character: ancient burial site; site of Christian foundation in the Isle of Man; military stronghold and workplace; simultaneously a place of communal and national identity and site of periodic resistance to religious control and conformity; iconic heritage monument; landscape as aesthetic ideal. It is a palimpsest of overlaid and transitional cultural and belief practices. To cite Kong again, 'Place is often multivalent and requires an acknowledgement of similar, fluctuating and conflicting investment of sacred and secular meanings in any one site' (Ibid.). Regular local Roman Catholic (and recent ecumenical) pilgrimage to the isle reiterates the sacred status of St Patrick's Isle within the local Christian tradition; the liminal character is acknowledged not only by these groups, but also by those who recognize a more generalized spirituality of the locale, and/or identify with the area as site of pre-Christian rites. The making of contemporary vernacular memorials represents an episodic continuity of practice from the Bronze Age through to nineteenth century Roman Catholic and Nonconformist minority burial and memorial practices. They also reflect the Celtic Christian life view in which the sacred and everyday are interwoven, and the strong place attachment common in expressions of Manx identity. Undoubtedly, were contemporary memorials more generally allowed on St Patrick's Isle proper, there would be those who would want to identify with its sacred history as well as the aesthetics of the place and for those who have sought to establish memorials on St Patrick's Isle, the constraints of heritage regulations mean that Peel Hill is the closest substitute site. Evidence also suggests that Peel Hill can be seen as both adjunct to and contra St Patrick's Isle's formal historic (and revived) sacred status. Those erecting memorial benches around St Patrick's Isle are not a homogenous group, but all implicitly or explicitly relate to the liminal character of the place and simultaneously create their own collective and micro-sacred-liminal spaces through their meaning-making memorials and the storying of their loved ones.

Acknowledgements

I would like to express my gratitude to the British Academy for funding research on St Patrick's Isle and the Isle of Whithorn in Galloway (SG-45177), and for co-funding, with the RGS-IBG HPGRG, my attendance at the Second International and Interdisciplinary Emotional Geographies Conference, University of Kingston 2006. My thanks to the organizers of the latter for an opportunity to air early thoughts on this work and to Carol Lipman and Jude Hill for inviting me to present a related paper at the RGS-IBG 2007 conference session on Folklore. I am grateful to all those who helped with my research on the Isle of Man, especially Roy Baker, Frank Cowin, Yvonne Cresswell, Andrew Johnson, Patricia Kelly, Bishop Graeme Knowles, Peter Ledley, Mr Ledley Senior, Paul Jones, and Wendy Thirkettle. I am also grateful to the book editors Laura Cameron, Joyce Davidson and Mick Smith, and to Yvonne Cresswell of Manx National Heritage, for comments on drafts of this chapter.

References

Anderson, J. (2004), 'Talking whilst walking; a geographical archaeology of knowledge', *Area* 36: 254–261.

Anderson, K. and S. Smith (2001) 'Editorial: emotional geographies', *Transactions of the Institute of British Geographers* 26: 7–10.

Blunt A. and Dowling, R. (2006), *Home* (London: Routledge).

Bondi L., J. Davidson and M. Smith (2005) 'Introduction: geography's "emotional turn"', in J. Davidson, L. Bondi and M. Smith (eds.), *Emotional Geographies* (Ashgate, Aldershot), 1–16.

Bonanno, B. (2006), 'Culture and continuing bonds: a prospective comparison of bereavement in the United States of America and the People's Republic of China', *Death Studies* 30: 303–324.

Clark, J. (2006), 'Authority from grief, presence and place in the making of roadside memorials', *Death Studies* 30: 579–599.

Davies, D.J. (2002), *Death, Ritual and Belief* (London: Continuum, 2nd edition).

Dempsey, W.S. (1957), *The Story of the Catholic Church in the Isle of Man* (Billinge: The Birchley Hall Press).

Dewsbury, D.J. (2003), 'Witnessing Space: "knowledge without contemplation"', *Environment and Planning A* 35, 1907–1932.

Dugdale D.S. (1998), *Manx Church Origin* (Llanerch: Felinfach).

Francis, D., L. Kellaher and G. Neophytou (2005), *The Secret Cemetery* (Oxford, Berg).

Freke, D. (1995), *Peel Castle* (Douglas, Manx National Heritage).

Freke, D. (ed.) (2002), *Excavations on St Patrick's Isle, Peel, Isle of Man 1982–88. Prehistoric, Viking, Medieval and Later* (Liverpool: Liverpool University Press).

Garrard, L.S. (1989), 'Additional examples of possible house charms in the Isle of Man', *Folklore* 100: 110–113.

Gough, P. (2000), 'From heroes' groves to parks of peace: landscapes of remembrance, protest and peace', *Landscape Research* 25: 2, 213–228.

Hallam, E. and J. Hockey (2001), *Death, Memory and Material Culture* (Oxford: Berg).

Harrison, P. (2007), '"How shall I say it …?" Relating the nonrelational', *Environment and Planning A* 39: 590–608.

Hockey, J., B. Penhale and Disable (2005), 'Environments of memory: home, space, later life and grief', in J. Davidson, L. Bondi and M. Smith (eds.), *Emotional Geographies* (Aldershot: Ashgate), 135–146.

Jackson, P. (1989), *Maps of Meaning* (London: Routledge).

Johnson, N. (1994), 'Cast in stone: monuments, geography and nationalism, *Environment and Planning D. Society and Space*' 13: 51–65.

Jones, O. (2005), 'An ecology of emotion, memory, self and landscape', in J. Davidson, L. Bondi and M. Smith (eds.), *Emotional Geographies* (Aldershot: Ashgate), 205–218.

Joyce, T. (1998), *Celtic Christianity. A Sacred Tradition of Vision and Hope* (Orbis: Maryknoll).

Kinvig, R.H. (1975), *The Isle of Man. A Social, Cultural and Political History* (Liverpool: Liverpool University Press).

Klass D., R. Silverman and S.L. Nickman (eds.) (1996), *Continuing Bonds. New Understandings of Grief* (London: Taylor and Francis).

Kong, L. (1999), 'Cemeteries and columbaria, memorials and mausoleums: narrative and interpretation in the study of deathscapes in geography', *Australian Geographical Studies* 37: 1–10.

Kong, L. (2001), 'Mapping "new" geographies of religion: politics and poetics in modernity', *Progress in Human Geography* 25: 211–233.

Laurier, E. and Philo, C. (2006), 'Possible geographies: a passing encounter in a café', *Area* 38, 353–363.

Lorimier, H. (2005), 'Cultural geography: the busyness of being "more-than-representational"', *Progress in Human Geography* 29: 83–94.

Lukas, S.A. (2007), 'The themed space. locating culture, nation and self', in A.S. Lukas (ed.), *The Themed Space. Locating Culture, Nation and Self* (Plymouth: Lexington Books), 1–22.

Milton, K. (2002), *Loving Nature. Towards an ecology of emotion* (London: Routledge).

Moore A.W. (1900), *A History of the Isle of Man* (London: T. Fisher Unwin).

Morris, S.M. and Thomas, C. (2005), 'Placing the dying body: emotional, situational and embodied factors in preferences for place of final care and death in cancer', in J. Davidson, L. Bondi and M. Smith (eds.), *Emotional Geographies* (Aldershot: Ashgate), 19–32.

Petersson, A. (2007), 'The production of a proper place of death', paper presented at the 7th Death, Dying and Disposal Conference, University of Bath.

Riches, G. and P. Dawson (1998), 'Lost children, living memories: the role of photographs in processes of grief and adjustment among bereaved parents', *Death Studies* 22: 121–40.

Rohr, R. (2002), 'Grieving as sacred space', *Sojourners Magazine*, 31:1, 20–24. Available online at: <www.sojo.net/index.cfm?action> Last accessed 13 September 2006.

Ryden, K. (1993), *Mapping the Invisible Landscape. Folklore, Writing, and the Sense of Place* (Iowa City: University of Iowa Press).

Scott Howard W. (2003), 'Landscapes of memorialisation', in I. Robertson and P. Richards (eds.), *Studying Cultural Landscapes* (London: Arnold), 67.

Silf, M. (2005), *Sacred Spaces. Stations on a Celtic Way* (Oxford: Lion).

Thomson, G. (2006), 'Tombstone lettering in Scotland and New England: an appreciation of a vernacular culture', *Mortality* 11: 1–30.

Trench, Jellicoe R. (2002), 'early christian and viking age sculptured monuments', in D. Freke (ed.), *Excavations on St Patrick's Isle, Peel, Isle of Man 1982–88. Prehistoric, Viking, Medieval and Later* (Liverpool: Liverpool University Press), 282–292.

Walter, T. (1996), 'A new model of grief: bereavement and biography', *Mortality* 1: 7–25.

Walter, T. (1999), *On Bereavement. The Culture of Grief* (Milton Keynes: Open University Press).

Williamson, T. and Bellamy, L. (1987), *Property and Landscape* (London: George Philip).

Worpole, K. (2003), *Last Landscapes. The Architecture of the Cemetery in the West* (London: Reaktion Books).

Internet sources

http://www.expressandstar.com/2008/01/04/row-over-photos-on-family-grave/
http://www.peelheritagetrust.net/wordocs/corrin-tower
http://www.timesonline.co.uk/tol/news/uk/article785802.ece

Archives

Peel Castle: papers and picture archive, Manx National Heritage Library

Interviews

Interview, Roy Baker, Curator, The Leece Museum, Peel, 2007.
Interview, Frank Cowin, Manx National Heritage Trustee and Manx Historian, 2007.
Personal Communication, Frank Cowin, 2008.
Interview, Andrew Johnson, Archaeologist, Manx National Heritage, 2007.
Interview, Mrs Patricia Kelly, 2008.
Interview, Mr Ledley (Senior), Deacon of St Patrick's RC Church, 2007.
Interview, Mr Ledley (Senior), 2008.
Interview, Mr Ledley, Town Clerk, Peel, 2007.
Interview, Paul Jones, Coxswain, Peel Lifeboat, 2007.
Interview, Graeme Knowles, Bishop of Sodor and Man, 2007.

Chapter 3

Historicizing Emotion: The Case of Freudian Hysteria and Aristotelian 'Purgation'[1]

R. Darren Gobert

Stage acting requires that actors make manifest what is initially immaterial, physically embodying characters that are incorporeal, existing only on the page. Indeed, the immateriality is doubled, since it is more accurate to say that actors physically embody their own imaginative conceptions of the characters they are charged to represent. These conceptions, of course, centrally include the emotions experienced by the characters: cast in the role, every actor conceives of Creon's anger or Phaedra's love differently, since he or she conceives of anger or love differently. These conceptions may not be, and indeed are usually not, consciously examined; they inhere in – and have been naturalized by – the cultural location of the actor in question. Thus, the material representations to which these conceptions give rise emanate from and reflect broader cultural understandings. The history of 'emotion' on stage therefore illuminates shifts in the history of 'emotions' more generally.

One thread of this stage history that has been of particular interest to me concerns interpretations of Aristotle's *katharsis*[2] clause in *Poetics* (1449b) – in which he claimed that tragedy is defined by its ability to induce emotional *katharsis* through the incitement of pity and fear[3] – since this claim has proven so vital to how actors and playwrights conceive of stage emotion (and, indeed, of emotion in general). While the question of what *katharsis* means (and, relatedly, how to translate the classical Greek word κάθαρσις) has been declared solved at various points in the histories of theatre and dramatic theory, the solutions are inevitably prone to debunking at later moments: the views of and translations by Aristotle's

---

1   For ease of reading, I have quoted English-language translations of all foreign texts. However, each quotation was scrutinized against the original; where appropriate or necessary, or where no published translation exists, this original is included.

2   A note about usage: like other critics, I use '*katharsis*' in its strictly Aristotelian sense and 'catharsis' in its more general senses, including that theorized by Freud and Josef Breuer.

3   In G.M.A. Grube's translation: 'Tragedy, then, is the imitation of a good action, which is complete and of a certain length, by means of language made pleasing for each part separately; it relies in its various elements not on narrative but on acting; through pity and fear it achieves the purgation (catharsis) of such emotions' (Aristotle 1989, 12).

earliest Italian Renaissance commentators receded as they were supplanted by others in a process that has continued uninterruptedly in Aristotelian commentary until the present day. Both solutions and debunkings are signposts, directing our attention to the axiomatic assumptions that undergird them. Commentators in different historical moments have not had the same thing in mind when they have written of the 'emotions' in general and 'pity' and 'fear' in particular, although they usually appear to have been unaware that their presuppositions were historically constituted: after all, the impulse to historicize at all is of very recent vintage. Examining these signposts helps us better to understand the landscape, the genealogy, of 'emotion' as an historical concept. I here consider one particular moment of this genealogy: the emergence, after the 1880 publication of classicist Jacob Bernays's *Zwei Abhandlungen über die aristotelische Theorie des Drama* [*Two Essays on the Aristotelian Theory of Drama*], of the notion of *katharsis* as a 'purgation' on the part of the theatrical spectator.

Bernays's medical view of *katharsis* emerges as dominant in Western scholarship, I argue, only and precisely because of its consonance with the simultaneously developing discourse of medical psychology and in particular psychoanalysis: Bernays's theory is directly supported by the model of emotion offered by his niece's husband, Sigmund Freud. In other words, if Bernays's theory came to be regarded as the 'solution' to the problem of Aristotelian *katharsis*, it did so because of assumptions – articulated by psychoanalysis but products of their culture more generally – that legitimated its cultural authority. These assumptions comprise a particularly spatialized understanding of emotion, one embodied by the Freudian hysteric whose contemplation leads Freud to his earliest theorizing of affect. Here, I read the paradigmatic hysteric – Josef Breuer's patient 'Anna O.' – in order to demonstrate the relationship between her particular somatic performance of emotion and the immaterial concepts that this performance embodies. Connecting Freud's 'hysteria' and Bernays's 'purgation', I historicize one key moment in the conceptual history of the emotions. This moment, I argue, has both ramifications and metatheoretical lessons for our theorizing about the emotions in this book.

Bernays on *Katharsis*

Even the most careful theatre historian may forget that knowing precisely what Aristotle himself 'meant' when he spoke of dramatic *katharsis* is an impossibility, perhaps especially because *katharsis* concerns emotions. We cannot adequately access either Aristotle's culture or the symbols (even its language is dead) with which his culture constructed emotions and disseminated emotional meanings; and the available evidence in Aristotle's corpus is too self-contradictory: Aristotle cross-refers his discussion of emotion in Book 8 of the *Politics* to the *Poetics*; in turn, Book 19 of the *Poetics* makes explicit reference to the 'rhetorical' emotions that he theorizes in some depth in Book 2 of the *Rhetoric*. Yet the seemingly protocognitivist understanding of emotions in the *Rhetoric* is irreconcilable with the discussion in the *Politics,* which considers emotions as noncognitive physiological perturbations.[4] That Aristotle also theorized the emotions in *Nichomachean Ethics,*[5] *On the Soul,*[6] *Parts of Animals* and *Movement of Animals*[7] complicates rather than clarifies matters.

In coming to terms with the elliptical *katharsis* clause in the *Poetics,* we have restricted and contradictory data. We know from Book 4 that all men take pleasure in imitative representations and that this emotion derives from the enjoyment of learning; this capacity, Aristotle claims, is inherent to man. We know from Books 6, 7 and 14 that the means to stir emotion in the spectator or auditor resides in the play's plot – although the play's spectacle does stir emotions to a lesser extent. We know from Book 13 that we feel pity for someone who does not merit his misfortune, and fear for the unfortunate who are like ourselves. And most famously, we know from Book 6 that the goal of tragedy is to effect, by means of these two emotions, *katharsis* – although how Aristotle means to use the term,[8] and whether the object of this *katharsis* is meant to be emotions more generally or 'pity and fear' specifically,[9] would both seem to be philologically unverifiable.

Theatre and classics students today are likely to be taught that the term *katharsis* best translates as 'clarification'; this translation of the word entered the Liddell-

---

4 On the recurrent claim that the *Rhetoric* betrays a cognitivist understanding of emotion in general and pity and fear in particular, see, for example, Fortenbaugh (1979), Lyons (1980, 33–35) and chapter six of Halliwell (1986). Belfiore (1992) dissents, seeing the account in the *Rhetoric* as internally self-contradictory, since it marries physiological and wholly cognitive elements (184, 187).

5 Aristotle has cause to theorize particular emotions and feelings at various points in the *Ethics*; on 'fear' in particular, see Aristotle (1976, 127–30).

6 Aristotle spends much of Book I of *On the Soul* discussing the composition of emotion, which, he says – in a distinction that is keenly relevant here – is both 'matter' and 'form': 'the natural philosopher and the dialectician would give a different definition of each of the affections, for instance in answer to the question "What is anger?" For the dialectician will say that it is a desire for revenge or something like that, while the natural philosopher will say that it is a boiling of the blood and hot stuff about the heart. And one of these will be expounding the matter, the other the form and rationale' (1986, 129).

7 Aristotle's explicit physiological explanations of animal emotions in the *Parts of Animals* shed light on human emotional reactions (see, for example, 1984b, 1013). In *Movement of Animals*, he similarly explains the relationship between emotional responses and physiology (see, for example, 1984b, 1092–93).

8 White (1984) catalogues all instances of the Greek word *katharsis* and its derivatives in classical literature; Aristotle's entries total 156 (110–23). The extended discussion of the word and its cognates in Moulinier (1952) is still widely cited. Belfiore (1992) more narrowly surveys Aristotle's usage of the word (291–336); her total is 161 (292).

9 The debate over whether 'pity and fear' effects either the *katharsis* of '*these* emotions' or '*such* emotions' begins with the earliest Renaissance commentators and is without end. (Some influential commentators have gone further afield: Else [1957] famously offers 'those painful or fatal acts', for example [221] and Golden [1962] offers 'such incidents' [58].)

Scott-Jones *Greek-English Lexicon* for the first time in 1940, and the theory that
this was Aristotle's intended sense had become by the late 1980s the majoritarian
view, due to the influence of Stephen Halliwell's *Aristotle's Poetics* and Martha
Nussbaum's *The Fragility of Goodness* (both 1986) and, as I argue elsewhere, the
ascent of cognitivism in philosophical and psychological discourse (see Gobert
2006).[10] Older views of *katharsis*, such as 'purification', are similarly historically
traceable; as with 'clarification', if they became naturalized as 'what Aristotle
meant', the process was facilitated by the concurrent naturalization of a specific
understanding of emotion.

   Such is the case with the notion of *katharsis* as 'purgation'. This translation of
the Greek word, unlike 'clarification', goes back to Aristotle's early Renaissance
interpreters: Minturno, for one, used the idea of a medical purgation of illness
to understand Aristotelian *katharsis* as early as the mid-sixteenth century, in his
1559 *De Poetica* and 1564 *Arte Poetica*. In the latter text, Minturno writes that
'[a]s a physician eradicates, by means of poisonous medicine, the perfervid poison
of disease which affects the body, so tragedy purges the mind of its impetuous
perturbations by the force of these emotions beautifully expressed in verse'
(Spingarn 1924, 80). But such a narrowly medical view of *katharsis* failed to gain
cultural traction in the sixteenth century, since there was no basis in science to
understand how theatre could be conceived as, to quote the French classicist André
Dacier, 'truly a medicine'.[11]

   The sense of *katharsis* as 'purgation' that an earlier generation of Aristotle
readers was likely taught has a comparatively recent historical emergence:
Bernays's 1857 treatise *Grundzüge der verlorenen Abhandlung des Aristoteles
über die Wirkung der Tragödie* [*Foundations of Aristotle's Treatise on the Effect
of Tragedy*]. The question is: How did Bernays's theory become so popular in
only a few decades as to be considered 'standard', in the words of one scholar,
and to have 'almost universal assent', in the words of another? (Sparshott
1983, 15; Barnes 1979, viii). Bernays writes that Aristotle is 'explained' if we
'understand' *katharsis* in its medical sense (1979, 159). *Katharsis* can only be thus
understood, however, in an historical context that views unpurged emotions as a
medical problem with physiological consequences, in a culture with appropriately
resonant assumptions about emotion. Bernays's 1857 publication was well
regarded by fellow philologists, but it did not have such resonance, as suggested
by Bernays's defensive forestalling of his detractors: 'Let no one primly wrinkle
his nose and talk of a degradation of aesthetics to the rank of medicine' (1979,
159). However, when he republished the treatise as *Zwei Abhandlungen über die
aristotelische Theorie des Drama* in 1880, he was read and discussed outside

---

   10   The first theory of Aristotelian *katharsis* as clarification appears in Golden (1962),
although it is adumbrated earlier in the century (see Haupt 1915). For Halliwell's most
succinct articulation, see 1986, 200–201; for Nussbaum's, see 1986, 391.
   11   'la Tragedie est donc une veritable medecine, qui purge les passions ... Mais c'est
une medecine agreable, qui ne fait son effet que par le plaisir' (Dacier 1692, 83).

academic circles. Indeed, the book and its central concept achieved something like popular success, especially in Vienna.[12] Evidence of this success is provided by Bernays's contemporary, Willhelm Wetz, who grumbled in 1897: 'Bernays could have supplied us with ten times more insight into tragedy than his research on ... catharsis ... actually does: [but] who doubts that, in contrast to his hundred admirers, he would have found only one?'[13]

Wetz's point is that Bernays's popularity reflected a trendy fascination not with dramatic theory or with Aristotle but rather with medical psychology, the discourse that was then revolutionizing cultural understandings of the emotions and, therefore, the emotions themselves. For example, between the two Bernays publications (that is, between 1857 and 1880) August Ambroise Liébeault's *Du Sommeil et des états analogues* [*Induced Sleep and Analogous States*] (1866) and Daniel Hack Tuke's *Influence of the Mind upon the Body* (1872) had both become influential, hypnosis had been resuscitated as a medical practice and psychological healing had come to be associated with neurology clinics instead of asylums in Western European practice.[14] It is in this context – the medicalization of human psychology that in the late-nineteenth century took 'lively ferment', to quote Stanley Jackson (1999, 7) – that Bernays's success must be located.

Bernays situates his analysis of the *katharsis* clause in the *Poetics* against the more specific usage of the term in the *Politics*, in which Aristotle had stated that music confers many benefits on its listeners, among them 'cathartic purposes'. These, he says in an unkept promise, will be treated 'more fully in [his] work on *Poetics*' (1981, 473). Since Aristotle himself connects the *Politics* to the *Poetics*, Bernays augments one brief discussion with another. Aristotle's remarks in the *Politics* figure musical *katharsis* as a means of alleviating excitement:

> Any feeling which comes strongly to some souls exists in all others to a greater
> or less degree – pity and fear, for example, but also excitement. This is a kind
> of agitation by which some people are liable to be possessed; it may arise out
> of religious melodies, and in this case it is observable that when they have been
> listening to melodies that have an orgiastic effect on the soul they are restored as

---

12 See, for example, Swales (1998, 28) who notes the 'surge of interest' in catharsis, which became a 'very fashionable topic of discussion among the fin-de-siècle Viennese *haute bourgeoisie*'. Ellenberger (1970, 484) also notes the 'widespread interest' that followed, calling catharsis 'the current topic of conversation in Viennese salons'. Sulloway (1979, 56–7) notes that 'In Vienna, as elsewhere, this whole subject ... assumed for a time the proportions of a craze'.

13 'J. Bernays hätte uns in selbständiger Forschung zehnmal mehr Aufschlüsse über das Wesen des Tragischen verschaffen dürfen, als seine Deutung des Katharsisprocesses uns thatsächlich lieferte: wer zweifelt daran, daß er für hundert Bewunderer seiner meisterhaften Katharsisabhandlung im andern Falle höchstens einen gefunden hätte?' (Wetz 1897, 30).

14 On the discourse and practices of medical psychology in the 1880s, see Jackson (1999), especially chapters one and six. Jackson's examples are drawn from France, Austria, Germany, England, Holland and Switzerland.

if they had undergone a [*katharsis*]. Those who are given to feeling pity or fear or any other emotion must be affected in precisely this way, and so must other people too, to the extent that some such emotion comes upon each. To them all inevitably comes a sort of pleasant [*katharsis*] and relief (1981, 473–74).[15]

In light of Aristotle's emphasis on orgiastic 'feeling', which Bernays views as pathogenic, Bernays argues that it is wrong-headed to assume, as commentators since the Renaissance frequently had, that the goal of *katharsis* is principally, or even consequently, moral: 'Aristotle's primary example of catharsis, which is drawn from the Greek experience of ecstasy, is pathological; and it is that which leads him to consider the possibility of a similar cathartic treatment for all other emotions' (1979, 158). Thus, the problem of *katharsis* in the *Poetics* should be approached from the same standpoint that Aristotle adopts in the *Politics*, which, as Bernays succinctly puts it, is 'a *pathological* standpoint' (1979, 158).

Bernays assumes that emotions are 'pathological' and 'pathogenic' for Aristotle, noting that Aristotle's paradigm in the *Politics* derives from the 'realm of psychopathology' (1979, 158). His failure to historicize this understanding provides a case in point about the very phenomenon that I interrogate in this chapter: Bernays's unexamined, naturalized assumption is generally symptomatic of prevalent late-nineteenth-century European ideas about the emotions and their association with irrationality.[16] Since Bernays assumes that the emotions to be relieved by music or tragedy are pathological, it follows that their *katharsis* should be constituted by their removal from the organism. This removal brings healing in the same way that a doctor's care brings healing: 'ecstasy turns to calm', Bernays writes, 'as sickness turns to health through medical treatment ... Thus the puzzling piece of *emotional* pathology is explained: we can make sense of it if we compare it with a pathological *bodily* reaction' (1979, 159). The negative, 'orgiastic effect on the soul' (Aristotle 1981, 474) that Aristotle finds in men possessed by ecstasy can be relieved by carefully administering these same effects, by means of intoxicating melodies, so that the effects can swell, diminish and eventually pass, leaving the sufferer in an improved state. This homeopathic model of musical *katharsis* Bernays applies to the *Poetics*, in which, he claims, Aristotle intends to theorize tragedy as a means of therapeutically inducing, through scenes of suffering, the emotions of pity and fear with sufficient force that pity and fear in the spectators' systems will be expelled: '*katharsis* is a term transferred from the physical to the emotional sphere, and used of the sort of treatment of an oppressed person which seeks not to alter or to subjugate the oppressive element but to arouse it and to draw it out, and thus to achieve some sort of relief for the oppressed' (Bernays 1979, 160).

---

15   T. A. Sinclair's translation, which I quote, renders Aristotle's two usages of the word '*katharsis*' as 'curative and purifying treatment' and 'purgation'.

16   For this insight, I am grateful to Mick Smith.

Psychoanalytic Emotion

The purgation theory of *katharsis* emerges as dominant when it does precisely because of its consonance with the model of emotion provided by the concurrently developing discourse of medical psychology – a discourse whose dominant voice would be provided by Freud. There is a symmetry, in other words, between purgative *katharsis* and Freudian psychoanalysis as 'the first instrument for the scientific examination of the human mind', in James Strachey's phrase (1957, xvi). Indeed, there is a very clear symmetry: in theorizing psychotherapeutic 'catharsis' in his 1895 *Studien über Hysterie*, written with Josef Breuer, Freud never recognizes a debt to the *Poetics*, but he seems to have cribbed substantially from his wife's uncle's treatise on Aristotle.[17] His recurrent references to plays by Goethe, Schiller and Shakespeare may reflect significantly more than the doctor's celebrated fondness for literature (Breuer and Freud 1957, 87, 192, 206, 229, 245 n.2, 250). Tragedy is never displaced from the heart of the theory of catharsis, even in medical practice – an explanation, perhaps, for Pedro Laín Entralgo's suggestive claim that in addressing the emotions '[t]he psychotherapist ... turns out Aristotelian rhetoric without knowing it' (1970, 181).

Tragic – or, at any rate, traumatic – events are central to Freud's earliest theoretical work on the nature and structure of emotion. This work takes place alongside his development with Breuer of the 'cathartic method', although Freud's most explicit formulations of emotion occur in his later writings,[18] after he had repudiated his mentor. Freud writes in his *Introductory Lectures on Psycho-Analysis* that emotions or affects are comprised of physiological changes together with the subject's registering of these changes in feeling: '[a]n affect includes in the first place particular motor innervations or discharges and secondly certain feelings; the latter are of two kinds – perceptions of the motor actions that have occurred and the direct feelings of pleasure and unpleasure which, as we say, give the affect its keynote' (1963, 395). Freud continues his definition by explaining: 'But I do not think that with this enumeration we have arrived at the essence of an affect. We seem to see deeper in the case of some affects and to recognize that

---

17   Bernays died in 1881, after Freud's engagement to Martha but before their wedding. Freud and Breuer's debt to Bernays has gone largely unrecognized by critics: for exceptions see Swales (1998, 28), Mitchell-Boyask (1994, 28–29) and Jean Bollack (1998, 54). There is surprisingly little discussion of the relationship between the two men, which Hirschmüller (1989, 157) characterizes as 'special'. He also speculates about Breuer's familiarity with Bernays's work, reminding us of the coincidence that Bernays died during Anna O.'s treatment, and of the many obituaries that popularly circulated at the time (Hirschmüller 1989, 156–57).

18   Hence Breuer's note in the *Studies* that '[n]o attempt will be made here to formulate either a psychology or a physiology of the affects' (1957, 201). Freud's clearest articulations of the structure of emotion can be found in the 'Anxiety' chapter of his *Introductory Lectures on Psycho-Analysis*; while many elements of psychoanalysis had by then radically evolved, Freud's conception of an affect had not.

the core which holds the combination we have described together is the repetition of some particular significant experience. This experience could only be a very early impression of a very general nature' (1963, 395–96). Freud thus locates the cause – the core, the essence – of the emotional response not in the context of the perceiver's immediate experience (say, witnessing a sad event) but in an earlier, unconscious experience, in a forgotten, even inherited, memory. (While Freud tends to focus on repressed memories from an individual's experience, he presumes that certain memories precede birth, a presumption that, Strachey notes, is possibly based on Charles Darwin's explanation of the emotions as evolutionary relics [Freud 1963, 396 n.1]. For example, Freud mentions Darwin's account of tail-wagging in dogs as an analogue for emotional reactions like screaming in his narrative of Emmy von N. [Breuer and Freud 1957, 91].) Elsewhere, in *Inhibitions, Symptoms, and Anxiety*, Freud reiterates this model by describing emotional states as having 'become incorporated in the mind as precipitates of primaeval traumatic experiences, and when a similar situation occurs they are revived' (1959, 93). Therefore, as William Lyons summarizes well in his important book *Emotion*, '[t]he actual emotion is a resurrection of the original traumatic emotional state triggered by some present event which stirs that memory' (1980, 26).

Emotions in the Freudian view are unavoidable, as many of them stem from traumas inherited or inevitable, like the primal scene. Secondly, emotions are healthy, since they serve as a safety valve through which people discharge unconscious drives. The failure to perform this discharge emotionally – a performance that Freud and Breuer term 'abreaction' – is in fact psychically and, therefore, physically unhealthy and potentially calamitous: failure to abreact causes neurosis, a particularly acute form of which is hysteria. Freud theorizes the structure of emotion inductively in light of his theory of hysteria, hence his formulation in the *Introductory Lectures* that 'an affective state would be constructed in the same way as a hysterical attack and, like it, would be the precipitate of a reminiscence' (1963, 396). Neurotic people whose blocks prevent abreaction will suffer from potentially severe physical symptoms; these symptoms can be alleviated only by removing the block and purging the emotion through the cathartic method. Psychoanalytic catharsis, therefore, aims to purge specific emotions that are impeding the psychic life of the analysand because of their failure to leave the unconscious; such emotions are said to be 'inadmissible to consciousness' or *bewusstseinsunfähig* (Breuer and Freud 1957, 225).

Despite Strachey's plaint that the term 'leaves much to be desired' because of its ambiguity (Breuer and Freud 1957, 225 n.1), the multivalency is richly productive: constructed out of *Bewusstsein* ('consciousness') and *unfähig* ('unable' or 'incompetent'), the coinage recalls its analogue *hoffähig* ('admissible to Court'). Breuer's neologism thus stresses both the imperial nature of consciousness, which determines admissibility, and the incapacity of the emotion, which lacks competence; it also makes clear the spatial terms in which he and Freud conceive of the emotions. In neurosis the inadmissible/incompetent emotion remains – like the originating memory in which it is tied up – in a repressed state, and in its failure to be abreacted properly, the emotion is made manifest only in strange symptoms of physical illness, bodily manifestations of suppressed memory. The analyst's task is to read the symbols of suppressed memory (which occur in the case studies primarily as bodily perturbations) and help ease the memory into consciousness.

Anna O.

The first of Breuer and Freud's hysterics, Anna O. (Bertha Pappenheim[19]), fell sick in Vienna in 1880, the same year that Bernays re-published his commentary on Aristotle. A 'markedly intelligent' young woman of twenty-one, possessed of 'great poetic and imaginative gifts' and guided by a 'powerful intellect' and 'sharp and critical common sense' (Breuer and Freud 1957, 21), Anna O. suffers symptoms that include diplopia (seeing double), a severe squint, hallucinations, episodic muscular paralyses, somnambulism and a strange aphasia that eventually culminates – shortly after her father's unfortunately timed death in April 1881 – in a complete inability to speak or understand her native language, German (Breuer and Freud 1957, 22–25).

Most strikingly, Anna O. comes during the time of her illness to suffer from the condition that figures in the *Studies on Hysteria* as the paradigmatic hysterical symptom: the dissociation that Breuer and Freud call 'splitting' or, using French, '*double conscience*', wherein the sufferer experiences two different ontological states or states of consciousness, one 'normal' and the other 'hallucinatory' – a '*condition seconde*'. (The term *condition seconde* is borrowed from the clinical work of Jean-Marie Charcot, a French neurologist to whose work on hysteria Breuer and Freud are indebted; as Jackson notes, Charcot is another important contributor to, and product of, the medicalization of psychology in the 1880s [1999, 84–85, 250–55].) The two states exist side by side, an ontological analogue for Anna O.'s diplopia: 'being double' in addition to 'seeing double'. Of her double consciousness, Breuer notes that 'though her two states were thus sharply separated, not only did the secondary state intrude into the first one, but – and this was at all events frequently true, and even when she was in a very bad condition – a clear-sighted and calm observer sat, as she put it, in a corner of her brain and looked on at all the mad business' (1957, 46). This dissociative state, Breuer tells us, is structurally similar and possibly causally related to her lifelong tendency to day-dream: in both cases, Anna O. indulges in what she calls, instructively, her 'private theatre' (1957, 41).

In spite of her intelligence and remarkable capacity for self-insight, Anna O. is ignorant both of the precipitating event of her bodily perturbations and the causal connection between them, since the traumatic event that has occasioned the

---

19 Pappenheim would later become as famous for her pioneering feminism and social work as she is infamous for being the first 'hysteric'. Melinda Given Guttman's biography *The Enigma of Anna O.* (2001) is the most extensive.

blocked emotion is similarly repressed from consciousness. As Breuer and Freud put it, 'the nature of the trauma exclude[s] a reaction' (1957, 10). The physical symptoms of hysteria are the emergent effects of the forgotten memory, which 'acts like a foreign body which long after its entry must continue to be regarded as an agent that is still at work' (Breuer and Freud 1957, 6). In this sense, hysterical trauma is unlike our universal, unconscious primal experiences described in the *Introductory Lectures*, which are also repressed but whose generated emotions are allowed to be expressed. Catharsis thus involves not the purging of an inherited affect-source but rather the purging of a block incurred in lived experience.

Since the psychoanalytic model of emotion links all emotion to memory, it is unsurprising that psychoanalytic catharsis yokes cathartic purgation to remembering. Once the memory has been retrieved and the emotion thereby abreacted, the illness ceases. As Breuer and Freud note, 'each individual hysterical symptom immediately and permanently disappeared when we had succeeded in bringing clearly to light the memory of the event by which it was provoked and in arousing its accompanying affect, and when the patient had described that event in the greatest possible detail and had put the affect into words' (1957, 6). The necessity of putting the 'affect into words' highlights two important features of Breuer and Freud's cathartic method. First, since the unearthing of the trauma happens through the questioning of a trained physician, the analysand is compelled to explore his or her[20] unconscious memories through language – a compulsion that Jacques Lacan would later seize upon in his revision of Freud, which denies the possibility of experience outside language.[21] In Breuer's succinct formulation, '[t]elling things is a relief' (1957, 211). This formulation lies at the foundation of the psychoanalytic method, which emerges out of his visits with Anna O.:

> The stories were always sad and some of them very charming ... As a rule their starting-point or central situation was of a girl anxiously sitting by a sick-bed. But she also built up her stories on quite other topics. – A few moments after she had finished her narrative, she would wake up, obviously calmed down ... If for any reason she was unable to tell me the story during her evening hypnosis she

---

20   Usually 'her', of course – in keeping with an association between woman and emotion that has been particularly pronounced in the West since the seventeenth century. While the name 'hysteria' (from the Greek word for womb) etymologically ties the disorder to woman (Veith 1965, ix), the possibility of male hysteria does arise incidentally in the nearly 4,000-year history of the disease. A cultural history is provided by Ilza Veith's *Hysteria: The History of a Disease*.

21   Lacan logically extends Freud by noting that language structures all human existence and experience. It is thus misguided to assume that any analysand can have an experience (after his or her infant 'mirror stage') that exists outside of language, or exclusively in the realm of biology, as Freud and Breuer assume. Hence Lacan's dictum that the unconscious is structured like a language: 'l'inconscient est structuré *comme* une langage. Je dis *comme* pour ne pas dire, j'y reviens toujours, que l'inconscient est structuré *par* un langage' (1975, 46–47).

failed to calm down afterwards, and on the following day she had to tell me two stories in order for this to happen (Breuer and Freud 1957, 29).

Thus, psychoanalytic catharsis is the 'talking cure', a term of Anna O.'s own that Breuer and Freud retained (Breuer and Freud 1957, 30).[22]

It is therefore telling that, in Anna O.'s case, the deterioration of her health can be directly mapped onto her faltering ability to communicate: word retrieval problems soon lead to an inability to conjugate verbs ('she used only infinitives', Breuer notes [1957, 25]) and a disregard for articles. At her worst she is virtually unintelligible, 'almost completely deprived of words ... put[ting] them together laboriously out of four or five languages' (Breuer and Freud 1957, 25). Eventually, she loses even her ability to understand German, which leaves her unable to communicate with her nurse; strangely, she regains expressive and receptive language skills in English, so that among her caregivers Breuer alone can communicate with her.

Conversely, in recovery Anna O. regains her linguistic abilities and can continue her talking cure. But in her seemingly endless capacity for producing 'imaginative products' lies Breuer's strongest evidence of her illness (1957, 30). Anna's inability to end her stories testifies to her emotional block: she cannot have closure, either narratively or emotionally. Just as *katharsis* in Aristotle's *Poetics* is linked to the unravelling of the action, and thus *katharsis* effects (and is an effect of) the plot's dénouement, the cathartic method brings resolution to the pathogenic emotion by purging or exhausting it through language. Thus, there is a connection between well-abreacted trauma and well-plotted narrative, a connection usefully underscored by Breuer's assertion that Goethe dealt with distressing affects by discharging them in literary creation (1957, 206–07). Freud would similarly point out in *The Interpretation of Dreams* (reliably or not) that *Hamlet* was written 'immediately after' the death of Shakespeare's father, under the 'immediate impact of his bereavement' (1958, 265).

In order to help her achieve creative closure, then, Breuer guides Anna O. as she mines her history for forgotten traumas; in accordance with the etiology of hysteria, he reads the manifest, mnemic symbols of her hysteria and prompts her to give voice to latent narratives. For example, Anna's hydrophobia (which prevents her from drinking water even during the particularly hot summer of 1880) is connected to a memory of seeing a little dog drink out of a glass – a 'disgusting' sight that her politeness initially prevented her from expressing and which she unwittingly forced into unconsciousness, to her later detriment (Breuer and Freud 1957, 34). But once she remembers and narrates the event under hypnosis, the

22   Indeed, Anna O.'s coinage 'talking cure' – as well as her insistence on 'chimney-sweeping' (Breuer and Freud 1957, 30) by telling stories to relieve herself – led to Ernest Jones's declaration that she was 'the real discoverer of the cathartic method' (1953, 223 n. *b*).

disgust is released along with the memory, and 'thereupon the ... [hydrophobia] vanished, never to return' (Breuer and Freud 1957, 35).

Since 'every sense-perception calls back into consciousness any other sense-perception that appeared originally at the same time', Breuer traces strands of her hysteria back to their originating traumas and cures Anna O. one symptom at a time (1957, 208). His goal is to arrive at the most important causal factor (1957, 35). In Anna O.'s case, Breuer notes in his chronicle of her treatment that she once had a 'waking dream' in which she:

> saw a black snake coming towards the sick man [her father, before his death] from the wall to bite him. ... She tried to keep the snake off, but it was as though she was paralysed. Her right arm, over the back of the chair, had gone to sleep and had become anaesthetic and paretic; and when she looked at it the fingers turned into little snakes with death's heads ... When the snake vanished, in her terror she tried to pray. But language failed her: she could find no tongue in which to speak, till at last she thought of some children's verses in English and then found herself able to think and pray in that language (1957, 38–39).

This episode, particularly rich in mnemic symbols, typifies the patient's hysteria: in the private theatre of her waking dream – experiencing what Breuer and Freud elsewhere describe as 'the hallucinatory reproduction of a memory' (1957, 14) – she suffers physiological symptoms (a hallucination and paralysis) that signify an emotional block, which is in turn reflected in compromised language skills. On the last day of her cathartic method, Breuer re-stages this scene by 're-arranging the room so as to resemble her father's sick-room' (1957, 40). In it, Anna is able to 'reproduc[e] the terrifying hallucination ... which constituted the root of her whole illness. During the original scene she had only been able to think and pray in English; but immediately after its reproduction she was able to speak German. She was moreover free from the innumerable disturbances which she had previously exhibited' (Breuer and Freud 1957, 40). Thus, the most serious plot-strand of her narratively complex hysteria is resolved: its originating emotional moment at her father's bedside is revisited, and the attendant blocked emotions are released in (and their release is signalled by) their unproblematic narration in German. Anna O. thereafter leaves Vienna and eventually 'regain[s] her mental balance entirely' and 'enjoy[s] complete health' (Breuer and Freud 1957, 41).[23]

23　Freud would later summarize elements of Anna O.'s story in his *Five Lectures on Psycho-Analysis*, in which he describes her traumatic emotions as 'unusual somatic innervations' and 'physical symptoms of the case' (1957, 18). In this telling, Freud radically de-emphasizes Breuer's role in the process – he has Anna O. 'reproduc[ing] ... scenes' unaided – which reflects the breakdown of his and Breuer's relationship. Details of Freud's later representations of Breuer's relationship with Anna O. are provided by Forrester and Cameron (1999).

As Breuer describes them, Anna O.'s aphasic incapacity to create language and, even more strikingly, her subsequent inability to end her stories highlight neurotic hysteria's resemblance to a dramaturgical problem; her story thus again adumbrates the *Studies'* debt to Aristotle's *Poetics*, which links emotional *katharsis* to plotting. One of Breuer's verbs, *tragieren* ('to play tragically' or 'to tragedize'), clearly evokes the theatrical nature of medical catharsis; Breuer assumes that Anna O.'s emotional experiences and narration are shaped by concerns of aesthetics, narrative structure or genre.[24] Indeed, Breuer's devotion to Aristotelian closure – Book 7 of the *Poetics* stresses the completeness of the plot and its organically connected beginning, middle and end (Aristotle 1989, 15–16) – may partially explain if not excuse the tidiness of Anna O.'s chapter in the *Studies*: the heroine of Breuer's narrative leaves Vienna to enjoy perfect health after her abreactive closing monologue, while Bertha Pappenheim needed to be institutionalized after Breuer's treatment (Jones 1953, 225; Borch-Jacobsen 1996, 21).

Moreover, that Anna O.'s recovery is enabled by Breuer's *mise en scène* – he mimetically reproduces the scene of her father's sickroom – highlights the important relationship in which his and Freud's cathartic method situates spectacle and sufferer, as well as the crucially gendered relationship between patient and analyst, which has been helpfully highlighted by feminist critics.[25] Breuer exploits Anna O.'s *condition seconde* – her capacity for 'being double' – by positioning her both as an actor and as a spectator to a show with which she is made imaginatively to engage. He thus facilitates the abreaction of her repressed emotion in catharsis. Later Freud would usefully mobilize this insight to explain the function of all tragedy: in 'Psychopathic Characters on the Stage', he argues that tragic plots are limited only by the 'neurotic instability of the public' (1953, 310). Echoing Aristotle, who assesses potential dramatic actions in the *Poetics* by applying the test of probability, Freud suggests that the best plots will resonate with the probable emotional blocks of audiences (1953, 308). Staging these plots, the stage director acts as an analyst, alleviating repression and thereby bringing pleasure. Freud therefore answers a vexing paradox in the *Poetics* – why audiences experience pleasure by witnessing fearful or pitiable events – with a gloss on Aristotle. Clearly recapitulating his wife's uncle's understanding of *katharsis*, Freud writes: 'If, as has been assumed since the time of Aristotle, the purpose of drama is to arouse

---

24 Breuer: 'Da sie, diese Dinge [her hallucinations] durchlebend, sie theilweise sprechend tragirte, kannte die Umgebung meist den Inhalt dieser Hallucinationen' (1895, 20). ['Since she acted these things through as though she was experiencing them and in part put them into words, the people around her became aware to a great extent of the content of these hallucinations' (1957, 27).]

25 See, to cite only one example, Guttman, who remarks of Jean-Marie Charcot that he 'turned his clinic into a theatrical spectacle. Charcot did not allow his "hysterics" to speak, nor did he listen to them. To him the "hysteric" was an actress who, under Charcot's direction, played the role of "woman", a creature of frailty and despair, who permitted man to be the strong protector and the superb lion' (2001, 76).

"terror and pity" and so "to purge the emotions", we can describe that purpose in rather more detail by saying that it is a question of opening up sources of pleasure or enjoyment in our emotional life' (1953, 305).

Theorizing the Emotions

That Anna O. fell ill in December of 1880, just after Bernays's *Zwei Abhandlungen* was published, is a coincidence worth reiterating: Juan Dalma (1963) has also delineated the connection between Anna's 'cure' and the interest in catharsis spurred by Bernays's 1880 publication, a connection that Ellenberger supplements by intimating that Anna was herself familiar with Bernays's book (1970, 484; see also Sulloway 1979, 56–57). Pappenheim's biographer, Melinda Given Guttman, concurs (2001, 62). In his *Remembering Anna O.*, Mikkel Borch-Jacobsen surmises that Pappenheim may have faked her symptoms accordingly (1996, 85–86) and argues that Breuer described them selectively and creatively when he was reconstructing her case from his 1882 clinical notes for the 1895 *Studies* (1996, 52–61).

However, whether Pappenheim had read Bernays is finally irrelevant. In delineating this convergence, in demonstrating that Freud's cathartic method was deeply rooted in Aristotelian notions, in arguing that Freud fertilizes the cultural environment that allows Bernays's 'purgation' model to take root, I do not seek to 'debunk' Bernays, Breuer or this or that idea of Freud's. Rather, I seek to demonstrate how a new set of assumptions about emotions makes possible a new set of emotional experiences and expressions. A certain understanding of emotions undergoes historical ascent in such a way as to make a certain reading of Aristotelian *katharsis* both possible and intelligible. But so too does this understanding of emotion make Anna O.'s illness possible and intelligible. Like a stage actor, Anna O. unwittingly expresses in her emotional behaviour many of her culture's unexamined presumptions about emotions. To locate a moment of intelligibility and to mark it as different from another serves to throw into relief the inherent instability, transhistorically, of 'emotion' as a concept.

All feeling persons, like actors on a stage, reveal their presumptions about emotion in every emotional expression. Moreover, as agents of culture we play a role in the naturalizing of our culture's suppositions about emotion. The suppositions of 1880 or 1895 Vienna – many of them now attributed to Freud but more diffusely co-authored by his cultural context – have subsequently become naturalized in our cultural discourse: for example, the notion that the essence of emotions resides in the unconscious or in forgotten memories from early childhood experience, the notion of a relationship between emotional health and language that underlines all narrative medicine or the notion that we should not 'bottle up' our emotions. (Incidentally, for Freud the 'pleasure' of theatre resides precisely in its help in this regard: 'the prime factor', he wrote, 'is unquestionably the process of getting rid of one's own emotions by "blowing off steam" [*austoben*]; and the consequent enjoyment corresponds ... to the relief produced by a thorough discharge' [1953, 305].[26]) One can see such ideas recapitulated and reinforced in our culture's representations: in our theatre, on our television shows, in our therapists' offices, in our everyday language and experience.

These ideas, however, currently sit uneasily with newer cognitivist ideas: to cite my central example, the psychoanalytic understanding of emotion is difficult to reconcile with the now-standard reading of emotional *katharsis* as 'clarification'. The case of Freudian hysteria and Aristotelian purgation is offered, then, for its metatheoretical implications, since the case can only be seen clearly with hindsight, now that its axiomatic assumptions have begun to be historicized. To insist on the historicizing of commonsense ideas is merely to spot a danger when we are talking about the emotions, merely to highlight the difficulty of defining exactly what we are presupposing when we say this or that about 'emotional states' – and the difficulty, even impossibility, of not taking for granted our own precepts, those that inform our own emotional expressions. After all, if Bernays's critical machinery is inevitably limited by its cultural location, so too is ours. As Stanley Fish has articulated, critical analysis requires not only the 'demotion' of the cultural norms under scrutiny, but also an unexamined allegiance to new norms that will enable such analytic scrutiny. Fish makes an important – ultimately Hegelian – point, which we ignore at our peril: relativism is 'not a position one can occupy ... because no one can achieve the distance from his own beliefs and assumptions which would result in their being no more authoritative *for him* than the beliefs and assumptions held by others' (1999, 53).

Kenneth Bennett has criticized the ahistorical tendency of Aristotelian commentary, in which the *katharsis* question 'has become a window through which the critic can complacently view his own critical machinery at work' (1981, 207–208). A more profitable way of making his claim would be to acknowledge that any critical machinery brought to illuminate the problem must inevitably be limited by its cultural location. We can nonetheless marry our theorizing about emotion, first, to an awareness of the impossibility of historicizing the present moment and, second (and therefore), to what Paul Ricoeur calls a hermeneutics of suspicion. I invoke Ricoeur purposefully: like his, my interpretive approach also presumes that texts (in my case, performances) are themselves heuristic, providing a mode of learning reality. Like all cultural artefacts, they are always materially expressive (like the actor's embodiment of Creon's anger) and also mimetic – whether consciously or not, they are imitative of previous expressions, previous

---

26   In theorizing theatre in 'Psychopathic Characters', Freud here revisits a term (*austoben*) that Breuer had used to describe emotional discharge in the *Studies*: 'Mental pain discharges it [excitation] in difficult breathing and in an act of secretion: in sobs and tears. It is a matter of everyday experience that such reactions reduce excitement and allay it. As we have already remarked ordinary language expresses this in such phrases as "to cry oneself out", "to blow off steam", etc. What is being got rid of is nothing else than the increased cerebral excitation' (Breuer and Freud 1957, 201–202).

conceptions. In this sense, the complicated nexus of relationships from which an actor's emotion derives – between immaterial conceptions (page) and physical manifestation (stage) – suggests the equally complicated nexus of relationships from which 'real-life' emotion derives.

For this reason, an apparent slippage in my argument is meant to be theoretically productive: my conflation of the public stage and the 'private theatre' and of actors and Anna O., herself a literary representation not coextensive with Bertha Pappenheim. Like Breuer and Freud, who cite the 'mental process' of Lady Macbeth as evidence for their theory of neurosis (1957, 245 n.2), I finally presume that 'real-life' performances of emotions are not different in kind but deeply interrelated with 'staged' performances of emotion: indeed, it is richly suggestive that the once-aphasic Anna O. teaches herself to write again 'copying the alphabet from her edition of Shakespeare' (Breuer and Freud 1957, 26). I similarly presume that Breuer's narration of Anna O.'s case is no less significant than Pappenheim's illness for the historian of emotions who seeks to understand 'emotion' in 1880 or 1895 by looking at its representation. Pappenheim's emotions and Breuer's narration of them are both performative; both materially express the conceptions of the performer in question and his or her cultural context. Therefore, both Pappenheim's hysteria and Breuer's narrative are artefacts, signs, to be read – and to be historicized. In both cases, an agent or actor – both words, tellingly, share the root, *agere*, to do – manifests or embodies understandings.

In theorizing the emotions, we have as evidence to be read only such signs: emotional expressions, products of culture, whether words or tears. A hermeneutics of suspicion recognizes that the meaning of any such text is concealed. If this concealment is largely unwitting on the part of the text's author, it is partly because of the ahistoricity of the present that I described earlier. The various historical understandings of *katharsis*, and especially Bernays's now-displaced theory of 'purgation', provide crucial object lessons in this ahistoricity for theorists of the emotion like ourselves. This is not to suggest, of course, that the task's theoretical impossibility means that we must abandon hope of making stable claims about the emotions. After all, the word 'emotions' (like 'affects', 'feelings', 'passions') is itself a laden cultural product, and we nonetheless adhere to a necessary fiction that the word can neutrally signify a class of responses, of phenomena across cultures, that bear a family resemblance to one another. But as Clifford Geertz puts it in *Interpretation of Cultures*, we need to aim not at 'discovering the Continent of Meaning and mapping out its bodiless landscape'; rather, we should restrict ourselves to 'guessing at meanings, assessing the guesses, and drawing explanatory conclusions from the better guesses' (1973, 20).

In this chapter and my work more generally, I view emotions as emergent products of culture: constructed in and by social contexts, neither transhistorical nor transcultural. This is not to suggest that there are no important biological factors involved in emotion. However, since both our understanding and experience of emotions are inexorably shaped by culture in general and discourse in particular, it becomes impossible to separate physiological factors from the linguistic tools we use to understand and express those factors: we are returned, perpetually, to the heuristic function of our own experiences, our own representations. Therefore, I follow the work of recent cultural anthropology[27] in imagining that emotions cannot be ontologically prior to the cultural beliefs that give rise to them and the cultural behaviours – such as crying or acting – that express them. To presume there are culturally specific emotions that are constituted within and by particular cultures helps to undermine the universalist and essentialist views that have dominated so much of the concept's history, steering attention to the more profitable notion that emotions are, in the words of Michelle Rosaldo, '*embodied* thoughts' expressed in the behaviour of cultural agents (1984, 143; her emphasis). Such a presumption helps us to understand the role played by an agent's cultural context: to see more clearly why Bertha Pappenheim's emotional performance may have unfolded as it did, and why Anna O.'s emotional performance unfolds as it does in Breuer's narration. Importantly, however, the presumption also implicates each of us as theorists of emotion, who reflect and shape our cultural understandings with each word that we write. After all, the views of emotion that predominate in any given moment require and are constituted by their cultural recapitulations.

Acknowledgements

I thank Ross Arthur for his aid with the classical Greek texts. Uncredited translations are my own. For their research assistance, I gratefully acknowledge Nemanja Protic and Katie Fry.

References

Aristotle (1976), *Ethics* [*Nichomachean Ethics*], trans. J. A. K. Thomson (New York: Penguin).

— (1981), *Politics*, trans. T. A. Sinclair, Revised Edition (New York: Penguin).

— (1984a), *Movement of Animals*, trans. A. S. L. Farquharson, in Jonathan Barnes (ed.), *The Complete Works of Aristotle*, i (Princeton: Princeton University Press), 1087–98.

— (1984b), *Parts of Animals*, trans. W. Ogle, in Jonathan Barnes (ed.), *The Complete Works of Aristotle*, i (Princeton: Princeton University Press), 994–1086.

---

27 This work is much indebted to Geertz's methodology: for example, work on Utku Eskimos (Briggs 1970) or among the Ilongot of the Philippines (Rosaldo 1980). Studies more narrowly focused on culturally specific emotions include (most significantly) those of Catherine Lutz, whose work on the emotional life of the Ifaluk is seminal: see Lutz (1988, 1995).

— (1986), *De Anima (On the Soul)*, trans. Hugh Lawson-Tancred (New York: Penguin).

— (1989 [trans. originally published 1958]), *Poetics: On Poetry and Style*, trans. G. M. A. Grube (Indianapolis: Hackett).

— (1991), *The Art of Rhetoric*, trans. Hugh Lawson-Tancred (New York: Penguin).

Barnes, Jonathan et al. (eds.) (1979), *Articles on Aristotle*, iv (London: Duckworth).

Belfiore, Elizabeth S. (1992), *Tragic Pleasures: Aristotle on Plot and Emotion* (Princeton: Princeton University Press).

Bennett, Kenneth (1981), 'The Purging of Catharsis', *British Journal of Aesthetics* 21:3, 204–13.

Bernays, Jacob (1857), *Grundzüge der verlorenen Abhandlung des Aristoteles über die Wirkung der Tragödie* (Breslau).

— (1880), *Zwei Abhandlungen über die aristotelische Theorie des Drama* (Berlin).

— (1979), 'Aristotle on the Effect of Tragedy', trans. Jennifer and Jonathan Barnes, in Barnes et al. (eds.), *Articles on Aristotle*, iv (London: Duckworth), 154–65.

Bollack, Jean (1998), *Jacob Bernays: Un Homme entre deux mondes* (Paris: Presses Universitaires du Septentrion).

Borch-Jacobsen, Mikkel (1996), *Remembering Anna O*, trans. Kirby Olson (New York: Routledge).

Breuer, Joseph and Freud, Sigmund (1895), *Studien über Hysterie* (Leipzig).

— (1957), *Studies on Hysteria*, trans. and ed. James Strachey (New York: Basic).

Briggs, Jean L. (1970), *Never in Anger: Portrait of an Eskimo Family* (Cambridge: Harvard University Press).

Dacier, André (1692), *La poétique d'Aristote traduite en françois, avec des remarques* (Paris: C. Barbin).

Dalma, Juan (1963), 'La Catarsis en Aristoteles, Bernays y Freud', *Revista de Psiquiatría y Psicología Médica* 4, 253–69.

Ellenberger, Henri F. (1970), *The Discovery of the Unconscious* (New York: Basic).

Else, Gerald F. (1957), *Aristotle's* Poetics*: The Argument* (Cambridge: Harvard University Press).

Fish, Stanley (1999), 'Is There a Text in this Class?', in H. Aram Vesser (ed.), *The Stanley Fish Reader* (Oxford: Blackwell), 38–54.

Forrester, John and Laura Cameron (1999), '"Cure With A Defect": A Previously Unpublished Letter by Freud Concerning "Anna O"', *International Journal of Psychoanalysis* 80, Part 5, 929–942.

Fortenbaugh, William W. (1979), 'Aristotle's *Rhetoric* on Emotions' in Jonathan Barnes et al. (eds.), *Articles on Aristotle*, iv (London: Duckworth), 133–53.

Freud, Sigmund (1953), 'Psychopathic Characters on the Stage', trans. James Strachey, in James Strachey et al. (eds.), *The Standard Edition of the Complete Psychological Works*, vii (London: Hogarth), 305–10.

— (1957), *Five Lectures on Psycho-Analysis*, trans. James Strachey, in James Strachey et al. (eds.), *The Standard Edition of the Complete Psychological Works*, xi (London: Hogarth), 3–58.

— (1958), *The Interpretation of Dreams*, trans. James Strachey, in James Strachey et al. (eds.), *The Standard Edition of the Complete Psychological Works*, iv-v (London: Hogarth).

— (1959), *Inhibitions, Symptoms, and Anxiety*, trans. James Strachey, in James Strachey et al. (eds.), *The Standard Edition of the Complete Psychological Works*, xx (London: Hogarth), 77–178.

— (1963), *Introductory Lectures on Psycho-Analysis: Part III (General Theory of the Neuroses)*, trans. James Strachey, in James Strachey et al. (eds.), *The Standard Edition of the Complete Psychological Works*, xvi (London: Hogarth), 243–463.

Geertz, Clifford (1973), *The Interpretation of Cultures* (New York: Basic).

Gobert, R. Darren (2006), 'Cognitive Catharsis in *The Caucasian Chalk Circle*', *Modern Drama* 49:1, 12–41.

Golden, Leon (1962), 'Catharsis', *TAPA* 93, 51–59.

— (1973), 'The Purgation Theory of Catharsis,' *Journal of Aesthetics and Art Criticism* 31, 473–91.

Guttman, Melinda Given (2001), *The Enigma of Anna O.: A Biography of Bertha Pappenheim* (London: Moyer Bell).

Halliwell, Stephen (1986), *Aristotle's Poetics* (Chapel Hill: University of North Carolina Press).

Haupt, Stephan Odon (1915), *Wirkt die Tragödie auf das Gemüt oder den Verstand oder die Moralität der Zuschauer?* (Berlin: Leonard Simion).

Hirschmüller, Albrecht (1989), *The Life and Work of Josef Breuer: Physiology and Psychoanalysis* (New York: New York University Press).

Jackson, Stanley W. (1999), *Care of the Psyche: A History of Psychological Healing* (New Haven: Yale University Press).

Jones, Ernest (1953), *The Life and Work of Sigmund Freud*, i (New York: Basic).

Lacan, Jacques (1975), *Le Séminaire: Livre XX*, ed. Jacques-Alain Miller (Paris: Seuil).

Laín Entralgo, Pedro (1970), *The Therapy of the Word in Classical Antiquity*, trans. and eds. L. J. Rather and John M. Sharp (New Haven: Yale University Press).

Liddell, Henry George et al. (eds.) (1940), *A Greek-English Lexicon*, 9th Edition (Oxford: Clarendon).

Liébeault, August Ambroise (1866), *Du Sommeil et des états analogues* (Paris).

Lutz, Catherine (1988), *Unnatural Emotions: Everyday Sentiments on a Micronesian Atoll and Their Challenge to Western Theory* (Chicago: University of Chicago Press).

— (1995), 'Need, Nurturance, and the Emotions on a Pacific Atoll', in Joel Marks and Roger T. Ames (eds.), *Emotions in Asian Thought: A Dialogue in Comparative Philosophy* (Albany: SUNY Press), 235–52.

Lyons, William (1980), *Emotion* (Cambridge: Cambridge University Press).

Minturno, Antonio Sebastiano (1559), *De Poetica...libri sex* (Venice).

— (1564), *L'Arte Poetica* (Venice).

Mitchell-Boyask, Robin N (1994), 'Freud's Reading of Classical Literature and Classical Philology', in Sander L. Gilman et al. (eds.), *Reading Freud's Reading* (New York: New York University Press), 23–46.

Moulinier, Louis (1952), *Le Pur et l'impur dans la pensée des Grecs* (Paris: Klincksieck).

Nussbaum, Martha C. (1986), *The Fragility of Goodness* (Cambridge: Cambridge University Press).

Rosaldo, Michelle Z. (1980), *Knowledge and Passion: Ilongot Notions of Self and Social Life* (New York: Cambridge University Press).

— (1984), 'Toward an Anthropology of Self and Feeling', in Richard A. Shweder and Robert A. LeVine (eds.), *Culture Theory: Essays on Mind, Self, and Emotion* (New York: Cambridge University Press), 137–57.

Sparshott, Francis (1983), 'The Riddle of *Katharsis*', in Eleanor Cook et al. (eds.), *Centre and Labyrinth: Essays in Honour of Northrop Frye* (Toronto: University of Toronto Press), 14–37.

Spingarn, J. E. (1924), *A History of Literary Criticism in the Renaissance* (New York: Columbia University Press).

Strachey, James (1957), 'Introduction', in Josef Breuer and Sigmund Freud, *Studies on Hysteria* (New York: Basic), ix–xxviii.

Sulloway, Frank J. (1979), *Freud, Biologist of the Mind* (New York: Basic).

Swales, Peter J. (1998), 'Freud's Master Hysteric', in Frederick C. Crews (ed.), *Unauthorized Freud: Doubters Confront a Legend* (New York: Viking), 22–33.

Tuke, Daniel Hack (1872), *Influence of the Mind upon the Body* (London).

Veith, Ilza (1965), *Hysteria: The History of a Disease* (Chicago: University of Chicago Press).

Wetz, Willhelm (1897), *Shakespeare vom Standpunkte der vergleichenden Literaturgeschichte* (Hamburg).

White, Daniel R. (1984), 'A Sourcebook on the Catharsis Controversy' (Dissertation Florida State University).

PART 2

Understanding

Chapter 4

Understanding the Affective Spaces of Political Performance[1]

Nigel Thrift

Introduction

Affect is a term sprinkled through many recent writings. Though its addition to the analytical vocabulary of the humanities and social sciences is often criticized as contributing nothing much more than a mere cultural frisson or, even worse, a highly questionable means of choosing choice which omits, ignores or diminishes many of the negative and obvious links to the exercise of power (Hemmings 2005), that does not mean that affect can therefore be written off as just a passing intellectual fad. Why? Because affect merges two collections of analytical objects that have been conventionally kept apart – namely 'the social' and 'the biological' – and in so doing addresses real issues about our fundamental understandings of what constitutes the work of the world.

There is, it seems to me, a political diagnosis to be made. Most obviously, this is a time of great political passions on both right *and* left (Nelson 2006). More importantly, however, we are living in a time of greater and greater authoritarianism. But this is an authoritarian capitalism that relies on sentiment, media, and lack of attention and/or engagement to most political issues to hold sway, a capitalist socialism or, at least, a neo-authoritarian new deal whose main interest is in accelerating innovation. At this moment, the left has very little purchase on how to combat this post-liberal form, which privileges media (news) time and election time over historical time (Runciman 2006), not least because it has so little purchase on how this form is able to use alternative modes of affective intelligence to produce compelling political impressions.

Too often the left falls back on the orthodox politics of resentment, which has become an increasingly sterile political repertoire whose appeals to unity simply repeat the old terms of succession within a foreclosed 'radical' community intent on the pleasures of victimization (Amin and Thrift 2005). The only alternative to left moralism often seems to be a mystique of protest which can call forth 'a community of angry saints in which the fire of pure opposition burns', as Sloterdijk (in Funcke 2005) aptly puts it, which then provides, simply through its existence,

1   This chapter is extracted from *Non-Representational Theory: Space, Politics, Affect* by Nigel Thrift (2007), London: Routledge, with permission.

an apparent revolutionary justification. Interestingly, a good part of this community finds its practices confirmed through a kind of affective justification. However, this alternative seems close to the prevailing regime too in that it relies on an appeal which is too often simply an appeal to affective force as if that somehow validated the political analysis. In other words, we need to find other keys to organizing and enduring which can combat the motivational propensities now being diffused.

So, I am searching for another way of going on, a different kind of politicalness which has its roots in new intellectual-practical formations which have cried 'enough' to the usual knee-jerk left analyzes and are attempting to re-materialize democracy. Such formations do not rely either on a politics of resentment or on the kind of 'spiritualism' that too often emerges in its stead, and in their search for a political reanimation they take biology seriously as a key to thinking about the political as a part of a more general search for political forms that are adequate to current modes of being: forms of multi-naturalism rather than multiculturalism, if you like (Viveiros de Castro 2005). This search must be both an experimental and a vigilant activity. After all, and in opposition to those who simply want a changing same, 'a subversive political theory must reveal an empty place that can be filled by practical action. Any political theory worthy of the name must await the unexpected' (Virno 2006, 42).

This chapter is therefore in three main parts. In the first part, I outline the main ways in which affect has been defined before going on to address the kinds of thinking about biology that are currently found in the social sciences, broadly defined. These means of thinking about biology provide a platform for the second main part of the chapter which is concerned with understanding affective contagion: how it is that affect spreads and multiplies, most especially through imitation. Finally, there is a (very) brief conclusion, not least because so much remains to be done to bear out and amplify many of the statements I make. But, as the conclusion makes clear, that work has now started, mainly by peering through the lens of individual affects and the way that they intersect with politics.

Scenes of Effusion

Broadly speaking, affect is an attempt to avoid an easy sociologism or a psychologism or a biologism. Put most simply, it refers to complex, self-referential states of being, rather than to their cultural interpretation as emotions or to their identification as instinctual drives, although, to muddy the waters, it is clear that affect is not easily separated off from either emotion or drive and that a good part of the current confusion in the literature derives from the difficulty of making such easy dividing lines. But to short-circuit what is clearly a complex debate, we might say, in line with Griffiths (1997), that emotions are everyday understandings of affects, constructed by cultures over many centuries and with their own distinctive vocabulary and means of relating to others. In a sense, they are a kind of folkbiology, a set of continually emerging beliefs about efficacy. The central Western concept of emotion is unlikely to be of more use than this because it assumes that one process category underlies all human behaviours, and can somehow explain them. But there is no evidence to suggest that this is the case.

> There is no one process that underlies enough [human] behaviour to be identified with emotion. Emotion is like the category of 'superlunary' objects in ancient astronomy. There is a well-designed category of 'everything outside the moon' but it turns out that that superlunary objects do not have something specially in common that distinguishes them from other arbitrary collections of objects (Griffiths 1997, 21).

Drives, in contrast, arise out of basic biological functions, such as hunger, sex, aggression, fear and self-preservation. They are often viewed as the source of many affects but, unlike drives, affects are able to be transferred to a wide variety of objects in order to be satisfied.

What seems certain is that any consideration of affect has to involve merging 'the social' and 'the biological'. As Brennan (2004, 1) puts it:

> This is not especially surprising, as any enquiry into how one feels the others' affects, or the 'atmosphere', has to take account of physiology as well as the social, psychological factors that generated the atmosphere in the first place. The transmission of affect, whether it is grief, anxiety, or anger, is social or psychological in origin. But the transmission is also responsible for bodily changes, as in a whiff of the room's atmosphere, some longer lasting. In other words, the transmission of affect, if only for an instant, alters the biochemistry of the subject. The 'atmosphere' or the environment literally gets into the individual. Physically or biologically, something is present that was not there before, but it did not originate sui generis: it was not generated solely or sometimes even in part by the individual organism or its genes.

Notice here how Brennan does not assume that the transmission of affect is from individual to individual, contained within one skin and being moved to another like a falling set of dominoes. Rather, that transmission is a property of particular spaces soaked with one or a combination of affects to the point where space and affect are often coincident.

Thus, as I hope is clear in what follows, I will be following a broadly posthumanist agenda. I am not, on the whole, interested in individuals but rather in how particular hybrids attain and keep coherence, become bodies of influence, so to speak. Thus, my interest is in trying to answer questions like 'what would the study of affect look like if it did not focus on the subject and subjectivity?' 'how do political formations generate affect?' and 'to what extent is affect a political form in itself?' However, as I have made clear elsewhere, I would not want to take this agenda to its limits. I believe that singular bodies can make an inventive difference which is sometimes of a different order from other hybrids. In certain situations, these bodies can stand out of the crowd as monad-like nodes of performativity constructed by the mass before falling back into the mass, as certain individuals seem capable of achieving for longer or shorter periods (see Elliott and Lemert 2006). But, equally, all kinds of other bodies are possible. There are, in other words, no stable ways to be human being because "human" is not the name of a substance, in the Aristotelian sense of the term, but the name of a relation, of a certain position in relation to other possible positions' (Viveiros de Castro 2005, 155).

There are five widely accepted schools of thought about affect that populate modern social thought (Thrift 2004, 2007). The first of these is the 'affect program' theory, derived from a Darwinian interpretation of emotions which concentrates on short-term bodily responses which it is claimed are pan-cultural (that is, present in most human populations).[2]

> In its modern form, the affect program theory deals with a range of emotions corresponding very roughly to the occurrent instances of the English terms surprise, fear, anger, disgust, contempt, sadness, and joy. The affect programs are short-term stereotypical responses involving facial expression, autonomic nervous system arousal, and other elements. The same pattern of response occurs in all cultures and homologues are found in related species. These patterns are triggered by a cognitive system which is 'modular' in the sense that it does not freely exchange information with other cognitive processes. This system learns when to produce emotions by associating stimuli with broad, functional categories such as danger or loss. (Griffiths 1997, 23)

In other words, the affect program approach makes a case for the view that certain so-called lower order affects at least have some degree of cultural generality but are not therefore necessarily innate. It makes no such claim for so-called higher cognitive affects such as love, jealousy, guilt and envy.

The second school of thought on affect is that of William James and Carl Lange, often referred to as the James-Lange theory. This famous theory essentially argues that bodily responses give rise to affective states and is popularly rendered by statements such as 'crying makes us sad'. The primacy that James gave to bodily changes, following in part from Descartes' belief that emotions are passive perceptions of bodily motions, has become a crucial element of modern

---

2   This is a different claim from the one that these programs are part of a universal human nature which Despret (2004) and others have rightly criticized. Rather, the claim being made is the same as the claim that brown eyes are found in all or most human populations and are a product of human evolution but are not thereby part of a universal human nature: affect programs display similar heritable variation within populations (Griffiths, 1997). It would, in any case be difficult to argue that certain affects can be found in all human cultures when the ethnographic evidence suggests that such a viewpoint cannot be sustained. For example, anger does not seem to be universal: thus both Tahitians and Utku Eskimos appear not to possess this quality (though they recognise it well enough in other peoples).

experimental psychology, in that it makes affect a matter of visual and auditory observation and so focuses attention on physiological change. The theory is undergoing something of a revival through the work of Antonio Damasio as the notion that emotion-feeling is the perception in the neocortex of bodily response to stimuli, mediated through lower brain centres. At the same time, the James-Lange theory is recognized to have serious defects and not least its overly simple model of causality.

The third school of thought is that of Sylvan Tomkins, whose main concern was to differentiate affects from drives. Unlike drives, affects can be transferred to a range of objects so as to be satisfied: they are therefore adaptable in a way that drives are not. As Hemmings (2005, 551–552) explains, Tomkins not only demonstrated that 'affect can enable the satisfaction of a drive (excitement might prepare the body for the satisfaction of hunger) or interrupt it (so that disgust might interrupt that satisfaction if you were served a rotten egg to eat)', but he also addressed the issue of our affective relations with others.

> Tomkins asked us to think of the contagious nature of a yawn, smile or blush. It is transferred to others and doubles back, increasing its original intensity. Affect can thus be said to place the individual in a circuit of feeling and response, rather than opposition to others. Further, Tomkins argues that we all develop complex affect theories as a way of negotiating the social world as unique individuals (Hemmings 2005, 551–552).

The fourth school of thought is that of Gilles Deleuze. For Deleuze, affect stands for the unruly body's ability to go its own way which cannot be reduced to just social organization. For Deleuze, therefore, the focus is on bodily displacement, the movement between bodily states, the map of intensities. As Hemmings (2005, 552) again succinctly puts it:

> Deleuze proposes affect as distinct from emotion, as bodily meaning that pierces social interpretation, confounding its logic, and scrambling its expectations. In contrast to Tomkins, who breaks down affect into a topography of myriad, distinct parts, Deleuze understands affect as describing the passage from one state to another, as an intensity characterized by an increase or decrease in power.

The final school of thought is psychosocial. It can be traced from Aristotle through Hobbes to Leibniz and involves an emphasis on corporeal dynamics which is based on the Aristotelian rather than the Cartesian model. It describes what might be called a political economy of affect in that affects are seen as having a different distribution across bodies. In opposition to the idea that the passions are seen as something housed in a body and shared by all human beings equally, affect is understood as the contours of a dynamic social field 'manifest in what's imagined and forgotten, what's praised and blamed, what's sanctioned and silenced' (Gross 2006, 15). Passions are constituted between politically and historically situated agents. In turn, this suggests that it makes a difference 'not only what sort of passions are distributed to whom, but also how they are hoarded and monopolized and how their systematic denial helps produce political subjects of a certain kind' (Gross 2006, 49).

Significantly, each of these different schools of thought involves a substantial biological component. In each case, the body is given its own powers that are outside social organization, *sensu strictu*, although obviously, in practice, it is very hard to tell the difference. Whatever the case, it is clear that affect signals a number of challenges to social theory as currently constituted but, most especially, a challenge to any easy dividing line between the 'social' and the 'biological' and to the apparent roles of each.

The Matter of Biology

In the context of these five schools of thought, this chapter is stimulated by two main currents of work, both of which either set out from the biological or at least take the biological seriously as an important element of affect which cannot be either wished away, shoved into a box marked 'naturalism', or made secondary to the social. These associative currents of work subscribe to the view that the biological is not something different in kind from the social but is an integral part of the business of building collectives.

The first current of work is concerned with a direct revaluation of the biological, a revaluation, which has been going on for some time now. Led by an evolutionary momentum which derives from three chief sources – the evolving debate around the interaction between culture and evolution, the debate around the natures of animals, and the rediscovery of the process philosophy of A.N. Whitehead (Thrift 2007) – this work has been going back to first principles, and especially the intersection between evolution and culture, in order to discover what it means to be human if human is understood as process of situated flow within which human bodies are just one of the sets of actors. It is no longer possible to avoid this fulcrum of activity, with all its undoubted historical baggage, since it both speaks to a time in which biopower has become biopolitics (Lazzarato 2002) and because the questions now being raised by biology press on that knot of interests formerly known as the social.

The second current of work informing this chapter approaches the biological in a less direct fashion. It has tried to take up the idea that it is possible to derive a thinking that is not conditioned or compressed by time but is of what might be called an ethological nature, recognizing the prime importance of territory since an object like affect is no longer regarded as in some sense 'internal' but is regarded rather as a moving map of passions making their way hither and thither. In terms of current thought, such a processual move towards a *spatial thinking* (cf. Buchanan and Lambert 2005) has most often been associated with the work of Gilles Deleuze.

(Re)Thinking Spatial Relations

I want to turn briefly to three developments in spatial thinking. Though each of them has a relation to Deleuze, they can also be used to go beyond his work and to think about the nature of shared animated space more deeply. The first is the work of Gabriel Tarde (1962, 1895/1999, 2000). Tarde was a formative influence on Deleuze's thought on difference and repetition but he offers much more than an extended footnote. The second is the work of the Italian operaismo Marxists, Paolo Virno and Maurizio Lazzarato. Both Virno and Lazzarato have become increasingly influenced by both Deleuze and Tarde but they provide a diagnostic reading of capitalism that is only faintly present in Deleuze. Finally, there is the work of Peter Sloterdijk, who shares an interest in Deleuze but in his *Sphären* trilogy (Sloterdijk 1998, 1999, 2004) arguably demonstrates a much more acute sense of space.

Why would authors like Virno and Lazzarato and indeed lately Sloterdijk become interested in Tarde (Barry and Thrift 2007)? Tarde provides a gathering point for those who believe that the term 'society' and the models of socio-cultural inscription which are its main theoretical legacy are completely exhausted and who want to work instead with a form of associationism which is regarded as the only way of following the multiplication of collectives, and forming 'landing strips' (Latour 2005) for new ones.[3] Most particularly, Tarde's work suggests the possibility of resurrecting an epidemiological model which is based on processes of imitative contagion, not least because the spread of feelings (through gesticulation, bodily movements, motor co-ordinations and repetitions, as well as all the technologies of the body that now exist) is such fertile ground for thinking about mental contagion. Such epidemiological models of mimetic 'vibrations' held sway in good parts of the social sciences at the turn of the nineteenth century as means of explaining phenomena as different as crowd behaviour (as in the work of Le Bon, Trotter or McDougall, and subsequently luminaries like Freud and Bion) and the diffusion of different kinds of cultural objects (as in the work of anthropologists like Tylor), but they went into steep decline as means of explanation for a variety of reasons, not least their association with right-wing diagnoses of disorder as pathological and irrational based on ideas such as that crowds were likely to regress to lower levels of mental functioning, and/or were highly suggestible, in both cases leading to loss of personality. Of late, epidemiological models have begun to make something of a comeback. That comeback is based on the ability of these models to express

---

3   We need to do this not just on theoretical but also on political grounds as a means of understanding how the envelope of understanding develops as a series of choices about what makes a collective.

expression and to frame sympathetic induction. Such models provide a much better sense of how particular kinds of affective phenomena do their work. A good example is provided by the seminal work of Brennan (2004) who tracks affect as so many 'atmospheres'.

It is against this background that the recently rehabilitated work of Tarde makes the most sense. Tarde provided a series of models of that which flows out from (or perhaps more accurately, through) us unawares, models of *imitation* and *invention*. These models serve the present moment well, emphasizing as they do non-conscious perception, dissociation, suggestion and suggestibility, and social influence as forming a part of a stream of thought, rather than a threat to the boundaries of an individual. For Tarde, imitation was akin to these kinds of phenomena in that it involved dissociation within which suggestion could thrive.

> As he declared in a characteristically stylish formulation, 'I shall not seem fanciful in thinking of the social man as a veritable somnambulist ... Society is imitation and imitation is a kind of somnambulism'. And: 'The social state, like the hypnotic state, is only a form of dream, a dream of command and a dream of action. To have only ideas that have been suggested and to believe them spontaneous: such is the illusion of the somnambulist and also of the social man' (Leys 1993, 279).

From Tarde, we turn to Italian workerist (operaismo) Marxism. Such a variant of Marxism is often associated with the name of Antonio Negri and his gradual journey from a Marxism that was still faithful to the Grundrisse (and most especially to that small part of the Grundrisse which expounded the idea of general intellect) to a Marxism that seems to be more faithful to Spinoza (cf. Wright 2002). However, arguably more interesting developments have occurred in the work of Paolo Virno and Maurizio Lazzarato, who both started from some of the same ideas as Negri but have diverged subsequently, increasingly taking on neo-vitalist ideas as they track from capital-labour to capital-life in which capital takes on all the powers of the transindividual general intellect – 'knowledge, the subjective spirit of invention, invention-power' (Virno 2005, 21) – by tapping in to the full range of powers of the human body rather than just labour, making it into a preindividual thing that can be operated on.[4] 'In a way, labour is today truly productive (of surplus value and profit) only if it coincides with the human abilities that previously explicated themselves in non-labour' (Virno 2006, 38). So 'labour power has become invention-power' (Virno 2006, 37). In other words, the whole of human praxis, all of the generically human gifts, come to be included in the productive process, all those aspects of forms of life that exist outside its stringent rationality and that allow this expanded praxis to both produce, hasten and adapt to a nonlinear productive fluctuation. 'Contemporary capitalist production mobilizes

---

4   This is a theme I have taken up in a number of papers.

to its advantage all the attitudes characterizing our species, putting to work *life* as such' (Virno 2005, 35, author emphasis).

In particular, Lazzarato (2002, 2005, 2006) has taken up the gauntlet of Tarde's work on the power of invention, concentrating on the modes of production of communication and knowledge that lead beyond economy, as well as on Foucault's work on biopower as a new political economy of forces that act and react against each other. Lazzarato tries to forge a new synthesis, a description of the new modes of extracting a surplus of power from living beings, which harness 'freedom' to the production and government of new forms of life that are based on relationships of differentiation, creation, and innovation. This Tardean-Foucauldian synthesis which uses the presence of others as a transindividual resource faces the problem of strategic relations that can escape (or at least minimize) domination by being based on *sympathy* and not just on asymmetry (and therefore domination-resistance).

The third school of thought arises from the work of Peter Sloterdijk on the genesis of manufactured environments (Thrift 2005). In his *Sphären* trilogy, Sloterdijk takes Heidegger on dwelling as a point of reference but then spatializes his thinking by posing the question of being as the question of being-together: 'one is never alone only with oneself, but also with other people, with things and circumstances; thus beyond oneself and in an environment' (Sloterdijk 1998).[5] 'Being-a-pair' or a couple precedes all encounters[6]. In other words, Sphären is concerned with the dynamic of spaces of co-existence, spaces which are commonly overlooked, for the simple reason that 'human existence … is anchored in an insurmountable spatiality' (Sloterdijk 2005a, 229).

Sloterdijk suggests that what is needed is an 'air-conditioning project' that can sweep through the totally managed and domesticated spatial environments of current social entities. This project of the *ventilation* of the atmospheres of modern life (Sloterdijk 2006) will change the socio-spatial terms of trade by providing new environments in which novel doctrines of living can thrive, environments which will provide the breathing space with which democracy can be re-invented, just as the original Athenian democracy was critically dependent upon the city and 'the pre-logical or pre-discursive premises of the art of urban co-existence' that were able to be constructed from it, premises that were the result of 'the skilful application of anti-misanthropic procedures'(Sloterdijk 2005b, 947) that included, most especially, explicit affective engineering.

I want to combine these different but related bodies of work by focussing on the vexed topic of affect, understood especially as a function of capitalism. In particular, I want to consider how motivation itself is being changed by more and more explicit engineering of affect.

---

5   Thus Sloterdijk retains Heidegger's radical emphasis on the then recently discovered notion of the environment, as circumstances being adjusted to accommodate the entity in their midst.

6   In bringing forward this formulation, Sloterdijk is making a similar move to those approaches based on joint action that have become increasingly common.

Understanding Affective Contagion

What is particularly hard to cope with in writing about affect is not so much its insubstantial nature as finding a model that can encompass its powers. That model must be, in part, biological, which adds to the challenge. What seems clear now is that such a task must mean attempting to understand affective contagion, for affect spreads, sometimes like wildfire. This was a central concern of turn-of-the-nineteenth-century social science in the form of the study of imitation and suggestibility. Of late, imitation and suggestibility have been making a return. Within cultural theory, viral models of contagion have been posited as explaining the workings of a range of phenomena, including ideology, governance, self-cultivation, and even resistance but often in highly speculative ways that posit a kind of performative energetics but without usually specifying what the source or content or form of that energy might consist of. But there is no need for this (often convenient) opacity, as I hope to show in this section through more detailed examination of the grip of affect.

Let me begin by summarizing what we have gleaned about affect so far. To begin with that means understanding affect is in large part a biological phenomenon, involving embodiment in its many incarnations, a phenomenon that is socially produced but not easily captured via specular-theatrical theories of representation (Brennan 2004). Affect brings together a mix of a hormonal flux, body language, shared rhythms, and other forms of entrainment (Parkes and Thrift 1979, 1980) to produce an encounter between the body (understood in a broad sense) and the particular event. Then, affect is generally semiconscious, sensation that is registered but not necessarily considered in that thin band of consciousness we now call cognition. Further on again, affect is understood as a set of flows moving through the bodies of human and other beings, not least because bodies are not primarily centred repositories of knowledge – originators – but rather receivers and transmitters, ceaselessly moving messages of various kinds on; the human being is primarily 'a receiver and interpreter of feelings, affects, attentive energy' (Brennan 2004, 87).

In turn, this depiction points to one more important aspect of affect, the importance of space, understood as a series of conditioning environments that both prime and 'cook' affect. Such environments depend upon pre-discursive ways of proceeding which both produce and allow changes in bodily state to occur (Thrift 2006). Changes in bodily state require understanding that essentially autonomic hormonal and muscular reactions are continually transferring between people (and things) in ways that are often difficult to track. At the same time, they challenge the idea that the body is a fixed component of humanity. It might be more accurate to liken humans to schools of fish briefly stabilized by particular spaces, temporary solidifications which pulse with particular affects, most especially as devices like books, screens and the internet act as new kinds of neural pathway, transmitting faces and stances as well as discourse,[7] and providing myriad opportunities to forge new reflexes. Thus, concentrating on affect requires a cartographic imagination in order to map out the movement between corporeal states of being which is simultaneously a change in connectivity.

Now we can also add in what we currently know about imitation and suggestibility. Imitation has become a paramount concern of the contemporary cognitive sciences, and this work is worth exploring in more detail, since it contains many insights. In particular, imitation is now understood as a higher level cognitive function, mirroring both the means and ends of action, and highly dependent upon the empathy generated in an intersubjective information space that supports automatic identifications. For example, hearing an expression of anger increases the activation of muscles used to express anger in others. There is, in fact, only a delicate separation between one's own mental life and that of another, so that affective contagion is the norm, not an outlier.

At the same time, it is important to stress that imitation is more than mere emulation. Imitation depends upon an enhanced capacity for anticipation, so-called mind-reading (Thrift 2006). In particular, much of human beings' capacity for mind-reading (whether this be characterized as inference or simulation) develops over years of interaction between infants and their environments, and involves processing the other as 'like me', and the consequent construction of high-level hypotheses like deception. That is, it involves a form of grasping which is innately physical and non-representational since our privileged access is to the world, not to our own minds.

Whatever the exact case might be, most imitation is clearly rapid, automatic and unconscious and involves emotional contagion, in particular (down to and including such phenomena as moral responsiveness). People seem to be fundamentally motivated to bring their feelings into correspondence with others: we love to entrain. What seems clear, then, is that human beings have a default capacity to imitate, automatically and unconsciously, in ways that our deliberate pursuit of goals can override but not explain. In other words, most of the time we do not even know we are imitating. Yet, at the same time, this is not just motivational inertness. It involves, for example, mechanisms of inhibition, many of which are cultural.

More generally, human imitative skills may be regarded as part of a widespread human capacity for mimesis. However, in contrast to many interpretations, for example those of Benjamin or Horkheimer and Adorno, I do not accept that the mimetic capacity has to be interpreted as somehow a primordial cognitive faculty which modernity has caused to decay. Many authors still argue that the main outpouring of mimesis now is in the play of children and is given up as the adult world approaches, or they consign mimesis to supposedly archaic categories

---

7 For example, it is possible to write about the history of facial expressions like the smile (Trumble 2003) because media have been invented which can transmit these expressions.

like magic. In contrast, I would argue that mimesis is in fact a perennial human imitative capacity, closer to a biological drive, and that it is an additive capacity for desire created out of a third term which René Girard calls the model or mediator and is therefore neither autonomous or innate. There is no exact copy so that our desires can never properly be ours (Fleming 2004). Rather, our desires are second hand and socially-oriented; we always desire what others desire, in imitation of them, and not under our own impetus. Take the case of modern advertising:

> advertisers rely on external mediation when they pay celebrities to use a product, and make use of internal mediation when they depict common people using common products. In the first case, they want our admiration for the celebrity to spark a desire for the product, and in the second case, they want us to 'catch' the nearby desire of someone like ourselves (Potolsky 2006, 147).

But, at this point, it is important to make a few observations about this imitative notion of affective corporeality which are important for the arguments in this chapter. To begin with, though much of the flow of affect can be described as a form of thinking, part of the ceaseless flow of mind-readings, it is not necessarily instrumental or knowledgeable, that is oriented towards determinate goals like comprehension, purpose or intention. This point is only underlined by the fact that a large part of corporeal life is simply oriented towards concern over the body's extreme vulnerability. And not surprisingly: bodies make mistakes, trip or fall, get toothache or migraine, can see only partially or not at all, get chronic diseases like arthritis – or simply drop dead. One of the remarkable facts about the recent interest in the social sciences in embodiment, practice and performance, and the body-in-action generally, is the lack of thought that has been given to the fact of this vulnerability and how it wraps back into mind-reading. The primary role in the theorization of embodiment across a number of theoretical perspectives is still played by intentional or auto-affective action. Vulnerability remains not only unthought-of but potentially un-thinkable within much current work on the body within Anglo-American social science. Yet much corporeal experience is based on bodily states that underline corporeal vulnerability, such as fatigue and exhaustion or pain and suffering or exposure to extreme cold or heat (Wilkinson 2005). In other words, corporeal life is inherently susceptible, receptive, exposed; open beyond its capacities to comprehend and absorb. One should not overdo this vulnerability, of course. For example, for some fatigue, weariness and general inaction can involve considerable psychic investments from hypervigilance to a certain kind of somatic self-regard (cf. Nunn 2005). But neither should one underplay it.

Then, it is also crucial to underline the role of things. Of late, the prosthetic impulse provided by the role of things has become a key theme in social sciences (Smith and Morra 2006) but, because of the social sciences' roots in interpretation, the emphasis still tends to fall on objects' meaning, on objects as cultural inscriptions. However objects do far more than represent. In terms of the argument of this chapter it is crucial to underline their other roles; 'the prosthesis is not a mere extension of the human body; it is the constitution of this body qua "human"' (Stiegler 1998, 152–153). In particular, objects form shields to human vulnerability by extending the body's circumference. They provide mental and physical resources to allow the body to be in the world, they add to what and how the body can experience, and they have their own agency, an ability to move bodies in particular ways. Thus clothes are not just ornamentation and display, they protect from the weather, provide resources for all kinds of specialist situations, and they produce particular corporeal stances. Similarly, houses provide a safe environment which wraps the comforting aura of familiarity around bodies. Thus, things redefine what counts as vulnerable. One of the key moments of affect is therefore an ability to produce controllable worlds.

Last of all, I want to stress – once again – the involuntary and precognitive nature of much of what is being described here. So, for example, feelings of vulnerability may not necessarily be expressly articulated though they may well be expressed in other ways, for example through bodily stances.[8] Thus we arrive at the subject of *automatism*. Generally speaking, affect is a semiconscious phenomenon, consisting of a series of automatisms, many of them inscribed in childhood, which dictate bodily movement, which arise from suggestion, and which are not easily available to reflection (Wegner 2002; Wilson 2002). These automatisms may often feel like wilful action but they are not and they have powerful political consequences, not least because they form a kind of psychic immune system which means that certain issues can be avoided or perversely interpreted as a matter of course (Wilson 2002; Milton and Svašek 2005). Equally, suppositions of causality may become firmly entrenched. For example, it is relatively easy to promote in populations feelings of responsibility over events for which they could not possibly have had any responsibility at all. Affect is, in other words, a series of highways of imitation-suggestion. As Wegner (2002, 314) puts it, we live in a 'suggested society' in which 'the causal influences people have on themselves and each other, as they are understood, capture only a small part of the actual causal flux of social relations'. In other words, societies are thought of, quite literally, as *en-tranced*, as only half-awake. I am not at all sure that current ways of thinking have come to terms with this vision. But without it, I am not sure we can even begin to understand what that which we glibly call 'the social' consists of. In conclusion, let me return briefly to Tarde.

Conclusion

Recall that Tarde paralleled imitation with invention. Moscovici (1985) is writing about another less mediated time but his work seems particularly relevant to the new mediated age of the imitative crowd we now inhabit, an age which might well be caricatured as mass mesmerism gone bad (Barrows 1981; Peters 1999;

---

8   In other words, intentionality does not have to be conscious (Brennan 2004).

Miller 1983). For Moscovici argues that political attempts to generate affective force often try to create an 'illusion of love' via a range of techniques – affective, corporeal, and psychological – aimed at maximizing processes of suggestion and imitation, including the use of symbols, images, flags, music, affirmations, phrases, speeches, and slogans, all jammed, as I see it, into the half-second delay between action and cognition. These are delivered through the hypnotizing use of repetition rather than didactic command and instruction. Thus, the population is touched in ways which might be non-conscious and may well instil the feeling that they are the originator of that thought, belief, or action, rather than simply and mechanically reproducing the beliefs of a charismatic other. The political challenge is to extend and reinforce 'mental touch'. In other words, waves of affect are transmitted and received, transmitted and received, constantly challenging the Lockean citadel of the consenting self (Barrows 1981) as they cook up a political storm.

We need to understand these kinds of systems. Currently we don't. But – and here is the optimistic moment – there is a wonderful confluence of all kinds of interests at present in feminist studies, literary studies, historical studies, media and cultural studies, aesthetic theory, and, of course, in geography (e.g. Abel 2007; Clough 2007; Gallop 2006; Ngai 2005; Stewart 2007; Thrailkill 2007), which is determined that we will do precisely that. Against those who want to cleave to the notion of a democracy of free thinkers freely thinking, these various contributors want to opt for a democratic politics that inherits other political traditions in which affect has clear purchase and can be understood as offering moments of political understanding and analysis in its own right. Recent studies (e.g. Berlant 2006, 2007, 2008) seem to me to have amply demonstrated that this ambition can be turned into a hale and hearty reality.

References

Abel, M. (2007), *Violent Affect. Literature, Cinema, and Critique after Representation* (Lincoln: University of Nebraska Press).

Amin, A. and Thrift, N.J. (2005), 'What's left? Just the future', *Antipode* 37, 327–338.

Arsic, B. (2005), 'Thinking leaving', in I. Buchanan and G. Lambert (eds.) *Deleuze and Space* (Edinburgh: Edinburgh University Press), 126–143.

Barrows, S. (1981), *Distorting Mirrors. Visions of the Crowd in Late Nineteenth-Century France* (New Haven: Yale University Press).

Barry, A. and Thrift, N. (2007), 'Gabriel Tarde: imitation, invention, and economy', Economy and Society 36, 509–525.

Berlant, L. (2006), 'Cruel optimism', *Differences* 17, 20–36.

—— (2007), 'Nearly Utopian, nearly normal: post-Fordist affect in La Promesse and Rosetta', *Public Culture* 19, 273–301.

—— (2008), *The Female Complaint. The Unfinished Business of Sentimentality in American Culture* (Durham, NC: Duke University Press).

Brennan, T. (2004), *The Transmission of Affect* (Ithaca: Cornell University Press).

Buchanan, I. and Lambert, G. (eds.) (2005), *Deleuze and Space* (Edinburgh: Edinburgh University Press).

Clough, P.T. (ed.) (2007), *The Affective Turn. Theorizing the Social* (Durham, NC: Duke University Press).

Despret, V. (2004), *Our Emotional Makeup: Ethnopsychology and Selfhood* (New York: Other Press).

Elliott, A. and Lemert, C. (2006), *The New Individualism. The Emotional Costs of Globalization* (London: Routledge).

Fleming, C. (2004), *René Girard. Violence and Mimesis* (Cambridge: Polity).

Funcke, B. (2005), 'Against gravity. Bettina Funcke talks with Peter Sloterdijk', Bookforum February/March. <http://www.bookforum.com/archive/feb_05/funcke.html>

Gallop, J. (ed.) (2006), 'Special Issue on Envy', *Women's Studies Quarterly* 34, Nos 3 and 4.

Griffiths, P. (1997), *What Emotions Really Are. The Problems of Psychological Categories* (Chicago: University of Chicago Press).

Gross, D. (2006), *The Secret History of Emotion. From Aristotle's Rhetoric to Modern Brain Science* (Chicago: University of Chicago Press).

Hemmings, C. (2005), 'Invoking affect: cultural theory and the ontological turn', *Cultural Studies* 15, 548–567.

Latour, B. (2005), *Reassembling The Social: An Introduction to Actor-Network Theory* (Oxford: Oxford University Press).

Lazzarato, M. (2002), 'From biopower to biopolitics', *Pli: Warwick Journal of Philosophy*, 13, 100–111.

— (2005), 'Introduction', in G. Tarde *Underground. Fragments of Future Histories* (Brussels: Les Maîtres de Formes Contemporains).

— (2006), 'European cultural tradition and the new forms of production and circulation of knowledge' (Unpublished manuscript).

Leys, R. (1993), 'Mead's voices: imitation as foundation, or, the struggle against mimesis', *Critical Inquiry* 19, 277–307.

Miller, J. (1983), 'Crowds and power: some English ideas on the state of primitive personality', *International Review of Psychoanalysis* 10, 253–264.

Milton, K., Svašek, M. (eds.) (2005), *Mixed Emotions. Anthropological Studies of Feeling* (Oxford: Berg).

Moscovici, S. (1985), *The Age of the Crowd: A Historical Treatise on Mass Psychology* (Cambridge: Cambridge University Press).

Nelson, D. (2006), 'The president and presidentialism', *South Atlantic Quarterly* 105, 1–17.

Ngai, S. (2005), *Ugly Feelings* (Cambridge, Mass.: Harvard University Press).

Nunn, C. (2005), *De La Mettrie's Ghost. The Story of Decisions* (London: Macmillan).

Parkes, D.N. and Thrift, N. (1979), 'Time spacemakers and entrainment', *Transactions of the Institute of British Geographers NS4*, 353–72.

— (1980), *Times, Spaces, Places: A Chronogeographic Perspective* (Chichester: John Wiley).

Peters, J.D. (1999), *Speaking into the Air. A History of the Idea of Communication* (Chicago: University of Chicago Press).

Potolsky, M. (2006), *Mimesis* (New York: Routledge).

Runciman, D. (2006), *The Politics of Good Intentions. History, Fear and Hypocrisy in the New World Order* (Princeton: Princeton University Press).

Sloterdijk, P. (1998), *Sphären I. Blasen* (Frankfurt: Suhrkamp Verlag).

— (1999), *Sphären II. Globen* (Frankfurt: Suhrkamp Verlag).

— (2004), *Sphären III. Schaume* (Frankfurt: Suhrkamp Verlag).

— (2005a), 'Foreword to the theory of spheres', in M. Ohanian and J.C. Royaux (eds.) *Cosmograms* (New York: Lukas and Sternberg), 223–240.

— (2005b), 'Atmospheric politics', in B. Latour and P. Weibel (eds.) *Making Things Public. Atmospheres of Democracy* (Cambridge, Mass.: MIT Press), 944–951.

— (2006), 'War on latency: on some relations between surrealism and terror', *Radical Philosophy* 137, 14–19.

Smith, M. and Morra, J. (eds.) (2006), *The Prosthetic Impulse. From a Posthuman Present to a Biocultural Future* (Cambridge, Mass.: MIT Press).

Stewart, K. (2007), *Ordinary Affects* (Durham, NC: Duke University Press).

Stiegler, B. (1998), *Technics and Time, 1: The Fault of Epimetheus* (Stanford: Stanford University Press).

Tarde, G. (1962), *The Laws of Imitation* (Gloucester, MA: Peter Smith).

— (1895/1999), *Monadologie et sociologie* [Monadology and sociology] (Paris: Les empêcheurs de penser en rond [Barriers to circular thinking]).

— (2000), *Social Laws. An Outline of Sociology* (Kitchener, Ont.: Batoche Books).

Thrailkill, J.F. (2007), *Affecting Fictions. Mind, Body, and Emotion in American Literary Realism* (Cambridge, Mass.: Harvard University Press).

Thrift, N. (2004), 'Intensities of feeling: towards a spatial politics of affect', *Geografiska Annaler*, Series B 86, 57–78.

Thrift, N.J. (2005), 'Movement space: the changing domain of thinking resulting from the development of new kinds of spatial awareness', *Economy and Society* 33, 582–604.

— (2006), 'Re-inventing invention: the generalization of outsourcing and other new forms of efficacy under globalization', *Economy and Society* 35, 279–306.

— (2007), *Non-Representational Theory. Space, Politics, Affect* (London: Routledge).

Virno, P. (2005), 'Interview with Paolo Virno', *Gray Room* 21, 26–37.

— (2006), 'Reading Gilbert Simondon. Transindividuality, technical activity and reification', *Radical Philosophy* 136, 34–43.

Viveiros de Castro, E. (2005), 'From multiculturalism to multi-naturalism', in M. Chanian and J.C. Royoux (eds.) *Cosmograms* (New York: Lukas and Sternberg), 137–156.

Wegner, D.M. (2002), *The Illusion of Conscious Will* (Cambridge, Mass.: MIT Press).

Whitehead, A.N. (1979), *Process and Reality* (New York: Free Press).

Wilkinson, I. (2005), *Suffering. A Sociological Introduction* (Cambridge: Polity).

Wilson, T.D. (2002), *Strangers to Ourselves. Discovering the Adaptive Unconscious* (Cambridge, Mass.: Belknap Press).

Wright, S. (2002), *Storming Heaven. Class Composition and Struggle in Italian Autonomist Marxism* (London: Pluto Press).

Wyatt, T. (2003), *Pheromones and Animal Behaviour. Communication by Smell and Taste* (Cambridge: Cambridge University Press).

Chapter 5

Environmental Aesthetics, Ecological Action and Social Justice

Jennifer Foster

When we imagine and experience nature, whether as untouched wilderness, agricultural fields or weeds growing through urban sidewalk cracks, we are deploying aesthetic interpretations of the world around and within us. Environmental aesthetics reflect the ways that we respond to and shape the world, including the ecologies and physical landscapes within which we are immersed. They are invoked by the sensory encounters of everyday life, encompassing routine as well as atypical and unexpected activities, and can call upon the entire repertoire of an individual's past experiences in relation to the sensation of a specific aspect, or even the whole, of one's present surroundings. In contrast to formal aesthetic readings of art, such as those associated with art criticism, environmental aesthetics infuse wide landscape interactions wherein everyday activities such as work, recreation, transit and home are situated (Berleant 1997; Leddy 2005). They are an inescapable feature of life. They have an affective potency at a personal, individual level, provoking diverse emotional responses to nature such as wonder, fear and curiosity that leave sensory imprints on us. But environmental aesthetics are also cultivated by the accretion of socially constructed knowledge: knowledge mediated by culture that accumulates over generations and is emotionally infused to produce explanatory stories of the world that shape visions of what is relevant, causal and possible, proper or egregious.

This chapter explores the interplay between the aesthetically affective and learned, how humans are figured by nature and nature is in turn socially produced, and how understanding nature may be both a highly personal matter and something conditioned by emotionally-based anticipatory structures related to particular histories and cultures. It identifies a need to consider the aesthetics of everyday life as important milieu for both emotional engagement on an individual level and as constitutive of the cultural politics of representation at a broader level, distinguishing an emotional basis for understanding aesthetic encounters as critical environmental determinants. To this end the chapter also explores a case study on the fringes of Toronto, Canada to illustrate some of the ways that aesthetics are emotionally formative of worldviews that may be enacted through environmentalism, suggesting a strong need to query the specific ecologies and social effects related to emotional dimensions of aesthetics.

An Emotional Basis for Environmental Aesthetics

Our worlds are inescapably both mediated by sensory emotional engagement and cognitively framed by knowledges that accumulate over time. Environmental aesthetics are the combinations of continuously changing spatial patterns and the subjective experiences of these spaces. They are experienced individually and collectively, are dynamic and multi-sensory. Encompassing the mundane to exceptional, commonplace to exotic, environmental aesthetics are influential in determining what is appreciated and how. Aesthetic ideals and preferences, meanwhile, are diverse and change over time.

From a social justice and environmental planning perspective, positive life experiences and overall quality of life are affected by environmental aesthetics (Beardsley 1982; Matilla 2002). Within the social and environmental justice literature, quality of life is typically treated in terms of spatial distribution of, and bodily exposure to, noxious elements. Allowing environmental aesthetics into this equation – weighing sensory experience as part of quality of life – enriches the terms through which social and environmental justice are considered by creating a broader and more nuanced picture of what constitutes an egalitarian society. Environmental aesthetics play a strong role in shaping and lending meaning to people's lives by prompting, stirring and congealing emotional interpretations of the world, its infinite complex relationships, and one's own place in the world. While there may be many positive facets of sensory engagement (development of empathies with the more-than-human stands out), aesthetic appreciation is also yoked by disciplining processes that reflect power relations across historical, geographic and cultural terrains. Although environmental aesthetics may be enriching and stimulate satisfying relationships with nature, they may also evoke negative emotional responses and/or ecologically unsustainable preferences and actions. Moreover, the social exclusions and erasures that are marked within, through, and as a result of environmental aesthetics reveal a great deal about how nature, as a concept and material outcome of cultural preferences, is equipped with power.

Aesthetics may be one of the most important forces shaping landscape and ecological patterns and structures. They are critical environmental determinants. Situating aesthetics as human practice, Emily Brady (2006) advocates acknowledging the importance of aesthetic values within environmental policy debates, not merely as decorative concerns to be addressed after serious scientific and economic matters have been dealt with, or simply as matters of subjective taste, but rather 'based on a critical set of activities that are practiced and developed in a public context' (279). Aesthetic values, she argues, are embedded in human relationships with nature. Building on arguments by Marcia Eaton (2001), Aldo Leopold (2000) and Holmes Rolston (2002), Brady also points out that aesthetic experiences can be visceral, powerful and enduring, contributing to moral values associated with nature that may inspire 'appropriate moral action' prompted by the perceptual sensitivity, imaginative freedom, creativity and emotional expression associated with aesthetic engagement.

Approaches to understanding aesthetic experience may be broadly grouped into engagement and cognitive interpretations. Ronald Moore (2007) refers to these as conceptualist and non-conceptualist views of nature. Focusing on immediate rapport between appreciator and objects of appreciation (incorporating an understanding of environment without disruption between life forms, perceptions, and place), Arnold Berleant (1992) captures the immersive, multi-sensory spirit of the engagement perspective when he observes that 'entering and participating in the landscape requires full sensory involvement' (6). Environments here comprise the interior worlds of individuals, as well as that which exists beyond the self. An engagement approach thus probes what is appreciated and the feelings and emotions of aesthetic encounters as subjective responses. This incorporates an understanding of aesthetic experience as tactile and ambient (Foster 1998), with an emphasis on aesthetic capacities for visceral sensory arousal (Carroll 1993), focusing on how places affect people in direct and individual ways, including the ways that feelings are kindled by surroundings and how moods, spirits and frames of mind are stimulated or dampened. The particular personal sensory effects of immersion in one's surroundings, unconditioned by social knowledge, are here appreciated as formative of human lived experience.

Cognitive approaches to environmental aesthetics are, on the other hand, more immediately concerned with how human responses are socially guided, how meanings are inscribed in people's imaginations in ways that route, enable, or delimit, aesthetic interpretations. This approach accentuates how knowledge affects environmental valuation and provides an interpretation of environmental aesthetics that helps illuminate how nature serves as material for the expression of social values. From this perspective environmental aesthetics disclose the historical and cultural situatedness that hermeneutic phenomenologists such as Gadamer (1989) claim prejudice all interpretation of the world. Cognitive views of aesthetics sometimes maintain that aesthetic experiences may be enhanced through learned knowledge of nature, for example, by absorbing the messages of sciences and natural history (Carlson 2003; Saito 1998; Rolston 1995). Others, such as Duncan and Duncan (2001, 2004) focus more directly on how aesthetics codify social order, how the advancement of landscape taste is figured to include and exclude different social groups. For instance, they describe the political history of environmental protection in Bedford, New York, by noting that 'the celebration of the natural environment, historical preservation, and the claimed uniqueness of a local landscape has often diverted attention from the interrelatedness of issues of aesthetics and identity on the one hand and social justice on the other' (2004: 7).

Allen Carlson (2007), a proponent of cognitive aesthetic interpretation, is dismissive of engagement approaches. As such, he embraces Ronald Hepburn's (1966, 1993) idea of serious beauty intuition, which classifies aesthetic appreciation along a continuum from passing, superficial, experiences to more challenging contemplation of nature that demands cognitive engagement as a requirement for adequate aesthetic appreciation of nature. Carlson is unenthusiastic about the significance of engagement experiences as aesthetically profound, explaining that

Multi-sensory immersion in nature, emotional arousal by nature, and ambient feeling for nature are interesting aspects of human experience and without doubt constitute some dimensions of our aesthetic appreciation of nature. However, each of such states is also at the easy end of the spectrum that runs from easy to serious beauty, and thus to make any of them the central core of aesthetic appreciation of nature is to disregard Hepburn's Serious Beauty Intuition (9).

But Carlson underestimates both the role played by aesthetic engagement in lending meaning to people's lives and inspiring actions, as well as the potential for combining engagement and cognitive approaches in order to appreciate the complex roles that aesthetics play in mediating human/nature relationships. Indeed, Moore (2007) proposes that the 'rivalry' between the two may be bridged through a syncretic approach that appreciates how they feed off each other. Brady (2003) also points to such a link, suggesting that caring attitudes toward nature may be cultivated toward aesthetically attractive as well as unattractive places through both sensory contact (engagement) and background ecological stories (cognition). Human imaginative capacities to shift between the perceptual and cognitive, she argues, enable us to appreciate nature by imagining creative connections that enrich experience with meaning. It is the process of imagining, however, that is largely figured by socially-relevant regimes of truth that, as Michel Foucault (1973, 1978) observes, shape the inconspicuous practices, techniques, and mechanisms determining how larger systems operate within a plurality of discourses by enabling and constricting action.

Combining the engagement and cognitive yields an understanding of how natures are both viscerally felt and learned in ways that cannot be ignored by those concerned with social justice and environmental sustainability. In particular, there is a need to consider the ways that environmental aesthetics motivate people to act 'environmentally' while generating social difference at scales ranging from the transcontinental to experiences and projections of and across individual bodies. Further, there is a need to reflect upon these in conjunction with the impacts of environmental valuation expressed through everyday decisions and actions that impinge on non-humans. This chapter attempts to contribute to such a conversation by considering both the affective and cognitive dimensions of environmental aesthetics as themes of social justice and environmental sustainability.

Prospects and Challenges of Environmental Aesthetics

As a medium through which the world is experienced, environmental aesthetics are formative of people's worlds and personal lives. Many environmentalists, for example, report that the lasting aesthetic imprint of childhood encounters with nature (ranging from remote protected areas to backyard streams and abandoned lots) inspire their ecological action. This portion of the chapter explores the affective and cognitive scope of environmental aesthetics through case study research that probes the goals and motivations of self-identified environmentalists. The specific group that is considered includes people who have made personal commitments to protect 'legacy landscapes' surrounding Toronto, that is, landscapes that are enshrined in popular historical accounts and provincial environmental legislation as worthy of enhanced valuation and preservation. The research participants are allies in political struggles to resist development in landscapes such as the Niagara Escarpment and the Oak Ridges Moraine (two linear geological features that enframe Toronto to the north and west), working in professional and/or advocacy roles to resist urban sprawl and its infrastructure spreading from Toronto. Ecological action among this group echoes many of the positive messages of the environmental aesthetics literature, but also suggests a number of challenges to social justice and environmental sustainability.

The focus here is specifically directed at environmentalists, on particular discursive formations of environmentally-identified landscape actors, as people who claim to privilege concerns for nature in their actions. The theme of landscape continuity serves as an analytical focus. This is a pervasive spatial and ecological metaphor drawn from the field of landscape ecology that suggests the patterns and degrees of similarity and harmony characterizing habitat networks. Within this field, continuity is characterized as the connections between patches of habitat across a landscape matrix, enabling ecological benefits such as species movement, diversity of landscape pattern, fulfillment of habitat and life cycle needs and ensuring genetic variance (see, for example, Turner 1987; Forman 1995). The concept is distinguished in the visions statements for multi-jurisdictional conservation legislation covering both the escarpment and moraine, and is an oft-cited trope in discussions and debates about the form and functions of environmental planning in these landscapes. Storylines that present historically and culturally situated explanations figuring social agency into accounts (from Hajer 1995) and environmental imaginaries that direct attention to future practices and relations between nature and society (from Peet and Watts 1996) are considered. They are articulated in this case through public hearings, municipal council assemblies, naturalist club meetings, documents circulated by environmental organizations, and a series of eighteen interviews (Foster 2005). This discourse coalition includes many people with personal and professional commitments to the scientific method and the conventions of rational comprehensive planning, for example wildlife biologists, ecologists, landscape architects and regional planners. Yet despite epistemic affinity with the precepts of rationalism, their depictions of landscape continuity are highly intuitive and socially-derived, presenting key aesthetic functions in the interpretations of the world they are protecting and fortifying the roles of conjecture, affect, emotion and sensation as legitimate means of detecting ecological phenomena. It could be argued by those critical of the constraints and disciplining of rational modernism that this is for the good. Recognizing and honouring experiential knowledge and feelings may certainly present a progressive challenge to the discursive control of conventional science's systematic knowledge of the material world. However, reliance on intuition also ignites the power of social ideals infused through landscape perception.

On the positive side, many evocatively describe how they have been deeply affected by youth experiences with nature, experiences that enrich their lives, help shape their identities and offer emotional affirmation of their sense of selves. These people locate their passion for ecology and environmental protection in childhood encounters with nature, where exposure to and immersion in natural worlds present seminal early life experiences. When asked to describe their entry point into environmentalism, many report a lifelong interest in ecology based on unstructured childhood time exploring local creeks, woodlots and fields. They report that this type of unsupervised time engendered both their curiosity about and empathy for the non-human world. When asked how she became an environmentalist, for instance, the head of a citizen's environmental group points to self-directed outdoor amusement with animals: 'I grew up on a farm. As a young child I spent a lot of time in the woods. Well, we always had many pets and we spent time throwing grasshoppers into the trout pond to feed the trout. I had a lot of time in the outdoors.' While such activity emphasizes manipulative play with nature (in this case at the grasshoppers' peril) these accounts evoke the ambient qualities of an engagement approach to environmental aesthetics, for example as summoned by a botanist who consults extensively along the escarpment and moraine. He talks about aesthetics as 'meaning in landscape', commenting on emotional attachments and personal values formed during his own 'ninety percent unsupervised, backwoods and old field loving' youth by simply stating 'if you live some place you begin to have certain connections and identity for that place that are real.' There is a palpable emphasis on the affective imprint of landscape interactions emanating from everyday activities and surroundings, and the formative relationships within nature forged during youth are considered by many to be their richest life experiences, often experienced in solitude. Part way through an extensive explanation of the biophysical attributes of landscape corridors, for instance, a hydrogeologist who leads hiking tours of the escarpment turns to the 'subtle influence' of early experiences 'solo wandering' as a basis for her own understanding of both the temporal and spatial dimensions of landscape continuity. These types of experiences made many feel settled and calm, with a comfortable sense of their own place in the world. They also made them curious about the world, and such curiosity was often later attended with scientific study or membership in outdoor-oriented groups such as 4H, an agriculturally-oriented youth leadership program. Many also convey discernable degrees of wistfulness and melancholy for a world perceived to have more opportunities for sensory stimulation and uncomplicated personal fulfillment.

As these environmentalists recount childhood experiences, they are also grieving for a lost world, a safe world and an ecologically rich world. An environmental planner who has guided the environmental land use plans for many of the municipalities surrounding Toronto, for example, reminisces about his own experiences and regrets that these encounters are inaccessible to today's youth. He also describes the introduction of a scientific instrument as a response to curiosity ignited by unaccompanied immersion in nature, suggesting links between perception, imagining and the culturally-produced scientific lens as a means of developing knowledge of nature.

> When I was a kid I would go out to a pond with a jar and a piece of string and get the water. And my mother and father had given me a microscope for my birthday and I'd take a look at it. Well now kids don't do that by themselves. You know they go to a science school. People need to experience things directly, even when it's just farming. I mean, how many kids know the difference between barley and wheat when they see it growing in a field or even know what it means?

The impact of childhood experiences with nature has been explored by numerous researchers, and has yielded interesting findings and observations. For example, Chipeniuk (1995) documents a direct relationship between childhood foraging and the acquisition of a sense of biodiversity in adulthood. Researchers in the field of significant life experience (SLE) have sought to understand the types of learning experiences that produce environmentalists, employing an autobiographical approach to understanding the environmental commitments of staff and directors of conservation organizations, environmental educators, and less frequently, environmental activists (for example, Palmer 1993; Corcoran 1999; Sward 1999). A study conducted by Kals et al (1999), meanwhile, uses multiple regression analysis to measure the linkage between emotional affinity toward nature as a motivational function of pro-environmental behaviour, finding that present and past unpleasant emotional experiences such as self-blame, indignation, and anger for environmental destruction combined with more pleasurable emotions such as senses of safety and love of nature stimulate a great deal of nature-protective behaviour. While much of this research has been heavily survey-based and statistical, there are also dimensions of the field that focus more qualitatively on emotions, such as Chawla's (2001) reflections on 'inner nature', memory and people's own constructions of their past. Other researchers, such as Wells and Lekies (2006), adopt a life course perspective to track the connections between childhood engagements with nature and pro-environment attitudes among the wider population beyond environmental professionals and activists. They find a direct relationship between experiences such as playing, hiking, camping, hunting and fishing in natural areas prior to age eleven and environmentally sensitive attitudes and behaviours. What these research themes do not explore, however, are the specific varieties of environmentalism, environmental attitudes and environmental behaviours that are expressed in relation to life experiences, particularly in relation to the social and cultural meanings that underwrite ecological action.

Environmental aesthetics serve as important conduit through which politicization of ecology congeals by fusing ecological knowledge and concepts of social order, often in ways that are not immediately discernable. While the ecological action of the environmentalists considered here may stem from, and is valorized by, engagement-type youth relationships with nature, it is also the case that emotionally-based anticipatory structures that presuppose interpretations as coherent, meaningful and whole (Gadamer 1989), and prefigure what constitutes satisfying, aesthetically particular landscape patterns in advance of sensory encounter, also play a strong role in adult motivation to act ecologically. This is where the positive, personally affirming dimensions of environmental aesthetics meet the more environmentally and equitably precarious ones, as people's conceptions of desirable and appropriate landscape aesthetics are not always environmentally sustainable, and they are often socially unjust. Discourses concerning what appear to be normal landscapes and natural ecology are also landscape-based expressions of beliefs or claims about reality that have profound political implications. They are sites where knowledge of ecological regimes intermingle with socially constituted beliefs. In the legacy landscapes of the Greater Toronto Area many environmentalists equate inhabitation of culturally-valued nature with colonial settlement patterns, and focus particularly on a sense of well-being associated with the expansive presentation of colonial aesthetic emblems. The former Executive Director of a prominent citizens' environmental advocacy group, for example, describes this ideal through the concept of landscape continuity:

> 'I [like] that checkerboard of ... farm fields and hedgerows and forested areas in the back of farms and farm houses and the configuration of the roads. I would talk about landscape continuity in that way, where if you leave the city and go to the rural area you start to picture this cultural landscape that has been imposed on the land, primarily by European settlers, then that would be the continuity I would look for.'

The aesthetic deployment of ecological ideals that reflect and reproduce world-views and preferences are to a large extent characterized by the presences and absences of different social groups within landscapes. This is a clear expression of what Iris Marion Young (1990) calls the 'injustice of cultural imperialism', where the experiences, beliefs, preferences and culture of dominant and privileged social groups are generalized for all groups. Histories of the escarpment, moraine and greenbelt are typically rendered with a celebratory emphasis on colonial encounters and achievements, prompting the removal of Others both materially and representationally from these landscapes. During the mid-1800s massive waves of immigrants came to southern Ontario from England, Scotland and Ireland, many of whom were offered land on the escarpment to establish agricultural homesteads, industries, rail lines and source fuels (McIlwraith 1997; CONE 1998). Yet archeological records suggest that 97 per cent of the history of human occupation of the escarpment is by First Nations people (CONE 1998), many of the descendants of whom now reside on or near the Nawash First Nation (at Cape Croker, on the Bruce Peninsula). Early human occupants of the moraine were Huron and Iroquois First Nations peoples who used the area for hunting and gathering as well as fresh water supplies (McQueen 1999). Until the arrival of European settlers, the forest cover of the moraine remained largely undisturbed, after which pressure for deforestation and land settlement was intense (Fisher and Alexander 1993). Extensive lumber harvesting, land cleared for agriculture and the erection of grist- and sawmills left the sandy moraine soils barren and unstable. By the early twentieth century, the moraine had lost ninety percent of its tree cover to agriculture (Wood 2000), causing turbid and irregular stream flows fluctuating between extremes of flood and drought (McQueen 1999), prompting broad-scale remedial action beginning in the 1920s (Oak Ridges Moraine Technical Working Committee 1994). While these landscape histories may tell stories of cultural displacement and ecological spoil, they are popularly presented as the awakening of crude and uncultured places to civilization. The environmentalist research participants themselves repeatedly employ metaphors that reveal high estimation such as 'landmark', 'anchors', 'unique', 'signature landscapes' and 'threatened' to evaluate these colonized landscapes.

An important social demarcation of 'insiders' and 'outsiders' in environmental planning and design processes is the differentiation of those who have rightful claims to inhabit the landscape and participate in land use decisions, including not only the spectacular disputes staged through hearings and court procedures, but more often the everyday aesthetic practices that fortify senses of belonging and incongruity. Aesthetics serve as an important channel for such regulation, where landscapes are simultaneously figured biophysically and through the production of subjectivities. Duncan and Duncan's (2001, 2004) observations about the ways that class and race codes are embedded in seemingly benign or universally beneficial environmental planning endeavours are germane here. They explain that 'collective memories, narratives of community, invented traditions, and shared environmental concerns are repeated, performed, occasionally contested, but more often stabilized and fixed in artifactual form' (29). In the fringes of Toronto, memory and identity expressions that do not conform to aesthetic ideals of the colonial picturesque settlement as a mosaic of agricultural fields, woodlots and hamlets are objectionable. Incongruity in housing style is a particular irritant for many environmentalists concerned with landscape continuity, who repeatedly rely upon this ecological concept to argue that aesthetically diverse homes and their inhabitants are out of place and disturb landscape continuity. Responding to an inquiry about how landscape continuity may be defined, a regional land use planner for the Ontario Ministry of Natural Resources explains his own aversion to 'discontinuous' housing styles, particularly those that are perceived as disruptive of the Anglo colonial style that dominated in the nineteenth and early twentieth centuries:

> If you drive about you'll see Mediterranean style houses, obviously built by immigrants from Italy or Spain or Mediterranean countries. And beside them you'll see Tudor style houses built by Brits. And you'll see houses that mimic New England clapboard. So, you don't have a whole lot of control over what

the housing looks like. So you get this mix which to me is not continuity. And if I go to Europe I would see every house with a tile roof and there's a continuity there and I can say 'yes, I'm in a particular area and sense the place.' In southern Ontario we're losing that or have lost it, so that even when you look at a landscape with the construction over the last fifty years there's a loss of continuity in the appearance of the built structures.

Responses to the question of how landscape continuity may be defined follow a distinct discursive pattern, with initial focus on biophysical descriptions of the non-human, incorporating themes such as connected tree cover, regional capacity for groundwater recharge or uninterrupted migration routes for birds. But the biophysical emphasis typically gives way to cultural considerations unrelated to biophysical concerns, for example objection to commercial signs printed in languages other than English (particularly Punjabi and Mandarin) or Jewish lawyers who are 'outsiders' representing developers seeking approvals and permits for land use change. A citizens' environmental organization leader, for instance, relates his frustration with an environmental land use tribunal process focusing on landscape continuity by explaining 'when I look around I think, and here I'll be racist again, these guys have got the best Jewish lawyer you can buy. This guy, I forget his name now, I just looked at him and you know, this is one of these guys with a giant brain, verbal powers, they can run circles around anybody. So how the process works, I don't respect the process very much. I know, it's democracy!' Melding an emotive sense of well-being based on youthful immersion in nature with signifiers of who belongs in an historically and culturally valued landscape creates particular landscape forms that are socially reproduced and, in this case, ecologically expressed through the landscape policy, planning and design ideals of environmentalists.

The rate of immigration along the outskirts of Toronto is remarkable, and this is the crux of a lot of enthusiasm for landscape protection that couches social exclusion. In the Region of Peel, perched between Toronto and the spine of the escarpment and moraine, 43 per cent of the overall population self-identifies as immigrants (Agarwal et al. 2007). The percentage of the population not born in Canada in suburbs lining the escarpment and moraine is far higher than the national foreign-born metropolitan average of 18.4 in 2001 and 19.8 per cent in 2006. For example, the percentage of population not born in Canada in Brampton is 47.5 per cent, in Mississauga is 51.6 per cent, in Markham is 56.5 per cent and in Richmond Hill is 51.5 per cent (Statistics Canada 2006). Population growth during the period between 1981 and 2001 was 118 per cent in Brampton, 513 per cent in Vaughan, 171 per cent in Markham and 249 per cent in Richmond Hill (Statistics Canada 2001). Yet, while aesthetically articulated conceptions of social belonging in these legacy landscapes excludes perceived newcomers, they also efface the presence and landscape histories of people whose ancestors predate colonial settlement of Ontario. This is aesthetically reflected in environmental storylines and imaginaries that efface native peoples' existence while envisaging the point of colonial encounter as a common-sense landscape ideal. Alternate interpretations of nature and landscape aesthetics are either invisible or ignored, and native people simply do not figure into these landscape representations.

Much of the distaste for perceived outsiders inhabiting these heritage landscapes is a desire for a sense of community that is in turn underwritten by complex emotional networks that foreground fear, what Leonie Sandercock (2000) identifies as fear of the other, the stranger, the foreigner, the outsider. While expressions of overt racism are unmistakable in the Greater Toronto Area– for example where visible-minority entrepreneurs face institutional discrimination in accessing business financing (Teixeira et al. 2007), racist epithets and white supremist graffiti appears on local university campuses (Sopinka 2008) or as anti-immigration groups such as Canada First (www.canadafirst.net) stridently claim to represent the non-immigrant 'dispossessed majority' – contemporary racism is typically more subtly articulated or even obscured. In the Greater Toronto Area concern is often expressed about the regional capacity to meet the service needs of newcomers (for example, Javed 2007; Kopun and Keung 2007), particularly in terms of the provision of culturally-specific services in ethnic enclaves (Agrawal et al. 2007). While such concern may reflect awareness of and attention to cultural diversity in urban planning, it also sometimes transmits xenophobic fear of Others taking over. In relation to a perceived strain newcomers place on cities, a senior fellow of the Fraser Institute (a conservative Canadian libertarian thinktank) opines that 'Canadians, according to surveys, think that there may be some major problems with immigration but they're constantly told that we need it anyway ... You don't really question immigration because you'll be a racist if you do' (Perkel 2007). Modern-day prejudgements about groups of people and discomfort with perceived outsiders are more commonly conveyed indirectly, perhaps often unwittingly. In this case environmental protection may be about more than mere ecological concern; it can also about the changing social profile of the landscape with introduction of a substantial non-white population, spatially expressed through the ways that recent immigrants, many of whom are members of visible minorities, are situated in and out of place. This is suggested by a regional planner's response to questions about how landscape continuity is perceptible, where biophysical criteria coalesce with his own unease about newcomers in Brampton. Along with an array of ecological processes, he identifies 'connection to place' as an important feature of landscape continuity, and believes that 'People moving into Brampton now, all the flow of immigration, don't have that same connection. They don't have it.' National census data reveal a 59.5 percent increase in foreign-born residents in Brampton between 2001 and 2006, and the largest proportion of the visible-minority population represented as South Asian (Statistics Canada 2006). Ahmed Iqbal, the executive director of the Brampton Multicultural Community Centre, explains that "For Indians, Brampton is Canada" and "They have heard about it and they know that it is a place where people 'look like us'" (Javed 2007). The following table illustrates the shifting demographic of Brampton concerning visible minority population data reported in the 2001 and 2006 censuses.

Table 5.1   Brampton's Visible Minority Population

	2006	% of Total Population	2001	% of Total Population
Visible minority population	**246,145**	**57.03%**	**130,275**	**40.20%**
South Asian	136,750	31.69%	63,205	19.50%
Black	53,345	12.36%	32,070	9.90%
Filipino	11,980	2.78%	6,965	2.10%
Latin American	8,545	1.98%	5,225	1.60%
Chinese	7,805	1.81%	5,445	1.70%
Southeast Asian	6,130	1.42%	3,005	0.90%
Arab	2,600	0.60%	1,850	0.57%
West Asian	2,875	0.67%	1,085	0.33%
Korean	580	0.13%	615	0.20%
Japanese	545	0.13%	535	0.20%
Visible minority, not included elsewhere	8,895	2.06%	8,180	2.50%
Multiple minorities	6,095	1.41%	2,110	0.70%
Not a visible minority	185,430	42.97%	194,120	59.80%

Source: Statistics Canada, 2001 and 2006 Census of Canada.

In Canada, researchers such as Kobayashi and Peake (2000) explore normative and relational positions that enrich the primacy of 'whiteness' in geographical imagination, reasoning that racialization is figured through both absence from as well as presence within particular places. Peake and Ray (2001) write about the interiorized and exteriorized landscapes of Canadian racism, explaining that 'construction of people of colour as outside the nation places them as negative disruptions of the Canadian landscape' (180). Work by Willems-Braun (1997), meanwhile, focuses on how native peoples are systematically made to 'vanish' from environmental and landscape accounts, a 'habit of thinking' that typifies the Anglo majority. The prevalence and imprint of 'whiteness' on Canadian landscapes such as those surrounding Toronto suggest the emotional disclosure of fear of identified groups through ecologies and landscape forms that often pass as natural and aesthetically appropriate. Transcription of xenophobia through landscapes is largely concealed by the supposed moral benevolence of environmental actions and pro-environment attitudes, particularly where aesthetic and ecological ideals converge through planning processes with goals that are intended to altruistically benefit nature and the non-human. Who gets to participate, how cognitive categories of social belonging and incursion are circumscribed and ecologically enacted, and how xenophobia is obscured through seemingly altruistic action are questions of how emotions are spatialized – through emotions ranging from curiosity, calmness and comfort to guilt, indignity and fear. Other emotional responses may also compel environmental action through affective dimensions of nature, but it is clear that these are enmeshed with conceptions of social order.

While most of the ostensibly positive features of sensory and emotional engagement with nature presented here relate to the affective dimensions of environmental aesthetics, the more worrisome ones are cognitive and socially-mediated. If emotional attachments based on childhood experiences with particular places and natures are valorized as legitimate bases for ecological action, how then might the experiences of those with geographically and culturally different experiences inhabit these same landscapes in more equitable ways? How might the more promising aspects of aesthetic awareness be linked to environmental sustainability and social justice, and the equitably vexing ones be re-channeled?

Linking Aesthetic Awareness to Environmental Sustainability and Social Justice

Emotions unquestionably play a strong role in the construction and advancement of environmental knowledge, in the everyday experiences of engagement with or cognition of aesthetic surroundings. People's feelings about their environments are at the core of their beliefs and environmental attitudes (Pooley and O'Connor 2000), and fear is clearly a critical emotional experience that helps mediate the ways that landscapes and ecologies are affected by human action and lived experience. Any strategy for dislodging xenophobia as it is aesthetically expressed, thus, must necessarily be about confronting fear, challenging prejudices and learning new aesthetics.

Writing about planning in polyethnic and multicultural societies, Leonie Sandercock (2003) proposes an approach to addressing fear amid urban social complexity through what she terms therapeutic practice. Often uncomfortable for those who have not confronted their fears and prejudices, this is an approach to social exclusion that moves beyond the rational discourse of communicative planning to emphasize emotions as an important interface for 'organizing hope, negotiating fears, mediating memories, and facilitating community soul searching and transformation' (153). Using techniques such as narrative through deliberative processes to allow people to recount their own stories and perspectives, Sandercock's therapeutic practice opens avenues for moving beyond dominant voices in landscape planning and design, where the emotions of aesthetics may be collectively explored, particularly where long-standing cross-cultural disputes are articulated biophysically through presumed commonality of place memory and history.

The progressive potential of Sandercock's therapeutic practice is further enhanced when linked to notions of aesthetic justice, particularly where this concept is incorporated into environmentalism as a whole. Appreciation of nature is, in many ways, performance of nature (Brady 2003; Fisher 2007), where we interact with expressions of nature through landscape practices and environmental narratives that reflect knowledge, values and beliefs. Aesthetic justice strategies must go beyond simple concerns of distribution of pleasing environments to also build aesthetic capacities that address the conception and production of aesthetically positive experiences generated through histories, emotional responses and performances of nature. This means situating planning and design as civic practices rather than professional and expert enterprises by encouraging and honouring aesthetic expression at broad societal scales. Just as Saito (1998) and Nassauer (1997) suggest that enhanced knowledge of science and natural history may boost ecological appreciation and further environmental sustainability goals, building awareness of the varied cultural and social contexts within which environments are experienced and performed is equally important.

As eminent cultural products that affect people deeply, nature, landscapes and ecologies are important sites for the inflection of social justice. Sensitivity to the ways that affective and cognitive aspects are intertwined – to mutually constitute emotionally based aesthetics helps generate clearer understanding of human/nature relationships. Socially constructed yet profoundly personal, the prospect of landscape discussions that build a public realm by generating reflexive, collaborative planning practices for inhabiting shared space might also help us understand our own ecologies, aesthetics, and emotions – to foster ecologies that are aesthetically responsive both in terms of individual experience and the cultural politics of representation.

References

Agarwal, S.K. et al. (2007), 'Immigrants' Needs and Public Service Provision in Peel Region', *Plan Canada* summer 2007, 45–9.

Beardsley, M. (1982), 'Aesthetic Welfare, Aesthetic Justice and Education Policy', in Wren and Callen (eds.)

Berleant, A. (1992), *The Aesthetics of Environment* (Philadelphia: Temple University Press).

Berleant, A. (1997), *Living in the Landscape: Towards an Aesthetics of Environment* (Lawrence, Kansas: University of Kansas Press).

Berleant, A. (ed.), (2002), *Environment and the Arts: Perspectives on Environmental Aesthetics* (Aldershot and Burlington: Ashgate).

Brady, E. (2003), *Aesthetics of the Natural Environment* (Edinburgh: Edinburgh University Press).

Brady, E. (2006), 'Aesthetics in Practice: Valuing the Natural World', *Environmental Values* 15, 277–91.

Carlson A. (2003), *Aesthetics and the Environment: Essays on Nature, Art and Architecture* (London: Routledge).

Carlson, A. (2007) 'The Requirements for an Adequate Aesthetics of Nature', *Environmental Philosophy* 4:1&2, 1–13.

Carroll, N. (1993), 'On Being Moved by Nature: Between Religion and Natural History', in Kemal and Gaskell (eds.).

Chawla, L. (2001), 'Significant Life Experiences Revisited Once Again: Response to Vol. 5(4) 'Five Critical Commentaries on Significant Life Experience Research in Environmental Education', *Environmental Education Research* 7:4, 451–61.

Chipeniuk, R. (1995), 'Childhood Foraging as a Means of Acquiring Competent Human Cognition About Biodiversity', *Environment and Behavior* 27:4, 490–512.

Coalition on the Niagara Escarpment (1998), *Protecting the Niagara Escarpment: A Citizen's Guide*, Coalition on the Niagara Escarpment, P.O. Box 389, Acton, Ontario, Canada, L7J 2M6.

Corcoran, P.B. (1999), 'Formative Influences in the Lives of Environmental Educators in the United States', *Environmental Education Research* 5:4, 207–20.

Duncan, J.S. and Duncan, N.G. (2001), 'The Aestheticization of the Politics of Landscape Preservation', *Annals of the Association of American Geographers* 91, 387–409.

Duncan J.S. and Duncan N.G. (2004), *Landscapes of Privilege: The Politics of the Aesthetic in an American Suburb* (London: Routledge).

Eckstein, B. and Throgmorton, J.A. (eds.) (2003), *Story and Sustainability: Planning, Practice, and Possibility for American Cities* (Cambridge, Massachusetts: The MIT Press).

Fisher, J.A. (2007), 'Performing Nature', *Environmental Philosophy* 4:1&2, 15–28.

Fisher, J. and Alexander, D. (1993), 'The Symbolic Landscape of the Oak Ridges Moraine: Its Influences on Conservation in Ontario, Canada' *Environments* 22, 100.

Forman, R.T.T. (1995), *Land Mosaics; The Ecology of Landscapes and Regions* (New York: Cambridge University Press).

Foster, C. (1998), 'The Narrative and Ambient in Environmental Aesthetics', *Journal of Aesthetics and Art Criticism* 56, 127–37.

Foster, J. (2005), 'The Social Construction of Landscape Continuity on the Niagara Escarpment and Oak Ridges Moraine: Whose continuity? Whose landscapes?' Ph.D. dissertation, York University, Toronto, Canada.

Foucault, M. (1973), *The Order of Things* (New York: Vintage Press).

Foucault, M. (1978), *The History of Sexuality: An Introduction, Volume 1* (translated from French by R. Hurley) (New York: Vintage Books).

Gadamer, H.-G. (1989), *Truth and Method* (J. Weinsheimer and D.G. Marshall, trans.) (New York: Crossroad).

Hajer, M. (1995), *The Politics of Environmental Discourse* (Oxford: Clarendon).

Hepburn, R. (1966), 'Contemporary Aesthetics and the Neglect of Natural Beauty' in Williams and Montefiore (eds.).

Hepburn, R. (1993), 'Trivial and Serious in Aesthetic Appreciation of Nature', in Kemal and Gaskell (eds.).

Javed, N. (2007), 'For Indians, Brampton is Canada', *The Toronto Star* [website] (published 5 December 2007) <http://www.thestar.com/News/article/282692>.

Kals, E. et al. (1999), 'Emotional Affinity Toward Nature as a Motivational Basis to Protect Nature', *Environment and Behavior* 31:2, 178–202.

Kemal, S. and Gaskell, I (eds.) (1993), *Landscape, Natural Beauty and the Arts*, (Cambridge: Cambridge University Press).

Kobayashi, A. and Peake, L. (2000), 'Racism out of Place: Thoughts on Whiteness and an Antiracist Geography in the New Millennium', *Annals of the Association of American Geographers* 90, 392–403.

Kopun, F. and Keung, N. (2007), 'A City of Unmatched Diversity', *The Toronto Star* [website] (published 5 December 2007) < http://www.thestar.com/News/GTA/article/282694>.

Leddy, T. (2005), 'The Nature of Everyday Aesthetics', in Light and Smith (eds.).

Leopold, A. [1949] (2000), *A Sand County Almanac, With Essays on Conservation* (New York: Oxford).

Light, A. and Smith, J.M. (eds.) (2005), *The Aesthetics of Everyday Life* (New York: Columbia University Press).

Matilla, H. (2002), 'Aesthetic Justice and Urban Planning: Who Ought to Have the Right to Design Cities?' *GeoJournal* 58, 131–8.

McIlwraith, T.F. (1997), *Looking for Old Ontario* (Toronto: University of Toronto Press).

McQueen, D. (1999), 'Oak Ridges Moraine', in Roots et al. (eds).

Moore, R. (2007), *Natural Beauty: A Theory of Aesthetics Beyond the Arts* (Peterborough, Canada: Broadview).

Nassauer, J.I. (1997), 'Cultural Sustainability: Aligning Aesthetics and Ecology', in Nassauer (ed).

Nassauer, J.I. (ed.) (1997), *Placing Nature: Culture and Landscape Ecology* (Washington: Island Press).

Oak Ridges Moraine Technical Working Committee (1994), A Cultural Heritage Resources Assessment Study for the Oak Ridges Moraine Area: Background Study No. 7 to the Oak Ridges Moraine Planning Study, Ontario Ministry of Natural Resources.

Palmer, J.A. (1993), 'Development of Concern for the Environment and Formative Experiences of Educators', *Journal of Environmental Education* 24:3, 26–30.

Peake, L. and Ray, B. (2001), 'Racializing the Canadian Landscape: Whiteness, Uneven Geographies and Social Justice', *The Canadian Geographer* 45:1, 180–6.

Peet, R. and Watts, M. (1996), 'Liberation Ecology: Development, Sustainability and Environment in an Age of Market Triumphalism' in Peet and Watts (eds.).

Peet, R. and Watts, M. (eds.) (1996), *Liberation Ecologies: Environment, Development, Social Movements* (London: Routledge).

Perkel, C. (2007), 'Newcomers Put Strain on Cities', *The Toronto Star* [website] (published online 4 December 2007) <http://www.thestar.com/printArticle/282348>

Pooley, J.A. and O'Connor, M. (2000), 'Environmental Education and Attitudes: Emotions and Beliefs are What is Needed', *Environment and Behavior* 32:5, 711–23.

Rolston, H. (1995), 'Does Aesthetic Appreciation of Landscapes Need to be Science-based?', *British Journal of Aesthetics* 35, 374–86.

Rolston, H. (1998), 'Aesthetic Experience in the Forest', *The Journal of Aesthetics and Art Criticism* 56, 157–66.

Rolston, H.R. (2002), 'From Beauty to Duty: Aesthetic of Nature and Environmental Ethics', in Berleant (ed.).

Roots, B.I. et al. (eds.) (1999), *Special Places: The Changing Ecosystems of the Toronto Region* (Vancouver: UBC Press).

Saito, Y. (1998), 'The Aesthetics of Unscenic Nature', *The Journal of Aesthetics and Art Criticism* 56, 102–11.

Sandercock, L. (2000), 'Negotiating Fear and Desire: The Future of Planning in Multicultural Societies', *Urban Futures Conference Proceedings, Urban Forum* 11, 201–10.

Sandercock, L. (2003), 'Dreaming the Sustainable City: Organizing Hope, Negotiating Fear, Mediating Memory' in Eckstein and Throgmorton (eds.).

Sopinka, H. (2008), 'Modern Intolerance: Bigoted Hate Acts on University Campuses: Systemsic, Isolated or Reflective of The City at Large?' *The Globe and Mail* [online] (published 22 March 2008) <http://www.theglobeandmail.com/servlet/story/RTGAM.20080321.wrace-sopinka0322/BNStory/lifeMain/home>.

Statistics Canada (2001). *2001 Census of Canada*. Ottawa, Canada.

Statistics Canada (2006). *2006 Census of Canada*. Ottawa, Canada.

Sward, L.L. (1999), 'Significant Life Experiences Affecting the Environmental Sensitity of El Salvadoran Environmental Professionals' *Environmental Education Research* 5:2, 201–6.

Teixeira, C., Lo, L. and Truelove, M. (2007), 'Immigrant Entrepreneurship, Institutional Discrimination, and Implications for Public Policy: A Case Study in Toronto' *Government and Policy* 25(2), 176–193.

Turner, M.G. (ed.) (1987), *Landscape Heterogeneity and Disturbance* (New York: Springer-Verlag).

Wells, N.M. and Lekies, K.S. (2006), 'Nature and the Life Course: Pathways From Childhood Nature Experiences to Adult Environmentalism', *Children, Youth and Environments* 16:1, 1–24.

Willems-Braun, B. (1997), 'Buried Epistemologies: The Politics of Nature in (post)Colonial British Columbia' *Annals of the Association of American Geographers* 87:1: 3–31.

Williams, B. and Montefiore, A. (eds.) (1966) *British Analytical Philosophy*, (London: Routledge).

Wood, J.D. (2000), *Making Ontario: Agricultural Colonization and Landscape Re-Creation Before the Railway* (Montreal: McGill-Queen's University Press).

Wren, M.J. and Callen, D.M. (eds) (1982), *The Aesthetic Point of View: Selected Essays: Monroe C. Beardsley* (London: Cornell University Press).

Young, I.M. (1990), *Justice and the Politics of Difference* (Princeton, New Jersey: Princeton University Press).

Chapter 6

Learning from Spaces of Play: Recording Emotional Practices in High Arctic Environmental Sciences

Richard C. Powell

> The learning of an emotional vocabulary is one of the essential skills of an ethnographer. To survive as a competent social being, the fieldworker must learn how to interpret, if not actually feel, the finer shades of anger, pity, or whatever the host population specializes in.
>
> (Beatty 2005a, 23)

Emotional Geographies and Affectual Anthropologies

Discussions amongst geographers of science have begun to highlight the potentialities of ethnographic methods (Powell 2007).[1] The critical dynamic of ethnography is the emotional immersion of the practitioner within the social life of a community. However, as of yet, geographers of science have rarely engaged with debates occurring in geographies of emotion and affect. In what follows, I attempt to meet these challenges through an emotional ethnography of environmental science in the Canadian Arctic. In doing so, this chapter attempts to derive in return, through ethnographic practice, some intimations for discussions in emotional geography.

Emotional geographies are concerned with the spatial mediation and articulation of emotions (Anderson and Smith 2001; Bondi et al. 2005). This has resulted in a focus upon the spaces through which emotional geographies are enacted. Such work has directed attention to the emotional intersubjectivity of social interaction and, specifically, to 'issues of relationality which are so profoundly embedded in our everyday emotional lives' (Thien 2005, 453). For Thien, if emotions are relational consequences of human interaction, then their embedding must relate to other positionings within the social order. This insistence on the relationality of emotions is in direct opposition to other work that has assumed the primordial status of affectual response (McCormack 2006; Anderson and Harrison 2006).

---

1 On geographies of science, see Livingstone (1995), Powell (2007) and Powell and Vasudevan (2007).

Such geographical research on affect has been criticized for ignoring the important power-geometries that impact upon human lives (Thien 2005; Tolia-Kelly 2006). As Tolia-Kelly argues:

> It is thus critical to think plurally about the capacities for affecting and being affected, and for this theorization to engage with the notion that various individual capacities are differently forged, restrained, trained and embodied (Tolia-Kelly 2006, 216).

I should stress that rehearsing and adjudicating between these different epistemic standpoints is not my concern here. Rather, I want to suggest that these debates amongst geographers have tended to operate in parallel to those that have taken place across anthropology during recent years (Beatty 2005a; 2005b; Josephides 2005; Svašek 2005; Tonkin 2005).[2] In this chapter, I draw from these related literatures to help provide some supplementary thoughts for geographers of emotion.

The central premise within anthropologies of emotions is the necessity of ethnography. As many scholars have argued, the role of field experience within anthropological discussions is conceived almost mythically. Fieldwork forms a disciplinary 'rite of passage' that all ethnographers must successfully undertake in order to become successful interpreters of human interaction. It is important to note that, as such, this understanding of ethnography is significantly different to that often supplied by geographers, where ethnographic methods are often seen as an important adjunct to other qualitative research techniques, especially in-depth interviews.

In what follows, I argue that emotional geographers could benefit from participation in debates about the ethnography of emotions, most evident in work by anthropologists. Indeed, the centrality of relations between emotions and modes of social organization within anthropological traditions has led Andrew Beatty to claim that 'to talk of social structure *is* to talk of emotions' (Beatty 2005b, 55, my emphasis). As Kay Milton puts it, '[a]nthropologists are professional observers of human situations, and emotions are present in everything they observe' (Milton 2005a, 215).[3]

There are, of course, invidious echoes of colonial praxis in this anthropological understanding of ethnography. This has been much discussed by scholars influenced by the general crisis of faith in representational practices during the

---

2   Beatty (2005a) provides an effective review of recent anthropological writing on emotion and affect. For a comprehensive survey of anthropological work that investigates the range of devices that convey affective meaning, see Besnier (1990).

3   Milton argues that rather than being solely social in origin, 'emotions are essentially ecological phenomena' (Milton 2005b, 25). In so doing, Milton outlines an argument that emotions are mechanisms via which humans learn *through* their environments (Milton 2005a; 2005b). Emotions are central to this learning process.

1990s (Clifford 1988; Sanjek 1990; Wolf 1996; Gupta and Ferguson 1997; Passaro 1997). As Joanne Passaro puts it in an enervating essay, even if anthropologists are now reticent to romanticize an 'exotic' other, they are still able to 'nonetheless continue to romanticize the "young" ethnographer and his/her ethnographic project' (Passaro 1997, 147). The epistemological anxieties that have accompanied this ethnographic crisis, as I will demonstrate in this chapter, can never be satisfactorily resolved.

The lesson to be drawn from the anthropology of emotions, I want to argue, is the necessity of the practice of fieldwork. The very experience of fieldwork, or 'being there', involves a series of emotional encounters (Tonkin 2005; Josephides 2005).[4] During fieldwork in Java, for example, the ethnographer is 'lacking an explicit set of rules or taboos, one must, as it were, *feel* one's way' (Beatty 2005b, 64, original emphasis). As such, developing awareness of emotional codes is part of the process of *becoming* in the field for any successful ethnography. For Beatty, it becomes critical to maintain 'heightened attention to what happens in the field' (Beatty 2005a, 18). This would require the emplacement of ethnographic fieldwork at the very centre of the geography of emotional practices. By focusing on the 'varying pragmatic contexts' of social life, ethnographers can make a central contribution to the understanding of emotions therein (Beatty 2005a, 34). As Beatty argues:

> This would entail a shift away from worrying about affect, and from inner/outer quandaries, to what I have called *emotional practices*. It would also mean a more exact reporting of what pass as emotional episodes in the field, for from such material we weave our theories (Beatty 2005a, 34, my emphasis).

After Beatty, then, I use *emotional practices* to indicate those displays and namings of social emotions that can only be grasped through their recording in ethnographic context. This recording of emotional practices can only be undertaken through intensive fieldwork. But as this chapter will investigate, the actual recording of such practices can be, at best, problematic, not least for the emotional experiences of the ethnographer.

---

4   There is, of course, a literature on the ethical challenges of emotional fieldwork in geography (Laurier and Parr 2000; Widdowfield 2000; Powell 2002; Bondi 2003). Much of this research has, however, concentrated on the emotional consequences of conducting qualitative interviews rather than participant observation. Drawing from object relations psychoanalysis, for example, Bondi (2003) discusses the utility of the concept of *empathy* for qualitative interviews. Although this provides for stimulating opportunities to rethink the research relationship, I would argue that ethnographic access is negotiated at the community-level and thus can provide for different encounters. For a comprehensive review of debates around the epistemic and ethical dimensions of geographical fieldwork, see Powell (2002).

Arctic Play

This chapter contributes to such debates on emotional practices by revisiting a
tradition of Arctic ethnography – that of *play*. However, it does so by widening
out the communities of practice involved in Arctic social life to include various
emotionally competent beings, such as environmental scientists, bush pilots,
and, especially here, base support staff. This theoretical mission is accomplished
by examining the ethical and epistemological consequences of observing, what
I term, *spaces of play*. In his philosophical account of the centrality of play in
culture, Johan Huizinga emphasizes that play is intertwined with ritual practices
and notions of the sacred (Huizinga 1950). There are two important elements of
Huizinga's discussion that I wish to stress for our present purposes. First, Huizinga
sees the observation of Arctic sociality as a critical dimension of his theory of
play. Indeed, as Huizinga puts it, Greenlandic Inuit society provides 'one of the
most cogent arguments for the intimate connections between culture and play,
namely the drumming-matches or singing-matches' (Huizinga 1950, 84–85). His
discussion of these Inuit practices illustrates the connections between play, humour
and the relieving of societal tensions. Second, Huizinga places great emphasis on
the *spatial boundedness* of play. For Huizinga's theorization,

> one of the most important characteristics of play was its spatial separation from
> ordinary life. A closed space is marked out for it, either materially or ideally,
> hedged off from everyday surroundings. Inside this space the play proceeds,
> inside it the rules obtain (Huizinga 1950, 19).

In his critique of Huizinga, French philosopher Roger Caillois stresses further
this domain of play (Caillois 1961). As Caillois argues, the 'laws of ordinary
life are replaced, in this fixed space and for this given time, by precise, arbitrary,
unexceptionable rules that must be accepted' (Caillois 1961, 7). These rules both
create the space of play and govern performances within it.

Through ethnographic research at a scientific base in the Canadian High
Arctic, I argue that the spaces that compose such an organization are contested
and thus involve competing, but always interacting, communities of practice.
Within these communities, social interaction is undertaken between field scientists
and logistical support staff, or between indigenous and non-indigenous groups.
This has a number of consequences for the individuals involved, not least, the
development of an emotional register of reactions to an imposed hierarchy of
importance for scientific and logistical practices. Moreover, these reactions result
in dissent being *performed* in and through spaces of play. This chapter interrogates
the possibilities and limitations that are involved in trying to trace the marking

out of those very spaces. Whether the emotional geography of such spaces *should be recorded at all*, then, becomes one of the guiding questions of this chapter.[5] In order to understand the implications of these statements, it is crucial to have a grasp of the role of play in the Arctic ethnographic tradition.

In a review essay, David Riches has argued that Inuit studies have been hampered by an overemphasis on the documentation of tradition, at the expense of embedding discussions within theoretical debates in social and cultural anthropology (Riches 1990).[6] However, this argument does not hold with respect to anthropological discourses of emotion and affect. Indeed, there has been a long tradition in Arctic ethnography of the anthropology of emotions, best represented in the work of Jean Briggs on the 'patterning of emotion' in Inuit family life in her path-breaking *Never in Anger* (Briggs 1970, 7). As Briggs showed through sensitive explorations, emotional interaction affects the ethnographer as much as the agents under investigation. This provides critical insight for geographies of emotion. As Briggs puts it, 'cross-cultural studies of emotion have so proliferated in recent years that to list a few authors is to offend many. Relatively few of these, however, have focused on the *socialization* of affect' (Briggs 1998, 267). Understanding the social embedding of affect, then, requires ethnographic experience.

Briggs's contribution to the development of an emotional anthropology is best evidenced in her engagement with the classic trope of Arctic ethnography of 'play'. Briggs argues for 'the emotional power of the games' within Inuit society (Briggs 1987, 13). It is through learning from these games that Inuit come to maintain feelings and values with respect to aspects of social life. In her evocative ethnography of Chubby Maata, a three year old Inuit girl, Briggs (1998) illustrates the importance of playful games for the development of emotional motives behind the experience of everyday life.[7] And as Nicole Stuckenberger demonstrates in her excellent ethnography of Qikiqtarjuat, play is still central to theorizations of Inuit social life (Stuckenberger 2005). For Stuckenberger, within the seasonal morphology of Inuit social life,[8] games carry a broader ritual significance. During

---

5   Eric Laurier and Hester Parr, in a stimulating discussion of emotional intersubjectivity during fieldwork, stress that 'further work should clarify whether it is possible, or indeed desirable, to have an 'ethics' of emotional research' (Laurier and Parr 2000, 101).

6   See Balikci (1989) for an attempt to place controversies within Arctic ethnography together with corresponding debates in social theory.

7   Briggs (1987) has noted that any connections between her field research and psychoanalytic theory emerged retrospectively – ethnographic observation of social interactions resulted in her interpretations, rather than prior reading. It is interesting to note that Briggs's person-centred ethnographies, developed through field experiences, appear to have important epistemic parallels with those psychoanalytical and psychotherapeutic approaches to emotional geography developed by Bondi (2005a).

8   The seasonal morphology of Inuit society was first outlined by Marcel Mauss in his *Essai sur les variations saisonnières des sociétés Eskimos. Étude de morphologie sociale* (1906). Mauss noted that a winter/summer distinction was evident in the cultural and sacred practices of Inuit social life (Saladin d'Anglure 2006).

religious festivals, such as 'Christmas',[9] participation in games allows structured competition and cooperation and thus the energetic development of social relationships. As Stuckenberger concludes her study:

> Thus the notion of play, so fruitfully developed by Jean Briggs, acquires a special meaning: it constitutes a modality which allows the community to connect itself to transcendental agencies, and to establish a sense of community and cooperation between the participants. All involved are well aware, though, that these relations are fragile and can only be fully realized in play (Stuckenberger 2005, 213).

Play, then, is a central trope in ethnographic accounts of social life in the Arctic.

Playing with Scientific Practices

> Social scientists who go to the Canadian Arctic are often challenged by Euro-Canadians there to spend less time studying the Inuit and more time studying the whites (Riches 1977, 166).

This chapter deploys an unconventional stance towards Arctic play by focusing on the social ordering of a scientific research base. The site in question is the research base of the Government of Canada's Polar Continental Shelf Project (PCSP) at Resolute, Nunavut. The PCSP is a branch of Natural Resources Canada and provides logistical support for a vast range of environmental, biological, geological and glaciological field research in the Arctic. As an organization today, the PCSP facilitates science by providing crucial aid to researchers from Canadian Universities, and Federal and Territorial governments.[10] Such support is in terms of essential subsidies for flights of parties and supplies out to respective fly-camps in the field, the loan of instruments, vehicles, and equipment, and residence at scientific base camps. These research costs are over and above those provided for in NSERC (Natural Science and Engineering Research Council), or similar, research grants.[11]

The contemporary Polar Continental Shelf Project spans numerous sites, ranging from central administrative offices in Ottawa, major research bases in Resolute, Cornwallis Island, and Tuktoyaktuk, Mackenzie Delta, a building which

---

9   Christmas is translated as *quviasuvik* [literally 'time of happiness'] in Inuktitut (Stuckenberger 2005).

10   This support is also provided on a cost-recovery basis to non-Canadian university and government researchers, but not for any private or commercial interests.

11   I will not discuss in detail the process by which this support is gained, but it is important to note the PCSP's consequent role as a gate-keeper in Arctic science.

is intermittently staffed at Eureka, Ellesmere Island, and co-ordinated scientific activities and dispersed buildings spanning across the entire Canadian Arctic. The retrenchment of research funding for science in the Canadian Arctic in recent decades has made winter scientific fieldwork impossible: funds have not allowed PCSP research bases to be open year-round, and this has prevented over-wintering.[12] Short scientific field seasons during the diurnal daylight of the Arctic summer result in cycles of very-long working days punctuated by occasional breaks. In many ways, scientific practice in the Arctic bears the hallmarks of seasonality that characterize other quotidian activities.

My study of field scientific practices revolves around the performance of the everyday at the PCSP base. Jean Lave argues that the everyday does *not* 'denote a division between domestic life and work' (Lave 1988, 14). The everyday is instead formed from what people do in quotidian cycles of activity (Lave 1988). Such conceptualizations are important in ethnographies of field science, where demarcations between domesticity and labour are particularly difficult. Moreover, recent work in anthropology has begun to take seriously notions of 'play' (Gable 2002; Lury 2003). The proceeding discussion problematizes any simple distinction between work and play due to the spatially contiguous nature of domesticity, science and indigenous practices at an Arctic field station. This is developed through the deployment of the notion of 'spaces of play', that is, spaces where resistance and dissent are performed *through* games, the mimicry of more senior and powerful individuals, and other activities.

Furthermore, following Michael Lynch and Steve Woolgar, I wish to investigate what counts as scientific practice in *particular* cases (Lynch and Woolgar 1990).[13] Through observations of quotidian activities in the Arctic, it quickly becomes evident that a notion of 'scientific practice' cannot simply entail, say, the calibration of instruments, but must include the development of practical understandings and the structuring of social interactions between individuals. Such social interactions often occur as much in spaces of play as in workspaces. And it is precisely for this reason that I will focus here on 'facilitators of science' rather than 'scientists'.

This account is further influenced by Lave's and Wenger's notion of learning about situations in the field through '*legitimate peripheral participation*' in

---

12 The intended scientific field season for projects operating out of PCSP Resolute in 2002 was Saturday 23 March–Tuesday 10 September, but days at either end were lost because projects were cancelled. Each year the operational season tends to get shorter due to financial constraints on the PCSP budget. This gradual decrease in the length of the season has had serious consequences in terms of financial hardships for employees on seasonal contracts for whom each field season provides their sole employment for the full year. My participant observation was conducted during two full summer field seasons, in 2001 and 2002.

13 As Latour argues, this dependence on the empirical origins of theoretical arguments is also 'what science studies does best, that is, *paying close attention to the details of scientific practice*' (Latour 1999, 24, my emphasis).

communities of practice (Lave and Wenger 1991, 29, original emphasis). This puts reliance on the researcher to develop her learning through participation in the field situation. As sociologist of science Harry Collins argues,

> ...my own (partial) mastering [of] the practices of the communities I study is a very concrete achievement akin to what the members of those communities themselves achieve as they become members. ... [M]astery of a practice cannot be gained from books or other inanimate sources, but can *sometimes*, though not always, be gained by prolonged social interaction with members of the culture that embeds the practice (Collins 2001, 107, my emphasis).

In this chapter, I focus on the everyday lives, opinions, and values, of the base staff at PCSP Resolute,[14] which are often revealed most starkly at space-times of play during the frenetic routines of the base. The seasonality of labour practices in the High Arctic is such that these spaces of play tend to emerge infrequently and in reaction to local variances in environmental and social conditions, such as malign weather or personnel changes on base. This can mean that such occasional spaces, whilst ephemeral, are peculiarly carnivalesque.[15]

The Polar Continental Shelf Project

Resolute has a remarkable position in Canadian *High* Arctic research, in that all researchers pass through Resolute airport and all Canadians, and the vast majority of non-Canadians, usually then stay at PCSP before flying out to the respective field sites. This meant that by spending long periods at PCSP Resolute I was able to come into contact with a huge range of researchers from, for example, Canadian Universities, Danish Research Institutes and even NASA.[16] Moreover, Resolute is a staging area for almost all North American, British, and even Scandinavian expeditions to the North Pole, as well as for adventure tourism. The fact that Resolute is a place that is passed through as a means to an end, rather than an end

---

14   In this way, I contribute towards anthropologies of labour that take seriously the role of work in the life projects of human agents (see Corsín Jiménez 2003).

15   On the theorization of the carnivalesque, see Stallybrass and White (1986) and Turner (1987). For a more developed discussion of Arctic carnival, with respect to scientist-Inuit interaction, see Powell (2008).

16   The scientists supported by PCSP include those from Canadian universities, various research branches of the Government of Canada, and non-Canadian researchers from countries such as Denmark (Greenland), the US, Japan, and the UK. The PCSP co-ordinates over 120 field parties each year. These groups vary in size from two to three scientists to large parties of ten to twelve researchers. The continual arrival and departure of these scientists thus involves around three to four hundred individuals over the duration of each annual field season.

in itself,[17] meant that I could also observe the way in which many social groups, including different bands of scientists, often did *not* converse with each other. For a relatively isolated community, therefore, there is a remarkable degree of human traffic.

My general activities around the base included helping in the kitchen (although this was not allowed during the second field period due to new insurance regulations), talking to scientists, talking to employees, hanging around in the various parts of the base or conducting more formal interviews. I also spent significant amounts of time providing assistance in the field to various scientific teams operating out of PCSP Resolute, and to other parties based at field sites across the Canadian Arctic. In this chapter, I focus upon discussions with one group of actors, or community of practice, within PCSP Resolute: the base technical staff.[18] I attend to the important but often overlooked voices and emotions of these actors that have been revealed through observation of spaces of play.

The base technical staff, who are all seasonal contractors from Newfoundland, are probably the most critical element in the functioning of the PCSP. These employees work incredibly long hours. For example, when the annual sea-lift arrived ahead of schedule on 24 August 2001, the Newfoundlanders worked through the night to unload the sea-lift, until (at least) 3 o'clock in the morning. These members of staff returned to duty less than four hours later for the next day's shift at 7 o'clock. As well as a general seasonal division in Arctic logistics, then, the working day of base staff depends on daily circumstances. Rapid changes in current, or predicted, weather conditions means that duties and tasks are performed as opportunities arise.

Dislocated Labour: Learning through Spaces of Play

Initially, it was difficult to be accepted by these members of the PCSP hierarchy because of my association with other communities of practice, such as the scientists or the base management.[19] However, over the course of fieldwork, it became easy to talk to the Newfoundlanders during space-times of play. At such moments, in response to circumstances provided by weather conditions, or by

---

17 Some researchers, however, use PCSP Resolute for daily accommodation, making day-trips to field sites on Cornwallis Island.

18 All personal identities have been heavily disguised, such as by the use of pseudonyms. As well as undergoing ethical review to gain a Research Licence from the Nunavut Research Institute, Iqaluit, 'informed consent' was also negotiated with all participants.

19 This was due to the initial conditions of my presence at the base, in that I had myself applied successfully for PCSP support as a scientific researcher. Although conducting ethnography of science, as an individual granted support by PCSP, I was constructed as a 'scientist' like any other researcher on base.

personnel changes on base, such as when staff were arriving at, or departing from, PCSP Resolute, these spaces would emerge. Towards the end of a field season, for example, a space of play would emerge because staff believed that 'the season was over'. Another occasion arose when, although operations on the base were expected to last a further eight weeks, a senior member of PCSP management was arriving from southern Canada the next morning, so it was a final opportunity to celebrate without surveillance. Such occurrences usually involved alcohol, music, games, and discussion, which would include gossip about scientists on base or senior members of the PCSP hierarchy.[20] These spaces of play are therefore doubly carnivalesque, both as areas for jollification and in that they involve the subversion of authority (Stallybrass and White 1986).

Even the base management would stress the importance of the staff from Newfoundland. It created great difficulties for the planning of logistics, complained a base manager named Edward, when all the PCSP seasonal staff returned to Newfoundland for the winter. As well as the staff being effectively unemployed for most of the year, their embodied expertise would be useful for planning various logistical arrangements. As Edward put it, the staff possess crucial knowledge and experience, but the costs that would be involved for them to come to Ottawa or to convene some other sort of meeting during the winter would prove prohibitive.[21] As Edward argues:

> We are such a small organization it is difficult to do things. ... Polar Shelf works because of the people, the 'Newfies' and the cooks, up here, but they are getting older, and they are getting pissed off as they get older. They don't get paid enough for being up here and away from their families, and they basically get abused by the Government. We have been fortunate because we have had 'Newfies' and there are no jobs in Newfoundland. ... But we can't afford to lose the experience.[22]

The reason for these constant reminders of the importance of this labour in the PCSP, repeatedly mentioned by both managers and scientists, was that the Newfoundlanders perceive themselves to be marginalized from the institution.

---

20   Riches (1977) identified similar tendencies for the development of social cliques and rituals, involving festivities and drinking, amongst Euro-Canadians in settlements in northern Canada during the 1970s.

21   Conversation with Edward Freeman, Resolute, 24 August 2001, recorded in field notes, August 2001, Book 2.

22   Conversation with Edward Freeman, Resolute, 24 August 2001, recorded in field notes, August 2001, Book 2. It is necessary to note the use of the term 'Newfie' in this passage. This designation has a contested history within Canada and is generally deemed to be offensive. However, this term was used commonly as a descriptor, both by the Newfoundland staff and by others, at PCSP Resolute. I reproduce it here only to remain faithful to the records in my fieldnotes.

Towards the end of the 2001 season, for example, Edward was pleased that the sealift arrived ahead of schedule and had been unloaded within a few hours. For Edward, this meant that the base staff should be able to finish the season, and thus go home, early and, he told Barry, 'we should get a reward for getting everything done quickly'. 'What,' Barry replied, 'two boots in the arsehole?'. This exchange reveals an evident demotivation and lack of job satisfaction.

However, what I want to stress about these feelings of alienation is that they were most evident at space-times of play. The *personal* costs of working for PCSP were revealed at such moments. It is difficult for any individual to cope with long periods of isolation. Researchers have produced psychological studies of cognitive stressors and coping strategies involved in residence at polar bases (Barbarito et al. 2001; Mocellin 1988; Mocellin and Suedfeld 1991). However, such studies are unable to quantify the impact of events happening *away* from the bases with those whom base staff have personal relationships. The first season that Gary, a Newfoundlander responsible for building maintenance, was in the Arctic, he spent 14 consecutive weeks on the PCSP Ice Island.[23] Arriving on the Arctic Ocean during the uncomfortable environment of early February, conditions were made more difficult because 'that was the first time I had ever been away from my wife'. A conversation between the Newfoundlanders, during a space of play, is illustrative of this point:

> Gary: It takes a certain kind of *person* to work up here. You have to be able to handle the isolation. It was really bad on the Ice Island.
>
> Barry: It takes a certain kind of *wife*.

Both Edward and Barry, as Gary put it, 'are on their second wives'. As he was working away, Barry's child was six months old before he saw her for the first time.

Another factor that contributes to low morale is that, unlike scientists, the staff are often unsure of how long they will be kept on base. They are often sent 'south' early, cutting short their envisaged season of employment, without prior warning. When Andrew Walder, a senior manager, returned to PCSP Resolute in 2001 towards the end of the season, he had decided that he would send Edward, two Newfoundlanders, and a cook, home on the next flight south. This would have the consequence that only Andrew and two Newfoundlanders would remain until

---

23 The PCSP Ice Island was in operation from c.1986–1989, and was literally a floating base on the Arctic Ocean from which various scientific projects were undertaken. The Newfoundlanders made constant reference to those individuals who had been employed on the Ice Island, and for how long they had 'served'. This was partly because it was seen as the ultimate challenge in Canadian Arctic employment, but it could also be interpreted in a sometimes humorous way, as it was generally regarded amongst PCSP staff and scientists that the Ice Island episode was a huge failure. Although there had been envisaged a twenty-five year programme of research, the chosen island disintegrated within three seasons.

the closedown of the base for winter.[24] This was upsetting news for some staff. As these seasonal contracts effectively form the annual salary for the staff, such logistical decisions that are taken as PCSP suffers from overall budgetary pressures have massive ramifications for the Newfoundlanders and their families.[25]

All the Newfoundlanders thus feel isolated from the decision-making process of PCSP.[26] The marginalization of the Newfoundlanders would therefore be expressed through frequent, implicit, challenges to the authority of management during spaces of play. Gary would make jokes about the perceived lack of technical know-how of the Director. Barry argued that:

> We're the ones who make the real decisions. We never listen to Andrew when we're loading the planes. I just want to know where the group are heading and for how long.[27]

On another occasion, when angry over a particular decision, Barry was more forceful:

> We do the fucking work. In the office, half of what they tell you is bullshit.[28]

At other times, Barry would tend to be more circumspect about the situation:

> I'm happy. … The management always seem to be increasing their numbers over the years. Whereas we, the ones who do the work, have always stayed about the same. Though it's the same with any job I suppose.[29]

---

24   Conversation with Andrew Walder, Resolute, 25 August 2001, recorded in field notes, August 2001, Book 2.

25   I should note here that this policy of sending staff home early for financial reasons has always been a management strategy for PCSP. The first PCSP Co-ordinator, E.F. Roots, for example, sent a telegram to Lionel Laurin, the camp manager at Tuktoyaktuk in May 1969, that read as follows: 'If we are to start operations by Feb. 1 of next year it is imperative that you reduce your total base camp staff by six men months before August 15. This would mean [a] reduction of two men for three months or equivalent. We are also reducing casual employment in other areas'. E.F. Roots to L. Laurin, PCSP Tuktoyaktuk, Telegram, 1.30 p.m., 22 May 1969, Library and Archives of Canada, Record Group 45, Volume 322, File 5-1-1, Part 1, Radio communications – Miscellaneous, Part 1, 1968-1969.

26   See Doing (2004) for similar findings regarding the politics of labour in a recent ethnographic study of a physics laboratory.

27   Conversation with Barry Fuller, Resolute, 26 July 2002, recorded in field notes, June-August 2002, Book 2.

28   Conversation with Barry Fuller, Resolute, 21 August 2001, recorded in field notes, August 2001, Book 1.

29   Conversation with Barry Fuller, Resolute, 22 August 2001, recorded in field notes, August 2001, Book 1.

At the same time, I always had the impression that the Newfoundlanders were *proud* of the occupation of their lifetimes. As Barry noted:

> I could've wrote a book from what I've seen. I've been to the North Pole twice, not many men can say that. My job is to help the science and the scientists. They go out in the rain and the snow and live in a tent, and we decide how much fuel they need. Ninety percent of what scientists want is wrong.[30]

It was through learning to participate in these spaces of play therefore, that I began to find evidence of hidden stories and agendas within the PCSP. However, these spaces of play often involved 'loose talk', and this leads to anxieties for the analyst as to how to represent such observations.

The Ethics of Recording Emotional Practices

Participation in spaces of play, I argue, results in epistemological and ethical quandaries. Such spaces are emergent from resistance to the official narratives of scientific practice in the Canadian Arctic. It is from the observation of activities in such spaces therefore that fuller descriptions of scientific and logistical practices become possible. In short, material derived from learning in spaces of play gives better access to the communities of practice involved in constituting PCSP Resolute. Studying play reveals the actual cultures of Arctic field science.

However, precisely because such spaces are resistant to management practices and official narratives, they are usually inaccessible, and perhaps should always remain so. Spaces of play become accessible only by participating in play. And this results in ethical anxiety for the ethnographer. Agents would appear to be much more relaxed and informal with an ethnographer who is participating in a space of play. Indeed, some of the spaces that I have partially documented here are, due to local regulations regarding alcohol, problematic. How is previously negotiated 'informed consent' complicated during spaces of play? How should material derived from such spaces be represented? The ethnographer must negotiate between fidelity to the representation of practices and the articulation of often repressed voices and experiences, whilst anticipating the consequences of representing spaces of play for informants long after return from the field. In this chapter I have been careful to censor much discussion, as well as making the usual attempts to disguise identities. But are such efforts sufficient?[31]

---

30   Conversation with Barry Fuller, Resolute, 26 July 2002, recorded in field notes, June–August 2002, Book 2.

31   There have been few social studies that investigate the ethical implications of representing such spaces. Although, in organizational studies, Michael Rosen has undertaken important work (Rosen 1985; 1988). Through an analysis of the social drama of the Christmas party organized for all staff at Shoenman and Associates, Rosen argues

The presentation of this material has therefore required careful attention to ethical issues, because it has been difficult to represent voices and emotions without making evident the identity of the articulating individual. I should note that some of these opinions may be surprising, even disturbing, to those expecting some sort of celebratory account of scientific practice at PCSP. Needless to say, my recording some of them here does not mean that I fully subscribe to any or all of them. However, this effort at careful representation has been crucial both to making my argument and to fulfilling my ethical obligations as an ethnographer of practices. This raises an emotional dimension of the practice of ethnography that is all too often neglected – that of *guilt*.

Participating in spaces of play has resulted in guilt, that most powerful emotion, about my incomplete attempts to represent some of the results of my learning. This, of course, is not an uncommon emotion for emotional geographers. Rebekah Widdowfield outlines a familiar set of responses, experienced during fieldwork with lone parents in the West End of Newcastle, such as anger, distress and, ultimately, demoralization. The resulting sense of powerlessness, as she puts it, led her 'to question both my ability and desire to carry on with the research project' (Widdowfield 2000, 205). In an essay on the emotional responses of the researcher to the field, Liz Bondi illustrates the commonality of feelings of guilt and shame at the inadequacies inherent in the research process:

> I felt guilty because, despite my efforts to be honest, people were spending time with me and were telling me all about their experiences, perhaps at least in part, on the basis of a misapprehension about my capacity to offer anything in return (Bondi 2005b, 238).

In an account of fieldwork in Papau New Guinea, anthropologist Lisette Josephides argues that the 'strongest motivator which emerges from the ethnographic vignettes is the feeling of *resentment*' (Josephides 2005, 72, my emphasis). Through this formulation, Josephides argues that resentment for the ethnographer 'is the feeling of not being acknowledged' (Josephides 2005, 86). Rather than resenting the communities of practice encountered in the field, I feel that it is the feeling of being unable to represent spaces of play more fully that causes anxiety for this ethnography.

For Bondi, through the process of qualitative research, the practitioner must *suspend* decisions about feelings so as 'to reflect on emotions in their full richness and complexity' (Bondi 2005b, 241). It is not that emotions of shame are because of poor research design or incoherent thought about research ethics; rather, guilt is a necessary emotion during such research and must therefore be thought through

---

that alcohol holds an important social function. Enacting the 'play' of the party, for Rosen, suggests the importance of those moments for peculiar agency beyond normal controls that, at the same time, reproduce the very social structures of the organizational community (Rosen 1988).

and resisted, lest it become methodologically disenabling. It is within these discussions that emotional ethnographies should take central place.

Conclusion

I have argued that the particular emotional practices involved in undertaking science in Resolute, some of which have been outlined here, have wider significance. This account has stressed the importance of ethnographic experience to the study of emotions in social life. As fieldwork involves a set of affectual encounters between the ethnographer and other practitioners within competing and complementary communities of practice, doing such work can contribute to debates about emotional geographies. The corporeal rigours involved in ethnography, the need to understand, interpret and empathize with emotional codes, sensitize the analyst to allow sensible theoretical discourse regarding emotions. In attempting to convey the viscerality of everyday scientific practices, then, I hope to have contributed to the development of (emotional) ethnographies of embodied practice (Wacquant 2005). By outlining some of the emotional processes of scientific research in the Arctic, I have argued for the broadening of these intellectual encounters in both geographies of science and emotional geographies.

The practice of ethnography has always had a difficult relationship with the discipline of geography. By recovering the sense of ethnography as an immersive practice involving dwelling at particular sites, I have attempted to outline some possibilities for emotional geographies. A focus upon the recording of the emotional practices involved in Arctic social life reveals empirical potentiality for geographers of emotion. Attempting to study the emotional aspects of scientific fieldwork may well uncover aspects that are best *revealed* through analyses of spaces of play. But as the consequences of these descriptions are uncertain, I am still anxious as to whether I should have participated, and thus learnt from them, at all.

References

Anderson, B. and Harrison, P. (2006), 'Questioning affect and emotion', *Area* 38:3, 333–335.

Anderson, K. and Smith, S.J. (2001), 'Emotional geographies', *Transactions of the Institute of British Geographers* NS 26:1, 7–10.

Balikci, A. (1989), 'Ethnography and theory in the Canadian Arctic', *Études/Inuit/ Studies* 13:2, 103–111.

Beatty, A. (2005a), 'Emotions in the field: what are we taking about?', *Journal of the Royal Anthropological Institute* 11:1, 17–37.

Beatty, A. (2005b), 'Feeling your way in Java: an essay on society and emotion', *Ethnos* 70:1, 53–78.

Besnier, N. (1990), 'Language and affect', *Annual Review of Anthropology* 19, 419–451.

Bondi, L. (2003), 'Empathy and identification: conceptual resources for feminist fieldwork', *ACME: An International E-Journal for Critical Geographies* 2:1, 64–76.

Bondi, L. (2005a) 'Making connections and thinking through emotions: between geography and psychotherapy', *Transactions of the Institute of British Geographers* NS 30:4, 433–448.

Bondi, L. (2005b), 'The place of emotions in research: from partitioning emotion and reason to the emotional dynamics of research relationships', in J. Davidson, L. Bondi, and M. Smith (eds.), *Emotional Geographies* (Aldershot: Ashgate), 231–246.

Bondi, L. Davidson, J. and Smith, M. (2005), 'Introduction: Geography's "Emotional turn"', in J. Davidson, L. Bondi, and M. Smith (eds.), *Emotional Geographies* (Ashgate: Aldershot), 1–16.

Barbarito, M., Baldanza, S. and Peri, A. (2001), 'Evolution of the coping strategies in an isolated group in Antarctic base', *Polar Record* 37(201), 111–120.

Briggs, J.L. (1970), *Never in Anger: Portrait of an Eskimo Family* (Cambridge, MA and London: Harvard University Press).

Briggs, J.L. (1987), 'In search of emotional meaning', *Ethos* 15:1, 8–15.

Briggs, J.L. (1998), *Inuit Morality Play: The Emotional Education of a Three-Year-old* (New Haven and London: Yale University Press).

Caillois, R. (1961), *Man, Play and Games* (Urbana and Chicago: University of Illinois Press). [First published in 1958 as *Les jeux et les hommes*; reprinted by Illinois in 2001.]

Clifford, J. (1988), *The Predicament of Culture: Twentieth-century Ethnography, Literature, and Art* (Cambridge, MA and London: Harvard University Press).

Collins, H.M. (2001), 'What is tacit knowledge?', in T.R. Schatzki, K. Knorr Cetina and E. von Savigny (eds.), *The Practice Turn in Contemporary Theory* (London and New York: Routledge), 107–119.

Corsín Jiménez, A. (2003), 'Working out personhood: notes on labour and its anthropology', *Anthropology Today* 19:5, 14–17.

Doing, P. (2004), '"Lab Hands" and the "Scarlet O": epistemic politics and (scientific) labor', *Social Studies of Science* 34:3, 299–323.

Gable, E. (2002), 'Beyond belief? Play, scepticism, and religion in a West African village', *Social Anthropology* 10:1, 41–56.

Gupta, A. and Ferguson, J. (1997), 'Discipline and practice: "the field" as site, method, and location in anthropology', in A. Gupta and J. Ferguson (eds.), *Anthropological Locations: Boundaries and Grounds of a Field Science* (Berkeley and Los Angeles: University of California Press), 1–46.

Huizinga, J. (1950), *Homo Ludens: A Study of the Play Element in Culture* (Boston: The Beacon Press) [First published in 1938].

Josephides, L. (2005), 'Resentment as a sense of self', in K. Milton and M. Svašek (eds.), *Mixed Emotions: Anthropological Studies of Feeling* (Oxford and New York: Berg), 71–90.

Latour, B. (1999), *Pandora's Hope: Essays on the Reality of Science Studies* (Cambridge, MA: Harvard University Press).

Laurier, E. and Parr, H. (2000), 'Emotions and interviewing in health and disability research', *Ethics, Place and Environment* 3:1, 98–102.

Lave, J. (1988), *Cognition in Practice: Mind, Mathematics and Culture in Everyday Life* (Cambridge: Cambridge University Press).

Lave, J. and Wenger, E. (1991), *Situated Learning: Legitimate Peripheral Participation* (Cambridge: Cambridge University Press).

Livingstone, D.N. (1995), 'The spaces of knowledge: contributions towards a historical geography of science', *Environment and Planning D: Society and Space* 13:1, 5–34.

Lury, C. (2003), 'The game of Loyalt(o)y: diversions and divisions in network society', *The Sociological Review* 51:3, 301–320.

Lynch, M. and Woolgar, S. (1990), 'Introduction: sociological orientations to representational practice in science' in M. Lynch and S. Woolgar (eds.), *Representation in Scientific Practice* (Cambridge MA and London: MIT Press), 1–18.

McCormack, D. (2006), 'For the love of pipes and cables: a response to Deborah Thien', *Area* 38:3, 330–332.

Mocellin, J.S.P. (1988), *A Behavioural Study of Human Responses to the Arctic and Antarctic Environments*, Unpublished PhD thesis, Interdisciplinary Studies, Departments of Geography and Psychology (University of British Columbia: Vancouver).

Mocellin, J.S.P. and Suedfeld, P. (1991), 'Voices from the ice: diaries of polar explorers', *Environment and Behavior* 23:6, 704–722.

Milton, K. (2005a), 'Afterword', in K. Milton and M. Svašek (eds.), *Mixed Emotions: Anthropological Studies of Feeling* (Oxford and New York: Berg), 215–224.

Milton, K. (2005b), 'Meanings, feelings and human ecology', in K. Milton and M. Svašek (eds.), *Mixed Emotions: Anthropological Studies of Feeling* (Oxford and New York: Berg), 25–41.

Passaro, J. (1997), '"You can't take the subway to the field!": "Village" epistemologies in the global village', in A. Gupta and J. Ferguson (eds.), *Anthropological Locations: Boundaries and Grounds of a Field Science* (Berkeley and Los Angeles: University of California Press), 147–162.

Powell, R.C. (2002), 'The Sirens' voices? Field practices and dialogue in geography', *Area* 34:3, 261–272.

Powell, R.C. (2007), 'Geographies of science: histories, localities, practices, futures' *Progress in Human Geography* 31:3, 309–329.

Powell, R.C. (2008), 'Canada Day in Resolute: performance, ritual and the nation in an Inuit community', in D. Cosgrove and V. della Dora (eds.), *High Places: Cultural Geographies of Mountains, Ice and Science* (London and New York: I.B. Tauris), 178–195.

Powell, R.C. and Vasudevan, A. (2007), 'Geographies of experiment' *Environment and Planning A* 39:8, 1790–1793.

Riches, D. (1977), 'Neighbours in the "bush": white cliques', in R. Paine (ed.), *The White Arctic: Anthropological Essays on Tutelage and Ethnicity* [Newfoundland Social and Economic Papers, No. 7] (St. John's, Newfoundland: Institute of Social and Economic Research, Memorial University of Newfoundland), 166–188.

Riches, D. (1990), 'The force of tradition in Eskimology', in R. Farndon (ed.), *Localizing Strategies: Regional Traditions of Ethnographic Writing* (Edinburgh and Washington, DC: Scottish Academic Press and Smithsonian Institution Press), 71–89.

Rosen, M. (1985), 'Breakfast at Spiro's: dramaturgy and dominance', *Journal of Management* 11:2, 31–48.

Rosen, M. (1988), 'You asked for it: Christmas at the bosses' expense', *Journal of Management Studies* 25:5, 463–480.

Saladin d'Anglure, B. (2006), 'Introduction: The influence of Marcel Mauss on the anthropology of the Inuit', *Études/Inuit/Studies* 30:2, 19–31.

Sanjek, R. (1990), 'On ethnographic validity', in R. Sanjek (ed.), *Fieldnotes: The Makings of Anthropology* (Ithaca, NY and London: Cornell University Press), 385–418.

Stallybrass, P. and White, A. (1986), *The Politics and Poetics of Transgression* (Ithaca, NY: Cornell University Press).

Stuckenberger, A.N. (2005), *Community at Play: Social and Religious Dynamics in the Modern Inuit Community of Qikiqtarjuat* (Amsterdam: Rozenberg Publishers).

Svašek, M. (2005), 'Introduction: emotions in anthropology', in K. Milton and M. Svašek (eds.), *Mixed Emotions: Anthropological Studies of Feeling* (Oxford and New York: Berg), 1–23.

Thien, D. (2005), 'After or beyond feeling? A consideration of affect and emotion in geography', *Area* 37:4, 450–456.

Tolia-Kelly, D.P. (2006), 'Affect – an ethnocentric encounter? Exploring the "universalist" imperative of emotional/affectual geographies', *Area* 38:2, 213–217.

Tonkin, E. (2005), 'Being there: emotion and imagination in anthropologists' encounters', in K. Milton and M. Svašek (eds.), *Mixed Emotions: Anthropological Studies of Feeling* (Oxford and New York: Berg), 55–69.

Turner, V. (1987), *The Anthropology of Performance* (New York: Performing Arts Journal Publications).

Wacquant, L. (2005), 'Carnal connections: on embodiment, apprenticeship, and membership', *Qualitative Sociology* 28:4, 445–474.

Widdowfield, R. (2000), 'The place of emotions in academic research', *Area* 32:2, 199–208.

Wolf, D.L. (1996), 'Situating feminist dilemmas in fieldwork', in D.L. Wolf (ed.), *Feminist Dilemmas in Fieldwork* (Boulder, CO: Westview Press), 1–55.

PART 3
Mourning

Chapter 7

'What We All Long For': Memory, Trauma and Emotional Geographies

Anh Hua

Hannah Arendt writes in *Between Past and Future*: 'In the words of Faulkner, "the past is never dead, it is not even past". This past, moreover, reaching all the way back into the origin, does not pull back but presses forward, and it is, contrary to what one would expect, the future which drives us back into the past' (Arendt 1977, 10–11). Roger Simon also reminds us to rethink our relation to the past, to reappraise the ties between civil life, historical consciousness, and the pedagogical force of various practices of remembrance. Remembrance practices are implicated in questions regarding human sociality and the problem of human connectedness across different times and places (Simon 2005, 2). He asks us to think through and formulate a public historical consciousness central to a new democratic form of community, one that represents the voices of the oppressed (4). While remembrance practices can sometimes be culturally divisive, *democratic* politics require, as Arendt argues, a pluralistic space where individuals are able to express their different identities in public, differences composed, at least in part, by and through the temporal 'pull' and 'press' that re-collects individual and collective memories.

I therefore agree with Simon that democracy requires that one is able to reclaim the past as experienced by the oppressed, by those often written out of history, and to formulate those experiences as part of public records and debates. He argues that the task of working towards social transformation and justice is not to forget the past but to remember it. One way to achieve this public historical consciousness and historical remembering is to seriously ask the question: '…what it might mean to take the memories of others (memories formed in other times and spaces) into our lives and so live as though the lives of others mattered' (Simon 2005, 9). Following Simon's argument for the need to work towards social transformation by formulating a public historical consciousness, I am interested in how works of fiction as an emotional geography can be utilized as a site of public memory to commemorate and record voices repressed or suppressed by official history. Fictions, at least those that are explicitly concerned with sympathetically revisiting those past events which have proved important in creating marginalized individuals and groups' senses of identity, might be thought of as re-composing emotionally saturated histories in ways that offer a 'novel' kind of public space and record. Such fictional landscapes and characters allow readers the possibility of vicarious, but still intimate and emotionally intense experiences of (often traumatic) events that have had profound affects – affects to which these readers might otherwise have remained oblivious. These fictional characters, whose stories and memories are present(ed) as part of an imaginary geography, can serve to express and foster public understandings of past events whose repercussions remain very real, not just for those who have their own personal memories of these events, but for those whose own identities have been composed in relation to the re-telling of these memories, perhaps in terms of the life-stories of previous generations.

In particular, I will examine the 'fictionalized' memories, which the Afro-Caribbean Canadian feminist author Dionne Brand presents, and how they relate to broader questions of the role of memory in diasporic historiography. I argue that Brand engages in what I call 'practices of remembrance,' and I will explore what those practices are and how they might include works of fiction. Brand explores a history of marginalized communities and how this history differs from 'mainstream' history, including official and popular history. There is a relation between Brand's works of fiction, emotional geography, and emotional historiography, as some works of fiction, particularly diasporic fiction, can challenge dominant historiography with stories intended to invoke the emotional memories of those concerned and emotional responses in readers. In her novel *What We All Long For*, for example, Brand reveals how a work of fiction might attempt to translate the trauma of the Vietnam War, to reveal how the diasporic Vietnamese community has to contend with the emotional geography and emotional historiography of loss and mourning.

In this chapter, I will begin by examining the difference between personal and collective memories, history and the past, and the political struggle over the meaning of memory within Euro-American culture. In the second section, I will explore Dionne Brand's practices of remembrances, such as her engagements with the cultural traumas of slavery in her earlier novel *At the Full and Change of the Moon*, and the cultural traumas of the Vietnam War in her novel *What We All Long For*. In the third section, I will give a close reading of this novel. Here, Brand not only engages with war and family trauma, melancholia and mourning, and a sense of loss, but also the struggle by the second-generation migrant youths to achieve a sense of identity and place, love and desire, longing and belonging. The chapter ends with a section on the significance of works of fiction as emotional geography for diasporic historiography, as exemplified by Brand's remembrance practices.

Personal and Collective Memory, History and the Past

The past is not the same as material history. History, whether oral or written, is a partial interpretation of the past, all those events and occurrences that involve various historical subjects that unknowingly or knowingly affect our presents. Some historical subjects who directly or indirectly experience the past may not have their personal memories and personal feelings of the events recognized and recorded as publicly accessible memory or official history, while others are recognized and recorded. Much of this selective interpretive representation has to do with who has power, privilege, resources and voice. The selective elimination of the experiences of the oppressed from official history or public memory has strategic value, especially for the nation: no representation, no voice, means little ideological and political power for the oppressed, which furthers the oppression.

Memory is a construction or reconstruction of what actually happened in the past. Personal memories are often different from public accounts of the past, which may have little to do with what individuals actually recall but have to do with how the past is remembered or put together to serve certain purposes. Collective memory, then, is not simply an addition of personal memories. The production of a collective or group history requires a public re-interpretation of personal memories that are placed at the service of the collectivity. Hence, it always involves a particular form of forgetting, one more or less deliberately sanctioned for political purposes. The question is whose voice, experiences, histories and personal memories are being forgotten by the collectivity such as the nation-state or a given community.

Memory and history can also be differentiated, although the two are sometimes fused. Whereas memory is associated with the continuity between past and present, history may be associated with a difference between past and present. While memory is marked by irregular and uncertain boundaries, conventional history writing divides the sequence of centuries into fixed historical periods and reconstructs the past from a critical distance (Misztal 2003, 101). Just as individual memory often works below the level of consciousness, so too collective 'memory' might be regarded as operating 'ideologically' behind our backs and, because of this, is more dangerously subject to manipulation by time and by societies (LeGoff 1992, xi). In particular, the ruling class frequently dominate and manipulate collective memory for their own gains and interests: 'To make themselves the master of memory and forgetfulness is one of the great preoccupations of the classes, groups, and individuals who have dominated and continue to dominate historical societies. The things forgotten or not mentioned by history reveal these mechanisms for the manipulation of collective memory' (54).

Memory can also tell us a great deal about our individual and collective past. As Vijay Agnew suggests, 'Memories establish a connection between our individual past and our collective past (our origins, heritage, and history). The past is always with us, and it defines our present; it resonates in our voices, hovers over our silences, and explains how we came to be ourselves and to inhabit what we call "our homes"' (Agnew 2005, 3). Memories involve 'an active process by which meaning is created' rather than 'mere depositories of fact' (8–9). As Agnew observes, memories can be used to ignite our imaginations and allow us 'to recreate our recollections of home as a haven filled with nostalgia, longing, and desire' (10) or they can 'compel us, as witnesses and co-witnesses, to construct home as a site and space of vulnerability, danger, and violent trauma' (Ibid.). Memories can evoke imaginary homelands and places of origin or 'an antidote to the struggles of the present' (Ibid.).

There is thus a tension between the idea of memory as recollection or witnessing of a forgotten past and memory being used to imagine or to create a present or future community. This tension is linked to the tension between telling the truth (i.e. what history writing promises) and creating a fictional account (i.e. critical works of fiction representing historical events) that is supposed to tell something of that truth. Perhaps history writing promises to witness and to recollect a forgotten past, and critical works of fiction promise to both witness and recollect a forgotten past as well as imagine or create a present/future community. While there may be debates about which is more truthful, history/historical 'facts' or fictional accounts that may tell something of a truth that may be missing in historical accounts, it is necessary to posit an interpretive ethics that respects the past as the past and does not just make it up free style to serve the present community's needs, however that community is imagined. In other words, authors of critical fictions do have to pay heed to the historical truth in their fictional accounts to be accountable towards an interpretive ethics, to represent feelingly and imaginatively, for example, an emotional or traumatic event with respects to the experiences of those trauma survivors.

Besides the debates around truth and falsehood with regards to memory and history, memories can also be used to produce nationalism or ethnic group identities, similar to Benedict Anderson's notion of nation as imagined communities (Anderson 1991). Memories or nostalgia for homelands, or wounds of dislocations and dispossessions have become important political narratives or metaphorical tools to imagine identity and community and to rewrite the nation of both origin and of settlement. The memory of personal and group experience is essential particularly for oppressed groups. Oppressed groups such as ethnic minorities are often pressured to forget and let go of their personal and collective memories and histories in order to assimilate into the new nation of settlement. Yet as Amritjit Singh et al. remind us, the memory of ethnic minorities or oppressed groups '…proves to be a powerful insurgent force against a dominant historiography that fails to represent the whole picture'. Writers associated with ethnic minorities often have their own distinct ways of creating memory 'out of history, silences, fragments, documentation, hints, and denials' (Singh et al 1996, 14).

For ethnic minorities and women, to forget one's individual and collective histories is to accede to the erasure or distortion of marginalized historical experiences. Moreover, to remember is to partially aid in the process and construction of our changing identities, haunted by pleasures and difficult events of the past. Nicola King writes: 'All narrative accounts of life stories, whether they be the ongoing stories which we tell ourselves and each other as part of the construction of identity, or the more shaped and literary narratives of autobiography or first-person fictions, are made possible by memory' (King 2000, 2). Yet memory, as King notes, 'can only be reconstructed in time, and time… "catches together what we know and what we do not yet know"' (Ibid.). One's identity is constructed by and through narratives, 'the stories we all tell ourselves and each other about our lives'. Moreover, the identity or self, constructed by these narratives, is dependent upon memories and the kind of access memories give us about the past (Ibid., 2–3). Yet, as King observes, individual and collective experiences including war, migration, assault or serious accident may make 'the relationship between the self "before" and the self "after" much more problematic' (3). Moreover, what is significant is not only what and how individuals remember and how they represent their memories, but also 'what might be termed a cultural struggle over the construction and meanings of memory within culture' (5).

What does it mean for a culture to remember and to forget? Cultural memory can provide a cultural identity and a sense of the importance of the past. Cultural memory, or what Marita Sturken (1997) calls 'tangled memory', is bound up with complex political stakes and meanings. It defines a culture but also the means by which its divisions and conflicting agendas are revealed. Cultural memory is crucial to understand a culture because it shows collective desires, needs and self-definitions. Memory is a narrative, not a replica of an experience, which can be retrieved and relived. Cultural memories are constructed as they are recollected; memory is a form of interpretation. What we remember is highly selective. How we retrieve memory is as much about desire and denial as it is about remembrance (1–2, 7). In examining memory, then, there may be contexts in which it is appropriate to place less emphasis on its power to invoke the real and more on its integrated relationship with forgetting, construction, fantasy, invention, and re-enactment. All of which suggests that fictional accounts are actually well situated to speak to issues concerning the public historical record and contemporary emotional geographies.

In the next section, I will examine what memories the author Dionne Brand engages with, and how they relate to broader questions of the role of memory in diasporic historiography and diasporic works of fiction.

Dionne Brand and Practices of Remembrances

Referring to the Holocaust, James Young argues that historians mistakenly make a forced distinction between memory and history, where history is defined as that which happened and memory as that which is remembered of what happened. This leaves no room for the survivor's voice and the survivor's memory of the events, whose value is lost to the historians. Young is interested in bridging the gap between the survivor's 'deep memory' and historical narrative, how to remember the past as it passes from living memory to history (Young 2003, 276–277). Historical inquiry, for Young, is understood as 'the combined study of both *what happened* and *how it is passed down to us*' (283). Instead of creating a gap between what happened and how it is remembered, Young proposes that we can benefit more by examining 'what happens when the players of history remember their [pasts for] subsequent generations' (Ibid.). Like Young, Dionne Brand in her 1999 novel *At the Full and Change of the Moon*, imagines how the subjects of history, the slaves and their descendents, remember their past for subsequent generations. Brand attempts to understand what happened during slavery and how slavery is passed down to us (Hua 2005, 78–79). The cultural trauma of slavery is the collective memory that grounds the identity formation of a particular people – the Black Diaspora. Slavery is traumatic, a 'primal scene,' which can unite all diasporic Afro-Caribbean subjects, whether an individual remembers or deliberately or unconsciously forgets it. Slavery is the root of an emergent collective identity and an emergent collective memory. The history of slavery distinguishes the Black Diaspora from other diasporas and dispersed communities (Eyerman 2001, 1). Today, slavery exists as a cultural trauma or collective emotional memory rather than an individual experience or an institution. Ron Eyerman defines cultural trauma as follows: 'As opposed to psychological or physical trauma, which involves a wound by an individual, cultural trauma refers to a dramatic loss of identity and meaning, a tear in the social fabric, affecting a group of people that has achieved some degree of cohesion' (2). Slavery has effected a dramatic loss and tear in the social fabric of African slaves and their descendents. The cultural trauma of forced servitude and nearly complete subordination found in slavery no doubt has impacted the lives, memory, fragmented history, pain and survival of slave descendents (Hua 2005, 79–80).

In the novel *At the Full and Change of the Moon*, Brand demonstrates her remembrance practices by taking historical fragments, moments and characters she found in historical documents and everyday life to reconstruct cultural or collective memory via literary imagination. For instance, the character Marie Ursule is based on the historical figure Thisbe who in 1802 was hanged, mutilated and burnt, her head spiked on a pole, for organizing a mass suicide as a slave revolt by poisoning on an estate. At her death sentence, Brand's Marie Ursule repeats the words of the historical Thisbe: 'This is but a drink of water to what I have already suffered' (Hua 2005, 85). In this novel, Brand begins with a historical encounter; the archive is her point of entry. She finds a historical figure – the resistant slave Thisbe – and begins to rewrite history with fictional narratives. Hence, this novel moves between fiction and nonfiction, invention and reclamation, to produce memory as a way to disrupt and then ideologically change official historiography (Hua 2005, 5–6).

While archival research was important for the writing of *At the Full and Change of the Moon*, it is not the case for the novel *What We All Long For*, where literary imagination alone plays a greater role. While *At the Full and Change of the Moon* deals with the cultural memory of slavery, Brand's 2005 novel *What We All Long For* engages with the cultural trauma of the Vietnam War and the family trauma of losing a family member. Brand, in *What We All Long For*, takes the Vietnam War and its refugee exodus, consequent dislocation and traumatic affects, and brings it to light to memorialize narratives of pain within a Vietnamese Canadian family. Brand examines how war, migration, dislocation or traumatic events can alter the self, psychically splitting the self 'before' and the self 'after'. She explores the effects of imagined traumatic memories, how does one, for example, carry on without being consumed by loss and melancholia, including the loss of homeland and family members?

'What We All Long For': War Memory and Family Trauma

Within the dominant North American public and cultural sphere, there is still a silence around individual and collective war trauma of the 'boat people' from their perspectives; their stories are overshadowed by mass media images of the war, Hollywood movies about the Vietnam War, the American veterans' experiences, and other subsequent catastrophes such as Rwanda, 9/11, and the Iraq War. The traumatic stories of the Vietnamese refugees including their experiences on the boats during the exodus in post-Vietnam War, like other historical or cultural traumas, evoke a sense of paralysis, a narrative void, and a fragmented disorganized story in search of a voice. Brand presents this forgotten or displaced memory of the Vietnam War in *What We All Long For* to remind the readers of the role of memory in diasporic communities and diasporic emotional geographies, in this case, the Vietnamese Canadian community.

What We All Long For opens with a scene of Vietnamese families fleeing the nation as part of the refugee exodus in post Vietnam War in the late 1970s. In the midst of confusion, six-year-old Quy lost his grip on his mother's hand and follows a stranger into another boat. His family shortly arrives in Canada, but Quy, trapped in the refugee camps in Thailand, is lost to them. In the summer of 2002 in Toronto, Quy's parents still have hopes of finding him. Their daughter Tuyen, an aspiring lesbian Vietnamese artist who rejects her immigrant family's suburban home in Richmond Hill, lives in a rundown apartment on College Street with her friends. She and her three friends struggle with family complexities, love, desire, identity, and strive for independence. In the mean time, Quy is finding his way to Canada and to an unexpected encounter with his lost family. Through the four young friends – Tuyen, Carla, Oku and Jackie – Brand captures the energy, diversity, multi-ethnicity and polyphony of Toronto as a global city, including the cultural geography of Kensington Market, the celebrations of the world cup in Korea Town, the streetcar, the subway and the vitality of urban life.

What We All Long For intersects the stories of friendship, love, urban life, family and community building for a group of Asian, Black and mixed-race youths with two family traumas: the trauma of losing a son or brother during the refugee exodus in Tuyen's family, and the 'silent' trauma of Carla witnessing her mother Angie committing suicide by jumping off the balcony of their high-rise apartment building. Tuyen's parents, for example, must deal with the trauma of having lost their son Quy during their refugee exodus. While her parents are paralyzed by the memory of this loss, marked by their persistent insomnia, guilt and melancholia, Tuyen struggles between the urge to know and the need to deny, to forget and yet to understand this traumatic family memory. She creates an art installation in her home, from debris and ruins as a way of mourning and working-through. Her installation recalls Walter Benjamin's allegorical 'angel of history,' the image of history as one single catastrophe piling wreckage upon wreckage (Benjamin 1969). Tuyen calls her working installation a '*lubaio*' or signpost, made of two railway ties, on which she plans to have the audience post messages to the city.

What is common for all four friends – Tuyen, Carla, Oku and Jackie – is that they cannot identify with the homeland narratives of loss or nostalgia told by their parents or the first generation migrants. When they listened to narratives of 'back home', 'descriptions of other houses, other landscapes, other skies, other trees' (20), they were bored. 'Each left home in the morning as if making a long journey, untangling themselves from the seaweed of other shores wrapped around their parents' (20). Home for these youths is their birthplace – the city of Toronto: 'They were born in the city from people born elsewhere' (20). Unable to assimilate into what their parents called 'regular Canadian life' because they were not 'the required race', these youths build a community out of unbelonging and they attempt to escape from the narratives of loss and melancholia found in their parents' recall. 'No more stories of what might have been, no more diatribes on what would never happen back home, down east, down the islands, over the South China Sea, not another sentence that began in the past that had never been their past' (47).

If Tuyen and her family must struggle with loss and mourning for a loss of a son/brother, Carla's personal trauma is witnessing her mother Angie jump off their balcony, after Angie instructed Carla to hold her baby brother Jamal. I pay substantive attention to Carla's personal and family trauma here, because it is important for a feminist project to recognize the silent insidious traumas of an individual's life history, not simply cultural or historical traumas such as war, displacement and slavery. Her story also helps one to understand place memory, for her trauma occurred at a particular location. The place of trauma is at 782 Wellesley Street, Apartment 2116. Brand's reiteration of the address haunts the novel. Carla returns physically to the site to understand that childhood trauma, yet the traumatic site also produces memories of pleasure, love and loyalty as Carla relives the memory of her mother. In Carla's return to 782 Wellesley, Apartment 2116, she realizes that 'this was not a home where memories were cultivated, it was an anonymous stack of concrete and glass' (110). Carla fears that 'she had overused these memories, wrung everything from them. She could hold them on one page of a notebook. The only way she could make them last was to spool them in a loop running over and over in her brain' (110). 782 Wellesley today stands 'indifferent and inhabited by other lives, other worries, other dramas', yet Carla knows that 'love can make you remember' (110–11). In her memory and mourning for the loss of her mother, Carla bears witness to Angie's life. Carla raises a crucial question about the relation between life history and place memory: 'Doesn't a life leave traces, traces that can attach themselves to others who pass through the aura of that life? Doesn't a place absorb the events it witnesses; shouldn't there be some sign of commemoration, some symbol embedded in this building always for Angie's life here?' (112). 782 Wellesley becomes a site of memory and trauma, a site of loss and mourning, an affective and emotional geography, a witness to a life loved and lost.

782 Wellesley embodies what Edward Casey calls 'place memory' (Casey 1987, 187). For Casey, place can aid remembering: *'place is selective for memories'*, that is, 'a given place will invite certain memories while discouraging others'. Memories are also selective for place; they seek out certain places as their habitats. As Casey writes: 'places are *congealed scenes* for remembered contents; and as such they serve to situate what we remember.... Place is a *mise en scene* for remembered events...' (Casey 1987, 189).

In *What We All Long For*, Dionne Brand also makes the emotional geographies of longing tangible. Tuyen, for example, asks people 'What do you long for?' while she is working in her brother Binh's electronics store, and she includes these longings of private lives in a public installation. Tuyen's installation is a work in progress, where she grapples with issues of belonging and unbelonging as well as longing by the migrants and citizens in the city of Toronto, as a way to understand her place as a lesbian Vietnamese artist within the Canadian nation. Witnessing in an artwork encourages the audiences or readers to produce a deliberate ethical consciousness, prompting an ethical response to rethink justice. Through the protagonist Tuyen, Brand provides us with a witness; Tuyen witnesses the lives of her parents, her brother Quy, her friends, and others who come from diverse cultural backgrounds in the city. Through her art installation, she is able to capture the beauty of the city, its 'polyphonic murmuring' (149). Her art becomes 'the representation of that gathering of voices and longings that summed themselves up with a kind of language, yet indescribable' (Ibid.).

Dionne Brand attempts to answer this very question: what do we all long for? Through her characters – Tuyen, Carla, Oku, Jackie, Quy and their families – Brand suggests that we all long for home and belonging, a sense of identity, desire and love, a sense of community, and nurturing emotional geography. Brand imagines a cultural livable space, where gendered exilic subjects who have been expelled because of their ethnicity, nationality or sexuality, can claim community and belonging through creative agency, cross-cultural dialogue, as well as rewriting the nation in their everyday practices. The novel conveys that in the midst of claiming home, identity and community, one, including the urban youth who seems to embody the present and the now, is also pressed towards the past, the haunting of loss and historical traumas.

One of the themes explored by the novel is trauma and traumatized subjects. E. Ann Kaplan tells us that: 'trauma produces new subjects, that the political-ideological context within which traumatic events occur shapes their impact, and that it is hard to separate individual and collective trauma' (Kaplan 2005, 1). How one reacts to a traumatic event depends on 'one's individual psychic history, on memories inevitably mixed with fantasies of prior catastrophes, and on the particular cultural and political context within which a catastrophe takes place, especially how it is 'managed' by institutional forces' (Ibid.). Furthermore, Kaplan notes the importance of translating trauma, 'of finding ways to make meaning out

of, and to communicate, catastrophes that happen to others as well as to oneself.'
Art may be one method to translate trauma; the wound of trauma and its pain
may be worked through in the process of translating trauma via art (Ibid. 19).
Besides the larger historical traumas such as the Holocaust, slavery, colonization,
the Vietnam War, 9/11, the Iraq War and so on, 'family' trauma, that is trauma of
loss, abandonment, rejection, and betrayal, also warrant study. These traumas may
be what T. M. Luhrman refers to as 'quiet traumas' (2000, 158).

For E. Ann Kaplan, it is necessary for us to achieve empathetic sharing and
witnessing rather than what she calls 'empty empathy', that is, 'empathy elicited
by images of suffering provided without any context or background knowledge'
(2005, 93). One often encounters 'empty empathy' in media images of war,
including the Vietnam War and the Iraq War, within our 'culture of "wounded
attachments" and sentimentality' (Ibid.). This empty empathetic media reportage
encourages sentimentality by presenting viewers and newspaper readers with
images that are merely fragments of a large, complex situation in another nation,
which the audience may know little about and the reporters often omit (Ibid.). I
have in mind the news coverage of the Vietnam War and mass-mediated images
produced by Hollywood Vietnam War movies, where the Vietnamese people are
portrayed as the bystanders or the victimized rather than as historical subjects and
the witnesses. These Vietnamese refugees who later become citizens of the nation
of settlement are rendered silent, without voice and subjectivity; their life stories
are overshadowed by the mass mediated versions of the war. From these dominant
representations, it is hard to detect the Vietnamese subjects and their narratives,
agency and resistance, desire, emotive geographies and interior lives.

Brand's novel tells us differently; it is a story of translating trauma via art, of
empathetic sharing and witnessing rather than 'empty empathy'. Although Brand
is an Afro-Caribbean woman who writes about a Vietnamese diaspora trauma,
Brand's work of fiction may become one of many recollections of the experiences
and emotional geographies of the Vietnamese Canadian community, along with
the reflective works of some historians, academic researchers, and the actual
voices or stories of the diaspora Vietnamese subjects. In creating this work, Brand
imagines and writes the Vietnamese Canadian community into the Canadian
literary and cultural landscape. I do not intend to imply that Brand provides us
with an 'authentic' voice. In fact, there may be diverse reactions, both positive and
negative, to Brand's representation from the Vietnamese Canadian community.
What is important is not 'authenticity' or who has the right to present an 'authentic'
voice, but that Brand engages with remembrance practices and personal and
historical traumas, in particular, those of the Canadian Vietnamese community,
to encourage the diverse readers to rethink and rewrite the nation as a more
democratic historically conscious landscape. Recognizing how Asian and Black
women and their communities continually struggle to live with certain traumatic
pasts such as war trauma, slavery, colonization and systemic sexist racism, Brand
attempts in her work of fiction to challenge 'the desire for a pristine and forgetful
nation,' a desire for an empty space to be filled with a flag and an anthem and 'no

discernible or accountable past' often found within the Canadian official nation (Brand 1994, 80). She disrupts and rewrites the nation by challenging its national myth of belonging and conventional historiography.

Brand's historical vision reminds us of a sense of history as haunting, the importance of telling histories and telling stories with emotional memories, emotional historiography and emotional geographies. She does not write historical fiction, but her fictional contemporary landscapes are shaped by emotional and traumatic histories. In *Demonic Grounds*, Katherine McKittrick suggests that Brand writes the land, a map that 'does not easily follow existing cartographic rules, borders, and lines' (McKittrick 2006, ix). Brand writes 'a different geographic story, one which allows pavement to answer questions...' (Ibid.). Brand, notes McKittrick, 'refuses a comfortable belonging to nation, or country, or a local street, she alters them by demonstrating that geography, the material world, is infused with sensations and distinct ways of knowing: rooms full of weeping, exhausted countries, a house that is only as safe as flesh' (Ibid.). Brand's decision to disclose that 'geography is always human and that humanness is always geographic' suggests 'her surroundings are speakable' (Ibid.). What Brand's emotional and human geography speaks is the need to develop a counter-memory and a counter-history that attends to how individual stories and histories of African diasporic experience, and other ethnic minority experiences, such as the diasporic Vietnamese, have been erased or silenced from written historical records. Brand attempts to make the land or emotional human geography speak, to tell of unheard life stories, of loss and mourning, desire, belonging, and psychic and physical longing.

In the novel, the four ethnic youths stake their claim of Toronto as their multicultural polyphonic home. Their energy, bodies, desire, and friendships, their movements, struggles, hopes, and longings, their art, music and stories, their anticipation and eagerness to embrace the future – these are their unconscious strategies to stake their place of settlement, nation of birth, as home despite the limits, the systemic racism and sexism, the violence within and outside their ethnic communities, the melancholy and disillusionment of some first-generation migrants, and the intergenerational memory of difficult pasts. It is a kinesthetic, sensual, creative desire for home and community, for individual and collective voice. But it is a stake for home on the edge of society, as embodied by Kensington Market within Toronto, a refusal to fit comfortably to the middle-class suburban life. It is a claim not for the status quo, but a skin shine, a flesh comfort, a body at home, in movements or stillness, beings in element, in high noon when the sun pierces blaze and haze and quietly cools in the gloaming. These four friends attempt to carve a life or a community, a rhythm or a pulse, that has an under current, hidden rivers below the common terrain, gully marked by defiance and difference, underground art and graffiti. Brand writes of the possibility of working through and letting go of traumatic pain and loss, to engage, more importantly, in the human and emotional geography of the present.

Brand writes about longing in this novel, both heterosexual and homosexual longing, longing for home, for rooms and nations that no longer weep, longing for un-traumatized families, longing for love, longing for those we have lost, the possibility of return, a sense that we (uncomfortably) belong. That discomfort, as daily rupture, keeps us, at least those who are dissonant, on our toes to live more critically, to always question and yet still enjoy and rejoice. The passion never ends because of the discomfort, but the discomfort, the struggle – political, personal, reflective – feeds passion for daily life. We long for something or someone, or a political state or a freedom, because we do not have it. That longing propels us forward, feeds passion to our imaginary and interior lives. Brand teaches us to never stop longing, critically, politically, and aesthetically. Perhaps Brand's aesthetic longing can be summed up by what Oku says to Carla in *What We All Long For*: '…in the middle of the melody, … you find yourself longing for the dissonance' (228).

Emotional Memories, Diasporic Historiography, and Emotional Geography

What Brand has created through her practices of remembrance with her works of fiction is an emotional geography, using emotional memories and historical and family trauma to write diasporic historiography. 'An emotional geography, then, attempts to understand emotion – experientially and conceptually – in terms of its socio-*spatial* mediation and articulation rather than as entirely interiorized subjective mental states' (Davidson, Bondi and Smith 2005, 3). By exploring personal and collective memory and trauma, Brand examines more than the subjective interior states of her characters but also the socio-spatial mediation and articulation found within environments, space and place, including cities and nations. She explores 'how environments, variously conceived, are encountered as sources of distress, pleasure and commemoration, sometimes intensifying exclusion and sometimes fostering well-being' (8).

Brand uses remembrance practices and cultural or historical trauma as narratives or sites of witness, healing, commemoration and anti-racist anti-sexist resistance. She deploys personal memories and self-identities in her works of fiction to link to public histories and the imagination of new cultural identities over time. She constructs identity – personal, sexual, ethnic, or diasporic – and narratives of home, identity, longing and belonging – to work through and mourn various historical traumas, including slavery for the Black diaspora community and the Vietnam War for the Vietnamese diaspora community. With her narratives of emotional memories, Brand attempts to rewrite the nation and official historiography, by acknowledging and commemorating the voices and narrative past of the silenced and the marginalized – ethnic minorities, women and the queer community. Brand explores practices of remembrance, family and historical traumas, to understand the localized histories of the Black diaspora community and other ethnic diaspora communities. She produces, via works of fiction, counter-histories and counter-memories to challenge the painful amnesia that marks historical and contemporary Euro-America, reminding us that our geography is influenced and haunted by emotions, emotional memories, and affective histories. Brand attempts to return to the trauma to witness, to work-through, and to heal historical and personal traumas through her writing. She recognizes how the painful effects of the past influence the present, how history is lived and experienced as a wound for women and marginalized individuals and communities.

One of the practices of remembrance Dionne Brand engages with, particularly in her novel *At the Full and Change of the Moon*, is the cultural trauma of slavery, how slavery is passed down to us as transgenerational memory. In *What We All Long For*, Brand is exploring the memory and trauma of other diasporic communities, and how that affects the multicultural communities living in the global city of Toronto. Brand engages with the trauma of the Vietnam War and the family trauma of losing a family member either to death/suicide or war/refugee exodus. She notes that war, migration, dislocation or traumatic events can alter the self, psychically split the subject, and at times, the psychically split personal identities can be linked to the trauma experienced by the cultural identities of a community.

Brand reminds us of the importance of the role of memory or remembrance practices in diasporic communities and diasporic emotional geographies, how the Vietnam War, for example, still affects the interior lives and identities of both first- and second-generation migrants of the Vietnamese diaspora community. What Brand shows us is that works of fiction can translate trauma, to communicate and make sense of catastrophes that happen to others and ourselves. The personal and collective wound of trauma (whether it is slavery, war, refugee exodus, or the traumatic loss or death of a family member) may be worked through in the process of translating trauma via art. Brand engages with the practices of remembrance for her own community and other ethnic communities to encourage the readers to rethink and rewrite official histories. In sum, Brand demonstrates how a critical work of fiction may function as an emotional geography that helps to make the memories and lives of others matter.

References

Agnew, V. (ed.) (2005), *Diaspora, Memory and Identity: A Search For Home* (Toronto, London, Buffalo: University of Toronto Press).

Anderson, B. (1991), *Imagined Communities: Reflections on the Origins and Spread of Nationalism* (London: Verso).

Arendt, H. (1977, 1954), *Between Past and Future: Eight Exercises in Political Thought* (New York: Penguin Books).

Benjamin, W. (1969), *Illuminations,* trans. Harry Zohn, ed. Hannah Arendt (New York: Schocken).

Brand, D. (2005), *What We All Long For* (Toronto: Vintage Canada).

– (1999), *At the Full and Change of the Moon* (Toronto: Vintage Canada).

– (1994), *Bread Out of Stone* (Toronto: Coach House Press).

Casey, E. S. (1987), *Remembering: A phenomenological Study* (Bloomington and Indianapolis: Indiana University Press).

Davidson, J., Bondi, L., and M. Smith (eds.) (2005), *Emotional Geographies* (Aldershot, England; Burlington, VT: Ashgate).

Douglass, A. and T. A. Vogler (eds.) (2003), *Witness and Memory: The Discourse of Trauma* (New York and London: Routledge).

Eyerman, R. (2001), *Cultural Trauma: Slavery and the Formation of African American Identity* (Cambridge: Cambridge University Press).

Hua, A. (2005), *Memory and Cultural Trauma: Women of Color in Literature and Film* (Ph.D. thesis, York University).

Kaplan, E. A. (2005), *Trauma Culture: The Politics of Terror and Loss in Media and Literature* (New Brunswick, New Jersey, and London: Rutgers University Press).

King, N. (2000), *Memory, Narrative, Identity: Remembering the Self* (Edinburgh, UK: Edinburgh University Press).

Le Goff, Jacques (1992), *History and Memory*, trans. Steven Rendall and Elizabeth Claman (New York: Columbia University Press).

Luhrmann, T. M. (2000), 'The Traumatized Social Self: The Parsi Predicament in Modern Bombay', in Antonius C. G. M. Robben and Marcelo M. Suarez-Orozco (eds.), *Cultures under Siege: Collective Violence and Trauma* (Cambridge: Cambridge University Press), 158–194.

McKittrick, K. (2006), *Demonic Grounds: Black Women and The Cartographies of Struggle* (Minneapolis and London: University of Minnesota Press).

Misztal, B. (2003), *Theories of Social Remembering* (Philadelphia: Open University Press).

Robben, A. C. G. M. and M. M. Suarez-Orozco (eds.) (2000), *Cultures under Siege: Collective Violence and Trauma* (Cambridge: Cambridge University Press).

Simon, R. I. (2005), *The Touch of the Past: Remembrance, Learning, and Ethics* (New York: Palgrave Macmillan).

Singh, A. et al. (1996), 'Introduction', in Amritjit Singh et al. (eds.), *Memory and Cultural Politics: New Approaches to American Ethnic Literature* (Boston: Northeastern University Press), 3–18.

Singh, A. et al. (eds.) (1996), *Memory and Cultural Politics: New Approaches to American Ethnic Literature* (Boston: Northeastern University Press).

Sturken, M. (1997), *Tangled Memories: the Vietnam War, the Aids Epidemic, and the Politics of Remembering* (Berkeley, Los Angeles and London: University of California Press).

Young, J. E. (2003), 'Between History and Memory: The Voice of the Eyewitness', in Douglass and Vogler (eds.), 275–283.

Chapter 8

Ephemeral Art: The Art of Being Lost

Mary O'Neill

> The very act of thinking objectively about distress places us at one remove from
> the distress. But if dissociation is a necessary part of clear thinking it may also be a
> defence against thinking.
>
> (Bowlby 1988, 11)

I have divided this chapter into two sections; both cover the same ground but are written differently, to demonstrate the possibility of thinking in different modes. The dual texts also respond to the difficulty raised by John Bowlby of thinking objectively about distress and the risk that the detachment this requires may result in an avoidance of thinking about the subject at all. This is particularly significant when faced with works of art that ask us not only to think about distressing subjects, but to feel the distress they embody. The first section explores a group of ephemeral artworks related to grief through the lens of the academic literature on grief and mourning, revisiting some of the age old questions about the nature of art, the nature of its effect on viewers and about the existence of universal human experiences underlying different cultural manifestations. The second section explores grief and mourning in the form of a personal narrative.

I did not set out to research mourning. The main focus of my research is ephemeral art practice; that is, art that disappears over time, where the disappearance or decay of the work is intrinsic to how the work communicates with its audience.[1] At the outset, I thought that this subject was primarily internal to the art world and the art market, and was concerned with the dematerialization and decommodification of the art object. I assumed that the loss to the art world of precious, durable objects would raise issues of documentation and legacy. However, as my research developed I became aware that there were far broader and more complex issues involved. As opposed to being merely temporary or transient, there are ephemeral works of art that engage directly with issues of mortality, and in a quite specific way. They are not about death in general, nor are they *memento mori*, reminders of our own mortality. What they have in common is that they address experiences of death that are arguably more difficult to deal with even than the normal sequence of the generations, as for example the pain of watching someone we love die slowly before us, or the shock brought on by violent death or the death of the young. From places as diverse as Indonesia, Thailand and Australia, I found a

---

1   For a lengthier elaboration of this definition, see (O'Neill 2008).

number of artworks which decay as part of their communication of the extreme grief associated with untimely death.[2]

This realization meant a significant shift in my research from a theoretical study of contemporary art to one of mourning. I began a wide ranging interdisciplinary study of mourning, in an attempt to understand how these works expressed, communicated and evoked feelings related to mourning – but without avoiding the emotional engagement through analytical detachment. My study came to embrace how not just ephemeral art but even discussion of it, can provide an opportunity whereby those present are given permission to tell their narratives of grief, when culturally we are all too often encouraged to remain silent. At every conference at which I have discussed ephemeral grief works I have been offered life experience by members of the audience who recognized this quality in the work and explicitly confirmed my suspicion that these works facilitate discussion of bereavement.[3] The association between narrative and grief is well known. Indeed, it is the basis of many forms of therapy (Neimeyer and Mahony 1995; Crawford, Brown and Crawford 2004). The experiences individuals offer in response to my research provide an insight into the possibility that ephemeral art can facilitate these narratives, even in an arena that normally inhibits such forms of engagement – the academic forum. These unsolicited stories exist only in memory and testify, in the moment of telling, to the secret nature of the bereaved and to the power of art to elicit their stories. These stories are not recorded or presented here as documentary evidence as they were offered as a spontaneous gift in response to my presentation; the process of recording or reproducing would be inappropriate and would change the nature of my relationship with the audience.[4]

Storytelling plays an important role in my research and operates on several levels: stories told by the artworks, which are interpreted by the audience, members of which bring their stories to the work; stories by the artists, offered in texts set alongside the art; stories I offer to exemplify the experience of viewing; and stories told to me in response to my discussion of the works. Like ephemeral art, stories exist in the moment of telling and require an engagement from the viewer/

---

2   In the narrative section of this chapter I will focus on parental bereavement which is described as the 'most grievous of losses' (Braun and Berg 1994, 105) and is characterized by a sense of failure.

3   Ephemeral works are not alone in this capacity to facilitate a discussion of one's experience of loss. In his essay *Bearing Witness to Apartheid: J.M. Coetzee's Inconsolable works of Mourning*, Samuel Durrant explores the capacity of literature to provide a way of working through a collective grief (1999, 430).

4   There is a parallel between the re-presentation of narratives and ephemeral art. Both are dynamics forms of communication and are fluid in their retelling. This has significant consequences when they are captured within the confines of more permanent media, such as video or text. My telling of the story of live performances here is not meant to be a definitive account of that piece but a description of my experience of the work on that particular day. For an overview of the arguments relating to the documentation of performance see Phelan (1993) and Auslander (1999).

listener to come fully into being. These witness records presented in different forms contribute to a broad landscape of mourning from which we can learn not only about an aspect of contemporary art but also about what it is to lose others, and to be lost.

William Worden (1991) states that grief has four dimensions: feelings, physical sensations, cognitions and behaviours. The cognitions and behaviours are particularly significant in relation to my research for it is through these that the two other dimensions, feelings and physical sensations, are made manifest. Of the cognition and behaviour manifestations described by Worden those which I will focus on are: a sense of the presence and hallucinations of the loved one; and searching for and calling out to the loved one. In his description of the universal aspects of the cognitive behaviour of searching he states: 'Whatever the society studied in whatever part of the world, there is an almost universal attempt to regain the lost loved object, and/or there is a belief in an afterlife where one can rejoin the loved one' (Worden 1991, 9). He also highlights that mourning is not a state but a process and emphasizes that it is a task, which, like all tasks, involves work. A mourner must complete this task for a successful resolution of mourning. In grief there is an element of irrationality, and this can be seen in both the sense of presence and hallucinations and the searching and calling out. The bereaved person often feels as though death could not possibility have happened and attempts to deny its reality. This denial can lead to the belief that the dead person will return – hence the use of the term 'lost' rather than 'gone'. In loss there is a possibility of being found. Some who are bereaved may wish not to 'get over' their bereavement but continue to mourn as a form of fidelity. To recover from mourning would mean the complete loss of their loved one.

> Although we know that after such a loss the acute stage of mourning will subside, we also know that we shall remain inconsolable and will never find a substitute. No matter what may fill the gap, even if it be filled completely, it nevertheless remains something else (Freud 1961, 386).

On the basis of the insights offered by a number of psychological and psychoanalytical theories of mourning I will discuss the work of Dadang Christanto and Araya Rasdjarmrearnsook, suggesting that the task of grieving represented in their ephemeral works offers the possibility of moving beyond the abject, understood in Julia Kristeva's sense of that which 'cannot be assimilated' (1982, 1). I will also discuss the web-based work *1001 Nights* by Barbara Campbell. Like the works of Christanto and Rasdjarmrearnsook, this piece can be read as a form of searching for and speaking to the dead. All three works are ephemeral and accompanied by a narrative of bereavement. In the spirit of these works, the second section of this chapter is in the form of a narrative that will, I hope, illuminate some of the theories discussed.

The relationship between the longing for permanence of art and the desire for immortality is a truism, and I will not dwell on it here other than to note the link between the creation and accumulation of durable objects as a form of denying mortality and the significant role that the ideals of permanence, stability and order play in the Western worldview.[5] Why would an artist make ephemeral art when there is considerable cultural and economic pressure to make permanent art? I suggest that the answer to this question does not lie solely in the art world or in rebellions against the art market, but involves a crisis of meaning and a resulting value shift experienced as a consequence of bereavement, in particular to mourning untimely or violent death.

In the works of Christanto and Rasdjarmrearnsook, we are confronted with a performance that re-enacts or re-experiences loss. This is distinct from death as a representation, which Kristeva tells us is comprehensible: 'In the presence of signified death – a flat encephalograph, for instance – I would understand, react, or accept.. corpses show me what I permanently thrust aside in order to live' (1982, 3). Through the tenderness of these ephemeral works we are able to witness without repulsion. They allow us to work through abjection, which is a process not unlike grief. In these works the abject is performative; if we open ourselves to them, they do not leave us unchanged. But rather than causing us to vomit, to expel the knowledge or, alternatively, filling us with disgust, they can be cathartic and ameliorative. There is a close relationship between these works and purification rituals which are traditionally employed as a defence against the abject. Here, however, the abject is assimilated rather than dispelled. To understand how these works perform this function I will explore the necessity of speaking to the dead, and the role of ephemeral art as a form of grieving.

The need to mourn and the overwhelming need to have a corpse that is the focus of a mourning ritual is powerfully embodied in the work of Dadang Christanto, in particular his 2005 performance *Collecting Displaced Bones* (Romanoff and Terenzio 1998). The performance took place at the Australian National University, in the grounds of Canberra House. It was performed by Christanto and his eight-year-old daughter, the age Christanto was when his father disappeared in 1965. In this work, Christanto's daughter excavates the body of her father, echoing the relationship between Christanto and his father.

Christanto was born in Yogyakarta, Indonesia in 1957 and now lives in Darwin, Australia. His work predominately deals with issues of human rights abuses especially those that took place in Java following the 1965 coup, which brought Suharto to power. During this period, members of the PKI (Communist Party of Indonesia) and those who were considered sympathetic to their cause were systematically rounded up and massacred. Christanto had personal experience of

---

5   I am particularly indebted to the work of Zygmunt Bauman (1992), *Mortality, Immortality and Other Life Strategies* (Cambridge: Polity); Ernest Becker (1973), *The Denial of Death* (London: Free Press); and the elaboration of Becker's work by terror management theorists J. Greenberg, T. Pyszczynski, S. Solomon, A Rosenblatt, M. Veeder, and S. Kirkland (1990), 'Evidence for terror management theory. II: The effects of mortality salience on reactions to those who threaten or bolster the cultural worldview'.

these atrocities when his father disappeared. *Collecting Displaced Bones* begins with two men wearing improvised cloth masks carrying a stretcher through a wooded area. They stop at intervals to rest and regain their strength, which is obviously flagging under the weight of the burden they are carrying. There isn't a seating area provided for the audience nor is there any indication of where they should stand, and so they follow like interlopers, unsure where they are going. The men stop in a clearing and carefully place the stretcher on the ground. On the stretcher there is a figure wrapped in cloth. Christanto's eight-year-old daughter is in the clearing; initially she circles the body slowly and rhythmically and after a while begins to untie the cloth and to excavate the body of her father, which is covered in white chalk. This work powerfully highlights the ongoing nature of mourning. In terms of his daughter's involvement it points to a group of disenfranchized mourners, those who mourn people they have never met, or who have been violently deprived of the opportunity to meet. Mourning does not stop with those who have experienced the bereavement first hand but it can be passed on through generations.

Eventually the wrapped body rises from the stretcher and envelopes the child and carries her to a tree just beyond the circle created by the audience. The performance is over. As with the beginning, the positioning, and the duration, this performance is borderless. As an audience member I am not given the usual clues but have to follow the performance bodily. This adds an extra dimension of unease to the work. Occasionally, one could hear murmuring from the audience but rather than being a distraction from the event, the others articulate questions I am also asking, 'Where are they going?' 'Should we stand here?' We are not sitting back and allowing the event to unfold, we must actually decide to follow, to risk intrusion, to bear witness. These silent performers do not acknowledge our presence but we know that they need us, for it is our role to be followers and witnesses; this is not a passive role but one we undertake by choice, and, having witnessed, we are changed. Witnessing is a central element in Christanto's work and through our witnessing we participate. Even the decision not to watch is participation; every reference to those who bear witness also refers to those who turned away, who decided to be blind to crime and injustice, to say nothing, not to get involved. As a signature on a legal document, witnessing is a form of contract; it does not leave us without obligation. Here we are asked to witness the mourning, and the pain of others.

This may be the abject as truth as described by Hal Foster in his discussion of the abject in art in *The Return of the Real*; the 'special truth [which] resides in traumatic or abject states, in diseased or damaged bodies.' These performers unearth the evidence that is the violated body and they exist as the 'important witnessing to truth, of necessary testimonials against power' (Foster 1996, 123). We witness the consequences of power relationships, the evidence of abuses, which were enacted in the dark, forcing silence upon victims. Christanto brings this evidence out into the cold sharp Australian winter sun. The absence of sound in this performance is striking and reflects Christanto's preoccupation with the silence of those who suffer oppression. Although this refers to personally experienced events, it also addresses universal issues relating to abuses of power and the consequences for individuals who are caught up in destructive and violent political events.

The participatory element in Christanto's work takes a more sinister turn in the performance Litsus. This piece, part of an exhibition titled *Text Me, An Exploration of Body Language* was performed in 2005 at the Sherman Gallery, Sydney. In this work, Christanto and his son sit shrouded in black cloth while the audience flings flour-filled 'bombs' at them. Considering both this work and Collecting Displaced Bones together, a disturbing question arises: What is our role in witnessing and are we complicit in the acts of violence? In this work both Christanto and his son, and the performers that are the audience, become abject. The abjection of the self would be the culminating form of that experience of the subject to which it is revealed that all objects are based merely on the inaugural loss that laid the foundation of its own being (Kristeva 1982, 5).

Christanto is aware of the possibility of his work performing the task of accommodating mourning, and the ability of art to have a healing effect both on the individual and on society (Gray 2006). While the works reference specific horrors both historical and ongoing, he is conscious that the viewer brings their own experience of tragedy to the works and reads the works through these experiences. He is equally aware that an encounter with his work can be painful for viewers and carries the possibility of reopening old wounds. In his gallery based work – *For those: Who are poor, Who are suffer(ing), Who are oppressed, Who are voiceless, Who are powerless, Who are burdened, Who are victims of violence, Who are victims of a dupe, Who are victims of injustice, (1993)* and *They Give Evidence* (1996–1997) – Christanto provides an opportunity for the participation in the work to ameliorate these consequences. The element of gift-giving in response to works is particularly evident in the audience participation in Christanto's *For those: Who are poor* and *They Give Evidence*. The text that accompanies the work invites the audience to participate:

> To overcome the immediate pain and potential conflict of this encounter, Christanto offers the audience an opportunity to enter the work by making an act of remembrance and consolation; to offer acknowledgement within the architecture of the artwork. Though undeniably a memorial to the victims of human violence, this installation also offers that essential space for a redemptive personal gesture (McDougall 2006).

It is through this interaction between offering, invitation, and gift giving that the viewer is given an opportunity to work through their grief in response to these works. Without this generosity and sensitivity on the part of the artist and the exhibition gallery this work might just reawaken memories of pain and trauma, of the abject, and leave the viewer there, lost without resolution. This work is not displayed permanently but is exhibited regularly. Each time the objects, flowers and letters left by the public are disposed of, allowing the work to begin again each time it is exhibited like the retelling of a story. In response to my enquiries the curator Ruth McDougall wrote:

> The Queensland Art Gallery retains close contact with Dadang regarding the documentation of his work and in consultation with him; the ephemera left by visitors to the installation is not kept. The personal act of giving, not collecting is felt to be the focus of the work and as such the work invites visitors to respond anew, each time it is exhibited (McDougall 2006).

While Christanto's work powerfully enacts the searching associated with mourning, the works of Araya Rasdjarmrearnsook address the sense of presence and calling-out. In June 2005, Araya Rasdjarmrearnsnook represented Thailand at the Venice Biennale alongside the posthumous exhibition of work by Montien Boonma. The exhibition, entitled *Those Dying Wishing to Stay, Those Living Preparing to Leave* (curated by Luckana Kunavichayanont, Sutee Kunavichayanont, and Panya Vijinthanasarn) is an exemplar of one of the main arguments of this chapter: works made in response to untimely death and ambiguous loss are substantially different to works made in response to mortality awareness and, in particular, work made with the knowledge of the certainty of, or immanent death. It is not surprising that the choice of these two artists concerned with grief should come after the devastation of the Asian tsunami, and this exhibition contrasts with the joyful celebratory spectacle that the Thai contingent offered in 2003. Their works incorporate very specific Thai references – architecture and literature, but the responses are universal.[6] In a radio interview the novelist Israeli Amos Oz, referring to the Palestinian-Israeli conflict said 'our dreams are all the same'. Not only are our dreams the same but so too are our fears, those horrors that haunt our waking. From the mother in the refugee camp who cuts up her identity photo and pins it to aid workers' shirts in the hope her lost children will recognize the fragment and eventually locate her, to the child survivor of the Beslan massacre who draws detailed and elaborate pictures of her captors in order to ritually tear them up and burn them in an attempt to exorcise her horrors, there is a universal vocabulary of pain (MacLoed 2005).[7]

---

6   For a discussion of cultural differences and responses to the dead see Klass and Roberts 1999.

7   In a *Guardian* article written by Donald MacLeod Wednesday August 3 2005, 'Academic finds Beslan still traumatized by school siege', MacLeod discusses the work of Dr. Cerwyn Moore, a lecturer from Nottingham Trent University, U.K. who had spent in Beslan as part of his research on the motivation of suicide Bombers. Moore stated 'For the survivors of the siege, especially the children, the trauma continues and there has been little in the way of psychological help ... Many adults showed clear signs of post-traumatic stress disorder, depression and other conditions. The vast majority of people we spoke to had not come to terms with it psychologically'.

While the work of both artists is linked by their response to grief, the works demonstrate the different positions in relation to bereavement. Rasdjarmrearnsnook (1953-) lost her mother, father, grandfather, and half-sister. Montien Boonma died of cancer in 2000 shortly after the death of his wife from the same disease. For the purposes of this chapter I will focus on the work of Rasdjarearnsook.

The works of Rasdjarearnsook and Christanto refer to specific personal losses, however, when exhibited they often have a site specific element in that they are often linked with a local sense of loss. In 2003 Rasdjarmrearnsook exhibited *Lament* at the Tensta Art Gallery in Stockholm. On September 11th, Swedish Foreign Minister Anna Lindh was murdered. A photo of Lindh stood in the entrance hall of the gallery, adding to the poignancy of the Rasdjarearnsook's work. One of the works exhibited was *I am Living* (2002). This video installation showed Rasdjarearnsook placing brightly coloured clothes over the body of a dead girl laid out on the floor of a morgue. This could appear like a macabre form of dressing dolls, a horrific sight which could evoke the abject. However, rather than revulsion, this work seems to offer tenderness and through this act of caring we can look.

The work of preparing the dead for their journey, like the task of nursing people into the world, is, in many cultures, the preserve of women. The clothes that the dead wear testify to the life they once had. The sterile setting of the morgue and the cruelty of death rob these bodies of individuality and any memory of the life they once were. Rather than anonymous bodies Rasdjarearnsook returns 'personality' to them. Here we see evidence of the stage beyond the abject, the promised 'rebirth against abject' introduced at the very end of Kristeva's essay to offer hope after abjection (Kristeva 1982, 31). This work by Rasdjarearnsook is evidence that a rebirth is not the only possibility, but that love, caring and tenderness can allow a passage through to acceptance. Coming from a culture (Southern Ireland) where the professionalization and commercialization of the management of death is a relatively recent phenomenon, I am not shocked by the tending of a corpse in the way that viewers coming from societies that abdicated this task to strangers and professionals might be. It is common now for most people in the Global North to have no experience of bereavement until well into their middle age and even then to have no contact with the corpse. The frankness with which Rasdjarearnsook approaches her subjects is testimony to her lack of, or at least her attempts to confront, the fear of death.

In his foreword to the programme book that accompanied *Those Dying Wishing to Stay, Those Living Preparing to Leave* Apinan Poshyananda states:

> By questioning the function and validity of art in society, some Thai artists have explored the potential of contemporary art as a healing force for mental relief. Others have created work as a space of refuge, contemplation, and communication beyond death (Poshyananda quoted in Sukhsvasti 2005).

The works of both Christanto and Rasdjarmrearnsook assert the role of art in society as a political tool, a means of drawing attention to extraordinary atrocities that cannot be spoken about in other media, or certainly not in the same way. This also involves art in addressing the universal ordinary atrocity, death, about which we avoid speaking. They offer a possibility of healing, and in the case of Christanto the possibility of forgiveness. *They Give Evidence* affords the viewer an opportunity to participate, and through this participation, not only complete the acting works, but also create their own rituals. The responses to *They Give Evidence* are not literal responses about the work but because of the work. The slowed time of ephemerality offers the time of grief and gives permission to mourn in a time when we are encouraged to 'move on'. These works do not ask us to forget or to 'get over' pain but to accept it and to find a way of living with it. In Mieke Bal's discussion of the work of Andreas Serrano she evocatively describes Serrano's ability to transform the abject body; in these works we can also see this transformation. She describes how the abject dead body becomes a powerful monument, brought to life by the '..."maternal love" – the slowed-down look that grazes the object, caresses it, and surrounds it with care...' (Bal 1999, 60). This is not death infecting life, but death as the inevitable consequence of living and the pain as the inevitable consequence of loving; this is what lies beyond abjection.

In Barbara Campbell's web-based durational performance *1001 Nights* the narrative of bereavement and searching appears on the introductory page of her web site. It reads:

> In a faraway land a gentle man dies. His bride is bereft. She travels across continents looking for a reason to keep living. Every night at sunset she is greeted by a stranger who gives her a story to heal her heart and continue with her journey. She does so for 1001 nights (Campbell 2006).

The introduction also makes reference to keeping the 'project alive'. Each day Campbell scans the press coverage of events in the Middle East and selects a word or phrase that has 'generative potential'. She renders the phrase in watercolours and posts it on her site to act as a prompt for writers to submit stories. The same evening at sunset Campbell performs a work based on the submitted stories live on the web. This process will continue for 1,001 nights. When I first started following this work it was performed from Paris; since then, Campbell has performed from various locations around the world, which means that the viewer must follow the sunset to catch her performance. The link between the performer and Scheherazade may be obvious but it is no less powerful for this. Scheherazade told stories to save her life as the bereaved do. The telling of bereavement stories is part of the process of assimilating information that can be too difficult to believe. The narrator is not simply telling the story to others so that they may know but so that they themselves might be able to believe. The book in which Scheherazade appears is set in the East, like the news stories which give rise to Campbell's prompts. Speaking about *1001 Nights* Campbell states: 'The stories within will be set in 'the East' but you may never venture there yourself because this East exists in the kingdom of the fantastic that is called the imaginary. It is a land formed out of words and desires' (Campbell 2007). Interestingly, this performance also takes place in another land formed out of words and desires – on the Internet. This land, like the land inhabited by the bereaved, is disconnected.

These works highlight the difficulty of confounding the requirement for permanence and the myth of immortality, even for the makers of the works themselves. In some of these works we are deprived of the dissociation which museums can offer, even in the presence of deeply emotional work. The possibility of an institutional anaesthetic facilitated by architecture, gilded frames and plinths has been removed. In relinquishing the myth of permanence and immortality, they speak a truth that may be too hard to bear and be easily dismissed as 'sentimental'. Through their ephemerality, they offer us knowledge not just of art and the role it plays in our world view, but of mourning. They 'speak' about the unspeakable which renders us mute, they speak of loving and the pain of loss and of being left behind. What is significant about these works is that they are active. In this they differ radically from death works which are static; death masks capture the frozen moment of death; ephemeral works begin with death and are alive. Their very transience creates a new life, if only the surrogate life of art. If we sacrifice our need for permanence, if we can make the shift in value orientation that transience requires, we have an opportunity to appreciate the here and now, the changing work, alive in the moment and soon to be gone. Our lives, and the lives of those we love, are chancy and short. This is the lesson of these works of art – the knowledge of what it is to be lost.

Being Lost (Part 2)

> Sometimes death does not come suddenly but is a slow process of loss. The dying gradually lose those characteristics which define us as living – speech, sight, hearing, laughter, swallowing, digestion, excretion, wakefulness – until all that is left is the slightest neurological function that maintains the heart and lungs, just barely enough of those autonomic actions to sustain life. Despite the signs of extinguishing life, we hope for a recovery and when that is no longer possible, at least for a halting of the decay that progresses slowly towards death itself. At every increment of loss we beg for respite – what more can we cope with losing, allow us to keep this much, this much will do, this much life is enough. But, it does not halt, deaf and determined, the cancer grows, overpowering yet another vital nerve, another fine string connecting the brain to those organs we had taken for granted, the spleen, the pancreas, the liver, the kidneys. The moment of death is as slight as all those other moments, just one more electrical impulse fading. Yet, that one moment was the difference between all and nothingness, between hope and hopelessness. Here the threshold is infinitely fine – life and then death. Did we even see it or is it only known to us in hindsight when we can recall and

say 'that moment there – that was the last moment of life'. And the next, was that the moment of death or had we moved into a new form of time? Time now is not measured by change, for the dead remain fixed and unchanging. Is this the meaning of eternity? Now there is no chance of a reprieve, no possibility of remission, or of the longed-for plateau in the bleak descent into the abyss of dying.

In the insanity that follows there is no point in checking that there may have been a mistake, no need to look for a pulse, death is too final, completely sure of itself, it leaves no room for doubt. The old myths of redemption, of resurrection, provide no comfort and we must create another. Now we must find other stories to tell ourselves, new possibilities to hold on to. In the face of the confidence of death we attempt to create doubt, to create possibilities no matter how ludicrous. We press our lips to the stones in the graveyard but it is hard and cold. We curl up on the grave to be near them but it does not accommodate us. Instead, we find consolation in the sea. We become obsessed with the ocean and those who have been lost at sea. We envy their mourners, they have possibilities, they have a dreamland of sunny islands and shipwrecks. This we know is not real but it will do. It is no more unreal than our hope of a cure and so it provides a focus, a way of communicating with the dead, a means of keeping them alive and of preparing for their return. We do not look at photographs because that would be about the past and memorializing. We want a future for the dead, with us, so we dream awake of their survival on a distant shore. For months we make images of objects to send to those lost at sea – the things they may have neglected to take, not knowing that their journey was going to be a long one. We are mothers, we have responsibilities, we must remind people to wash their teeth, to eat well, to keep warm, to take their gloves and button up their coats. They will need fresh fruit, a hot water bottle, maps and directions. We think of all the things we forgot to teach them. They will need to know about the planets and how to navigate by the stars, they will need to learn to tie knots, make string, and make bricks to build houses. They will need to know about us, our culture, the things we have produced that exhibit those qualities of humans of which we are proud, lest they forget all this on that island where they have washed ashore. We will keep a place for them here for their return, we will talk about them so they know we haven't forgotten them, we will remind ourselves what they look like so that we will not fail to recognize them when we open the door and they are there, wet and tired from their journey.

Others tell us 'We are sorry for your loss', and we are relieved that they are colluding with us that our loved ones may be found again. We have all agreed this is the best way to proceed, to delude ourselves with the help of kindness, which does not confront us with the truth. No one says 'don't you remember the funeral? the coffin? the headstone?' those landlocked solid reminders of the truth. But we arranged those things for others, to comfort them, to trick them into believing we would survive, to show we could perform the dignified rituals our culture demands, without shouting or raising our fist to the air and screaming 'give them back', 'they are ours', 'we need them, you do not'. We protested with art, we read poems that said those things for us, which discreetly refuse to be dignified, to be accepting. For the poet can object passionately and then others can enjoy the transformed grief, the beauty of the language. But beauty was not what we intended – we wanted to give voice to our pain, our ugly untransformed grief. These poems are our language because they speak a truth that is too hard even to say and so we become mute except for rhymes, like Alzheimer's sufferers who can remember only things learnt by heart. We too only had our learnt by heart information but in the end it is of little use.

We know there is a limit to our comforting fantasy of loss and learn to keep it secret. After a time people start to speak of our lost ones in the past tense, they no longer write their names on Christmas cards, in phone calls they do not send their love, they do not send birthday cards or invitations to parties. The others know they are not coming back and hope that we do too. But we do not give up, we stand on the shore and hear the water rushing over the pebbles, clinking the little stones together before rushing out again and we wait and watch. Sometimes we swim far out to sea to be near then. We stay in the water too long until we are exhausted.

It is not the dead who are lost, who are on a distant island, it is we who are lost, it is we who have drowned and been washed up on the shore of a strange country. We do not recognize this land and we feel numb. When we walk in cities we hold hands because we are afraid that we will become more terribly lost. We are nervous crossing streets because the world has become unreliable and strange, we do not know where we are going. The dead have no need of knowing where they are but we are alive and adrift in a landscape of death where we feel marked. We have been infected with death and the knowledge of fragility it has taught is. In other places, they know this about us and are afraid of those who have witnessed death, who have had it in their home, and they send them away until it has left them. We do not need to be sent away for we have already left and we do not know if we will ever find our way back. We do not know if we want to go back for there is a comfort in being lost. Being lost now is our form of fidelity – without our dead ones we are lost and if we return to the world we may lose them again, lose the possibility that being lost gives us, we may find them gone forever. When you are lost you have a task, you leave breadcrumbs for others who may come this way, like others have done for you. We remember having seen these signs but did not understand them. We start to recognize others who are also in this world, their clothes are not torn, they do not wear a sign openly but you can spot them and you nod to each other. You gesture 'I know you', 'you are here too'. It may be years before we speak, before we share our stories of being lost at sea and washed up on this shore. We are surprised how the stories are the same, how we drowned and sank to the depths and were cold and numb and slept a deep sleep where eyeless fish eat our hearts in the dark and left us thin and hollow and didn't dream for fear of recalling happiness and waking to find it gone. Now we find pleasure in absence, because absence is all we have. We are not part of the world of others with their lofty towers. These are pointless and foolish; do they not know they will not save them? They think that by making them tall they will be able to see them from afar when they are lost, but they are deluded. In our country the rules have changed, it is not enduring monuments which help us find where we are. Instead we seek out the fleeting, the transient. Even among the ancient stones of the Alhambra, we know that with our feet, with our breath we are wearing it away. We fill our pockets with seeds and these give us hope. We have vomited, we have cried, we have screamed in an attempt to expel the knowledge we have learnt, but it will not leave us. Meaning has collapsed but we find new meaning in this place and the death holds no fear for us. For others the world is round but for us the world is flat and like explorers, we may fall off but we are not afraid.

Ognuno sta solo sul cuor della terra trafitto da un raggio di sole: ed è subito sera.

Each one stands alone at the heart of the earth, pierced by a ray of sunlight: and soon it is evening (Quasimodo 2004/1942, xix).

References

Auslander, P. (1999), *Liveness: Performance in a Mediatized Culture* (London: Routledge).

Bal, M. (1999), *Quoting Caravaggio: Contemporary Art, Preposterous History* (Chicago: University of Chicago Press).

Bauman, Z. (1992), *Mortality, Immortality and Other Life Strategies* (Cambridge: Polity).

Becker, E. (1973), *The Denial of Death* (London: Free Press).

Bowlby, J. (1988), 'Introduction to the First Edition', in Colin Murray Parkes, *Bereavement: Studies of Grief in Adult Life* (London: Penguin), 11–13.

Braun M.J. and Berg D.H. (1994), 'Meaning construction in the experience of parental bereavement', *Death Studies* 18:2, 105–129.

Cambell, B. "1001 Nights" <http://1001.net.au/> Last accessed 8 May 2006.

Durrant, S. (1999), 'Bearing Witness to Apartheid: J.M. Coetzee's Inconsolable Works of Mourning', *Contemporary Literature* 40:3, 430–463.

Foster, H. (1996), *Return of the Real: Avant-garde at the End of the Century* (Cambridge, Mass: MIT Press).

Freud, S. (1961) *Letters of Sigmund Freud, 1873–1939*, Ernst L. Freud (ed.) (London: Hogarth Press).

Gray, A. 'Dadang Christanto: A calling to account' <http://www.realtimearts.net/rt56/gray.html> Last accessed 5 July 2006.

Greenberg, J., Pyszczynski, T., Solomon, S., Rosenblatt, A., Veeder, M., Kirkland, S., and Lyon, D. (1990), 'Evidence for terror management theory: II. The effects of mortality salience reactions to those who threaten or bolster the cultural worldview', *Journal of Personality and Social Psychology* 58, 308–318.

Klass, D. and Goss, R. (1999), 'Spiritual bonds to the dead in cross-cultural and historical perspective: comparative religion and modern grief', *Death Studies* 23:6, 547–67.

Kristeva, J. (1982), *Powers of Horror: An Essay on Abjection* (New York: Columbia University Press).

MacLeod, D. (2005), 'Academic finds Beslan still traumatized by school siege', *The Guardian* 3 August. Available online at: <http://education.guardian.co.uk/higher/news/story/0,1559887,00.html> Last accessed 8 June 2008.

McDougall, R. (2006) Personal correspondence, 1 January.

O'Neill, M (2008), 'Ephemeral Art: Mourning and Loss', in J. Schachter and S. Brockman (eds.), *Impermanence: Cultures in/out of Time* (Philadelphia: Penn State University Press).

Parkes, C. M. (1986), *Bereavement: Studies of Grief in Adult Life* (London: Tavistock).

Phelan, P. (1993), *Unmarked: The Politics of Performance* (London: Routledge).

Romanoff, B. D., and Terenzio, M. (1998), 'Rituals and the grieving process', *Death Studies* 22, 697–711.

Sukhsvasti, U. (2005) 'Waking the Dead', *Bangkok Post*, 6 May.

Worden, W. (1991), *Grief Counselling and Grief Therapy: A Handbook for the Mental Health Practitioner* (London: Routledge).

Quasimodo, S. (2004/1942), 'Ed e Subito Sera', in R.L. Payne *A Selection of Modern Italian Poetry in Translation* (Montreal: McGill-Queen's University Press).

Chapter 9

'To Mourn': Emotional Geographies and Natural Histories of the Canadian Arctic

Emilie Cameron

In a recent study of encounters between the Alaskan and Yukon Tlingit and European explorers, scientists, and settlers, Julie Cruikshank (2005) asks the provocative question: 'do glaciers listen?'. Her book is a meditation on the entanglements of two very local cultures, between the knowledge and stories of the Tlingit, who have lived among the glaciers of the region for thousands of years, and the knowledge and stories of early European explorers and scientists, who had their own ideas about the landscapes and peoples they encountered. In Cruikshank's book, as in most treatments of encounter between Indigenous and European peoples, it is the Indigenous Tlingit that believe in an animate and personified nature, a world in which glaciers protect, punish, and listen to human conversations. The 1786 French expedition led by La Pérouse provides, in stark contrast, an example of Europeans who scoff at the possibility of glaciers listening; they understand the movements of glaciers through physical, chemical, geological, and meteorological principles. The differences between Indigenous and Western knowledge systems are consistently highlighted in studies of the history of science, exploration, and empire, and while Cruikshank troubles this familiar line of inquiry in important ways (refusing, for example, to answer her own question – *do* glaciers listen?), she makes clear the incommensurability of the storied, animate, responsive nature the Tlingit described and the factual, inanimate, scientific nature the European visitors strained to apprehend.

Several hundred miles to the East, in the tundra stretching north to the Arctic Ocean, another encounter story would unfold. In the summer of 1821, British Navy officer John Franklin, leading the first of his three Arctic expeditions, journeyed north along the Coppermine River. His party encountered several Copper Inuit groups as they explored the region and made scientific observations. But instead of confronting glaciers that listened, the Franklin expedition encountered a flower that mourned. In this case, though, it was not the Copper Inuit who told stories of a living and breathing world; there was no transfer of Indigenous knowledge to imperial centres of calculation. In this case, it was the *Europeans* who conjured a personified and storied landscape, a landscape that, in fact, runs counter to Inuit understandings of that place.

Senecio lugens (Figure 9.1), known commonly as the black-tipped groundsel, is a flower still found along the banks of the Coppermine River whose name

Figure 9.1   *Senecio lugens*
Source: Photo by author.

is derived from the Latin word 'lugeo', meaning 'to mourn'. Named by John Richardson, surgeon-naturalist with the Franklin expedition, the flower references the massacre of a Copper Inuit fishing party by a group of Chipewyan Dene along the shores of the Coppermine River fifty years before, in 1771. The massacre story had been passed on by the English Arctic explorer Samuel Hearne, who claimed to have witnessed the event. Richardson gave very little indication as to how a flower might be conceived of as 'mourning' in the botanical appendix to the expedition's travel narrative, only that it was collected at 'Bloody Fall', a place along the river 'where the Esquimaux were destroyed by the Northern Indians that accompanied Hearne, whence the specified name' (Richardson in Franklin 1823, 759). In this chapter, I would like to consider what the naming of *Senecio lugens* reveals about the emotional geographies of Arctic exploration and about the role of emotion and story in early nineteenth century scientific practice. I would like to consider how 'feeling' inflected imperial encounters in the Central Arctic and how particular places and things were (and continue to be) invested with emotional significance.

Even while recuperating some of the pain, anger, and grief that is usually written out of Arctic historiography, however, I would also like to problematize the production of Bloody Falls and the black-tipped groundsel as grieving and mournful. I will argue that *Senecio lugens* tells us much more about emotion, death, and scientific practice in nineteenth century Britain than it does about the flowers of the Central Arctic. Botanical names like *Senecio lugens* register some of the deeply emotional aspects of the Franklin expedition, a three year journey that resulted in starvation, murder, and the death of over half the original crew, but they fail to index centuries of other stories and feelings associated with Bloody Falls and with the wildflowers that bloom there. Rather than characterize *Senecio lugens* as yet another example of imperial appropriation and erasure of 'the local', however, I would like to consider how multiple 'locals' came together to produce a mournful yellow flower, and how yet still other 'locals' remain unaccounted for in this name. My interest, then, is in 'situating' knowledge (Haraway 1988) of an 'emotionally heightened' place like Bloody Falls to understand not only how 'social relations are mediated by feelings and sensibility' (Anderson and Smith 2001, 8) but also how particular feelings and sensibilities can occlude the expression and experiences of others. If, as Davidson et al argue, a 'genuine emotional geography... must try to express... a sense of emotional involvement with people and places, rather than emotional detachment from them' (2005, 2), then in this chapter I am additionally concerned with how emotional investment and involvement in the Central Arctic intersects with imperial and colonial power.

Emotion, Science and the Local

There is a small but significant literature addressing issues of emotion, gender, and race in the Canadian Arctic. Bloom (1993) and Grace (1997; 2002) have underscored the profound masculinism and racism of polar exploration, both as practice and as discourse, and have identified the gendered and racialized foundations of textual, visual, and other material representations of the Arctic in both academic and popular venues. Hulan (2002) additionally emphasizes the importance of 'experience' in representations of the Canadian North, outlining the ways in which 'being there' contributed to the 'epistemic privilege' (Powell 2005, 375) claimed by explorers, scientists, and ethnographers; that is, their authority and reliability was particularly tied to having 'experienced' the North, even if only briefly and superficially. Those who had been to the North buttressed their claims to 'know' the region through assertions of objectivity and neutrality, lending a particular emotional tone to scientific accounts and the strategic management of sentimentality, intimacy, and other threats to masculine detachment. Behrisch (2003) suggests that in response to these constraints, Arctic explorers in the mid-nineteenth century wrote poetry expressing 'the emotion inherent in the experience of discovery' (2003, 73), emotions that were not acceptable in their expedition reports, and Pálsson (2004; 2005) has devoted significant attention to the ways in which Vilhjalmur Stefansson, a prominent early twentieth century Arctic explorer and ethnologist, excised the emotional and intimate from his public persona. Behrisch's study of mid-nineteenth century exploration identifies an historical moment where the confinement of emotion to the personal, intimate, and poetic was resisted by explorers like John Ross, and it remains, to my knowledge, the only study of nineteenth century science, emotion, and exploration in the Canadian Arctic. Studies of the earlier decades of the nineteenth century remain to be done, however, and this chapter is a contribution along these lines.

By limiting her study to poetic expression, Behrisch tends to reinforce the notion of scientific prose as stripped of its emotional content. In this chapter, I argue that the emotional geographies of early nineteenth century exploration and science in the Canadian Arctic were not necessarily confined to the personal and private and that in fact they register in the scientific names assigned to a number of plants. Moreover, I argue that even while Richardson's scientific naming practices can be understood as part of a universalizing imperial project drawing heterogeneous 'local' knowledges, materials, and experiences into standardizing 'centres of calculation', Richardson also undermines and contradicts this explanation of his activities along the Coppermine River. The naming of *Senecio lugens* raises questions about how we understand the 'local' activities of traveling naturalists like John Richardson and troubles notions of science as a universalizing, detached, and emotionless pursuit. Richardson forces us to rethink the opposition of local and universal, objective and emotional, in accounts of scientific 'discovery' in the Canadian Arctic, particularly in the early decades of the nineteenth century.

A number of geographers have recently considered the meaning of the concept of 'local knowledge'. As Powell (2007, 312) points out, among science studies scholars in particular, the 'local' was initially proposed as a critical counterpoint to the idea of a universal body of scientific knowledge and truth. Drawing in part on Haraway's (1988) famous argument that all knowledge is 'situated', partial, and emerges from particular, inherently political contexts, scholars have emphasized locality as a way of calling attention to the contingency and partiality of scientific knowledge. Geographers have recently begun to problematize the 'local', however, as necessarily a political and critical concept, as in David Matless' (2003) consideration of the production of explicitly 'local' scientific knowledge of the Norfolk Broads, and in Bravo's (1999) examination of the ways in which 'local' knowledge was collected and rationalized by La Pérouse while he traveled through the Bay of Tartary in 1787. Bravo argues that the translation of knowledge about this North Pacific region into universalizing imperial frameworks was much more complex and cumulative than theories of imperial science-making reflect. Knowledge was produced gradually and through negotiation with those who knew the region best. It was not possible to 'collect' this knowledge without developing a set of ethnographic skills and transacting multiple exchanges, a process that calls into question the seemingly direct transfer of knowledge from the 'local' to the 'global'. Matless pays attention to the 'cultural charge' of the 'language of locality' (2003, 356) itself, pointing out that the claim of 'localness' on the part of scientists is worthy of study by scholars interested in the production of scientific knowledge. Indeed, as Hulan (2002) has pointed out from a different angle with regard to arctic exploration, claims of having 'local' experience in the Canadian Arctic are central to the production of authority about that region. Unlike in the Norfolk Broads, however, those who claim local experience in the Arctic are overwhelmingly from elsewhere, suggesting that there is more than one 'local' at play in the production of Arctic science and exploration. If, as James Secord has persuasively argued, we must do more than merely demonstrate that scientific knowledge is 'local' in the sense of being specific to a cultural-geographical context (2004, 659) and instead consider its 'connections with and possibilities for interaction with other settings' (664), then accounting for the multiple, conflicting 'locals' at play in the production of imperial science remains an important focus of study. In this case, the names Richardson assigned to a number of Arctic plants are best understood as emerging from multiple contexts, including shifting ideas of death, nature, and feeling among botanists, the Romantic sensibilities of the British public, and the specific culture of the Franklin expedition. 'Emotion' was expressed and understood differently in each of these contexts.

Senecio Lugens and the Emotional Geographies of Exploration

Here I highlight three key 'local' contexts within which to make sense of the naming of *Senecio lugens*. First, the naming of *Senecio lugens* can be understood as a form of tribute to God and to a theological understanding of plant morphology and botanical practice, tempered by the increasing secularization of botanical practice through the adoption of Linnaean taxonomy. Second, the plant name can be interpreted as a botanical equivalent to the popular genre of travel narratives and as an appeal to the Romantic sensibilities of the British public. Third, *Senecio lugens* can be read as an attempt on the part of Richardson to record the emotional experiences of the expedition itself, a journey fraught with hardships and personal tragedies.

The story of *Senecio lugens* begins in 1771, as Hudson's Bay Company employee Samuel Hearne undertook his third attempt to reach the Arctic Ocean by foot from Prince of Wales' Fort near present-day Churchill, Manitoba. Hearne was looking for copper deposits rumoured to exist along the Coppermine River, as well as any information that might assist in discerning a northwest passage. He made it to the Ocean, but only a few miles from the river's mouth, on the night of July 17, 1771, Hearne was detained by the interests of his Chipewyan Dene guides, who planned and executed the massacre of a group of sleeping Inuit, encamped at what is now known as Bloody Falls. In itself, this was a relatively unremarkable event; the Dene and Inuit were longstanding rivals and occasionally tensions erupted into killings of this nature (Fossett 2001; McGrath 1993). But Hearne, who allegedly witnessed the scene, was horrified. After completing his travels, Hearne went on to tell and retell his version of the massacre to his friends and colleagues, at London parties and scientific meetings,[1] and eventually published a sensational account of the attack in 1795. This story, full of writhing, eviscerated Inuit bodies, naked women begging for mercy, and dispassionate, ruthless Dene murderers enthralled the British public, and was foundational in establishing images of the Inuit as helpless, peaceable, and feminized, in contrast to the savage, marauding Indian (McGrath 1993).

Fifty years later, the Franklin expedition would be the second group of Europeans to travel overland to the Arctic Ocean, largely retracing Hearne's route. Charged by the Royal Society with the task of mapping the region and exploring new territories, Franklin was also urged to make a range of scientific observations. Second in command to Franklin was Surgeon-Naturalist John Richardson, a talented Scottish doctor with knowledge of botany and natural history, hoping to earn promotion and financial rewards for discovering and conveying specimens of Arctic flora and fauna home to Britain (Houston 1984a). Franklin was instructed to give 'every facility to [Dr. Richardson] in collecting, preserving, and transporting, the various subjects of Natural History, which can be allowed consistently with the primary object of the Exploration' (Lord Bathurst, cited in Levere 1993, 105). Identifying plants, in other words, was valued alongside activities like discovering deposits of precious metals, charting transportation routes, and accurately mapping the region.

Richardson was an ambitious and talented naturalist, if not particularly well-versed in Arctic flora before being hired for the expedition. Trained at the University of Edinburgh in botany, zoology, and geology, Richardson studied under naturalist Robert Jameson and associated with other leading scientists throughout his university and early naval career (Houston 1984b). In an 1819 letter to William Hooker, Richardson worried over his 'inexperienced eye' (Richardson 1819, 2b) and lack of botanical expertise (having spent the previous years working as a naval surgeon and not as a botanist), but also claimed to have a great store of energy and enthusiasm for the task ahead of him. 'The only quality I have in common with a Botanist is zeal for the advancement of the science' (1819, 3), Richardson wrote, and so began a long relationship with Hooker and other leading botanists of the day. The early decades of the nineteenth century were a fascinating and dynamic time to be practicing botany. Imperial exploration and trade had radically expanded the scope and volume of plants available for study, motivating the development of a more systematic means of classifying and naming plants.

---

1   Evidence of Hearne's oral tellings of the story remain circumstantial and are the focus of current research. Novelist Ken McGoogan alleges that Hearne was a 'minor celebrity' (2003: 263) while in London in 1783 and that he spoke at a number of meetings, even suggesting that Hearne spoke with a young Samuel Taylor Coleridge in 1791. Upon returning from his overland trek Hearne relayed the story of the massacre to Andrew Graham, Chief Factor of Prince of Wales Fort (Rollason Driscoll 2002), and it is reasonable to assume that Hearne also told the story to others throughout the many years he spent in Hudson's Bay, but details of his activities in London are not well established.

Although in use from as early as the mid-eighteenth century (Heringman 2003), the early decades of the nineteenth century saw the official adoption of the Linnaean system of plant and animal taxonomy by scientific bodies like the Royal Society, a decision that was not without controversy. The Linnaean system was meant to overcome the long and unwieldy scientific naming systems that preceded it and to universalize a diversity of local, Indigenous names. Botanists were directed to classify plants according to their visible sexual features, stamens and pistils, and assign binomial Latin names (*Genus*, followed by *species*) to the proliferation of plants pouring in from around the world. Linneaus was insistent that names be issued in Latin and that species names be assigned either in tribute to naturalists or prominent public figures, or as descriptions of the appearance or habitat of the organism. Names were not to reflect Indigenous names or languages, nor medicinal or other plant usage, nor were they to be biographical except in the form of attribution (Schiebinger 2004). The very efficiency of the Linnaean system was also a source of concern for some critics, who charged that it was overly mechanistic and sacrificed the evocative power of common plant names (Kelley 2003). Indeed, the adoption of Linnaean taxonomy can be understood as part of a broader move towards the rationalization and secularization of scientific practice, a process well underway before the publication of Darwin's *Origin of the Species* in 1859 (McOuat 2001; Rehbock 1983). Linnaean names did not so much challenge theologically-based understandings of the natural world as sidestep them, and it was this lack of explicit reference to the Divine qualities of plants that made some botanists uneasy.

The gradual secularization of botanical practice does not necessarily indicate a corresponding secularism among botanists themselves, nor even an immediate challenge to theological understandings of plant morphology, distribution, and origin.[2] Early nineteenth century botanical theory was deeply informed by theology and metaphysics. Belief in God as nature's designer was largely unquestioned at this time and the nascent sciences were understood, in part, as a process of revealing His work (Rehbock 1983). Indeed, natural philosophy – the precursor to the modern natural sciences – was a branch of knowledge devoted both to the study of the natural world and to philosophical ruminations on the origins and causes of natural phenomena. The study of plant life, in that sense, was not only a descriptive practice but also involved questions of meaning, cause, and origin. One of the ways in which the divine properties and origins of plants were discerned by botanists was through their own 'reactions' to the natural world. A sense of awe and wonder at the divine workings of God has been expressed by scientific figures

---

2   White (2003) has pointed out, moreover, that the religious beliefs of scientists may be of less importance in understanding nineteenth century conflicts between religion and science than an understanding of the ways in which various professional groups struggled to gain 'cultural authority' (101) against the Anglican church. Brooke (2001) also provides a much more nuanced account of the relationship between religious belief and scientific practice.

for centuries, as Brooke (2001) notes, and botanists are no exception. Allen argues that from the late eighteenth century onwards, naturalists were expected not only to record 'what one saw plainly and accurately' (1976, 54) but also 'to record one's reactions – and the livelier these reactions appeared, the more beneficial, the more exalting, the more "tasteful" the contract with nature was assumed to have been' (54). A kind of engagement and involvement with plant specimens was encouraged, and emotional engagement was particularly valued. Outside of the botanists' realm a similar notion circulated. Plants were sometimes personified during this era and attributed with both agency and profound symbolism (Kelley 2003).[3] Erasmus Darwin (1799) famously produced erotic poems describing the sexual exploits of various plants and, as Bewell (1996) argues, the explicitly sexual quality of botanical investigation raised concerns about the moral implications of closely inspecting and identifying plants. Plants, it was believed, could affect people, but it was also believed that plants were themselves effects, physical manifestations of God's mysterious universal design.

Richardson was devoutly Christian, filling his letters to his family and his colleagues with theological interpretations of everyday life (Richardson 1821; 1823c; 1823d; 1824). He was convinced that the Arctic landscape was 'inhospitable to atheists' and that only his Christian faith saved him from despair on the arduous journey. In a letter sent to Franklin in 1824, Richardson commented on a Mrs. Gracroft who was bearing the death of her husband

> ...with a resignation so truly Christian. Time, the grand physician of the mind will soften down the poignancy of her grief into a tender remembrance of the excellent qualities and benevolent disposition of her departed husband. In the mean time her pious frame of mind will enable her to contend with the anguish inseparable from so recent and severe a deprivation by directing her to the contemplation of the Almighty disposer of Events who orders every thing according to his pleasure and all for the best. It is in the power of him who hath taken away to blunt the sting of death, destroy the victory of the grave and in another and a better world bring about that reunion which is the hope to which a Christian most fondly clings when the ties of friendship are thus suddenly swept asunder (1824, 23).

Such passages were a regular feature of Richardson's letters to his wife, Mary, and his Christian faith was remarked upon by the expedition members with whom he traversed the Central Arctic (Franklin 1823). Richardson has also been memorialized as a generally emotional and sensitive man who stored up

---

3 Browne (1996, 155–156) makes the interesting observation that Linnaeus himself personified plants, describing plant fertilization as a form of 'marriage' involving husbands, wives, and a marriage bed. Richardson's personification of *Senecio lugens* as a grieving witness to the Bloody Falls massacre thus took its cue from Linnaeus as much as it subverted Linnaean naming protocols.

'the best poems and songs of Burns…, ready to be poured forth when his feelings were touched by either the pathetic or the humourous' (McIlraith 1868, 4). Such a characterization perhaps attests more to the Romantic climate within which this biography was penned than to Richardson's own sensibilities, but it nevertheless suggests that feeling and emotional engagement with the world did not diminish, and perhaps contributed to, Richardson's status as an upstanding man of science. He became a fellow of the Royal Society in 1825 (Royal Society 2007) and maintained close relations with the major figures of nineteenth century botany throughout his career; Richardson is celebrated even today as an important figure in the history of science (Houston 1984a). Daston and Galison (2007, 36) have recently pointed out that objectivity, understood as 'the suppression of some aspect of the self, the countering of subjectivity', only emerged as a model of scientific practice in the mid-nineteenth century. The scientific climate Richardson operated in was not only tolerant but to some extent supportive of emotional reaction and engagement with the natural world, whether in a religious or a more generalized Romantic register.

According to Rehbock (1983) and Brooke (1991), the botanical sciences in Britain remained thoroughly teleological until the 1830s; that is, plants were believed to be designed by God and outfitted with features that reflected specific meaning or purpose. It was God, in other words, who gave the leaves of *Senecio lugens* their black tips, and it was the job of naturalists like Richardson to decode God's purpose in fitting the groundsel with its own peculiar features, in part through scrutiny of his own emotional reaction to the plant. Because Richardson himself provided so little explanation about the naming of *Senecio lugens*, one can only speculate as to who, or what, he imagined the flower to be mourning. Certainly, in the botanical community within which Richardson operated, it would have been plausible to imagine that God's own grief at the slaying of the Inuit was manifested in the leaves of the black-tipped groundsel, and, moreover, that this grief was accessible to the close eye of the naturalist. The naming of *Senecio lugens* might therefore have been both a tribute to God and a mournful record of a savage act, reminding us that God was watching that day, as Richardson was certain He was. This tribute, moreover, does not explicitly violate Linnaean taxonomical guidelines. In one sense *Senecio lugens* references precisely the kinds of things Linnaean names were meant to overcome: its mournful name evokes a sensational event the British public was amply familiar with, it pays tribute to Indigenous history, and it indexes an emotional reaction to the natural world – all things that Linnaeus aimed to eliminate from plant naming practice. And yet the species name 'lugens' was also accepted among botanists as an appropriate reference to an organism's black colouration (Charters 2003–2005; Jaeger 1955, 145), colouration that evoked the mourning dress that was increasingly common in Europe from the late eighteenth century onwards. Known commonly as the 'black-tipped groundsel' for the black tips on its involucral bracts, *Senecio lugens* is, in that sense, named for its morphology as much as its poetry. It is a brilliant negotiation of early nineteenth century botanical naming debates.

The reference to mourning dress is not insignificant. Richardson was collecting and naming plants for a much wider audience than the network of botanists with whom he corresponded. Not only had plants like coffee, sugar, and cotton transformed imperial commerce, but from about the middle of the eighteenth century onwards natural history had also begun to capture the imagination of writers, artists, and the rising middle class (Fulford et al. 2004; Shtier 1997). By the turn of the nineteenth century, it was not unusual for middle class families to spend a day at the beach identifying local species or for ladies to wander the woods with botanical guides in hand. A handful of poems were written about Linnaean taxonomy (Kelley 2003) and the discovery of new species elicited interest from a much wider constituency than the network of botanists involved in the identification, circulation, and naming of plant specimens. Knowledge and enthusiasm for plants circulated so widely in part because a separate, specialized, scientific realm had yet to be constructed, and because knowledge of the natural world grew alongside and in relation to literary and philosophical thought. Although specialized knowledge and skills were certainly required of Richardson to carry out his assignment, his findings were meaningful to a broader public audience, and accessible to that audience through the publication of an expedition travel narrative.

Very soon after returning from North America, Franklin and his colleagues began work on the publication of their expedition travel narrative (Richardson 1823–1842). Beginning with the publication of James Cook's journals through the late eighteenth century, travel narratives had become an important and expected venue for the transmission of knowledge about Britain's growing empire and exploratory activities. As Driver points out, the authority of nineteenth century explorers 'depended substantially on the writing of a narrative of travel, either first or second hand' (2004, 77), and Franklin published his within months of his return to Britain (Franklin 1823). The narrative included a botanical appendix penned by Richardson, but Richardson also arranged for the production of separate bound copies of the appendix for independent circulation, one of which was presented to William Hooker at Kew (Richardson 1823a).[4] Richardson was well aware of the public's interest in the expedition (Richardson 1823b) and was keen to augment his own reputation through the publication of his work. The travel narrative itself was written to appeal to a broader, more public audience and *Senecio lugens* can be interpreted as a botanical iteration of this popularization effort.

Franklin and Richardson were writing for an audience increasingly influenced by Romanticism. This was the era of William Wordsworth and Samuel Taylor Coleridge; it was a time when picturesque and evocative landscapes were celebrated, sentimental language flourished, and outpourings of emotion became important forms of public display (Pinch 1996). There is a growing body of research into the links between Romantic literature and scientific practice (e.g., Fulford et al. 2004;

---

4   Thank you to Dr. D. J. Nicholas Hind at Kew Gardens for bringing this separately-paginated appendix to my attention.

Heringman 2003), suggesting that the lines between poetry, feeling and science were not so neatly drawn in this era. This literature has focused primarily on the ways in which scientific knowledge influenced Romantic writers, however, and has paid less attention to Romantic influences upon the production of scientific knowledge (although see Cunningham and Jardine 1990). Certainly, poets had greater license to express the emotions expected of Romantic subjects than their scientific colleagues, and Behrisch (2003) is right to identify the expressive limitations placed upon the men who explored and catalogued the Canadian Arctic throughout the nineteenth century. Nevertheless, I would argue that the naming of *Senecio lugens* can be interpreted as a gesture to the Romantic sensibilities of Richardson's public audience.

Julie Rugg argues that Romanticism was characterized by 'a movement away from concerns with the death of the self, and towards a stress on loss, and the death of the "other". At a time when "feeling" was paramount, the almost wild expression of grief at the loss of a family member was considered to be appropriate and laudable'(1999, 210). Performances of grief peaked in the Victorian era, but were certainly active in the earlier part of the nineteenth century. The British were quite simply obsessed with death and mourning, and along with a proliferation of mourning protocols, costumes, and lavish funerals, they began to grow concerned at the state of their cemeteries. In the early years of the nineteenth century the notion of the garden cemetery began to take hold in Britain. These bucolic landscapes were meant to replace the 'gloom of the grave' (Curl 2000, 70) and provide an eternal resting place where the dead could be 'eternally present, immortal, joined indissolubly with the living' (70). Full of picturesque scenes and evocative ruins, garden cemeteries also included graveside flowers. Aside from the symbolism of individual flowers (the lily, for example, was used as a symbol of virginity), as a whole the seasonal rebirth of flowers signaled the ongoing life of the dead. Flowers were seen as happy companions to buried souls but also directly symbolized the possibility of rebirth and reunion in the afterlife (Curl 2000).

A garden cemetery to mark the death of the Inuit at Bloody Falls seems an absurd notion, but the Franklin expedition did their best to effect such a place. Consider expedition artist George Back's illustration of the site (Figure 9.2). Note the skulls and bones in the foreground, rendered in a typical picturesque style. Contemporary scholars insist that it is next to impossible that these bones were actually on site due to periodic flooding of the clearing upon which they are depicted (McGrath 1993), and argue, moreover, that the massacre actually occurred 1 km downstream (St. Onge 1982). Niptanatiak (2006) also questions Back's placement of the massacre site and argues that the Inuit would more likely have been camping on a flat, gravely stretch downstream known as Onoagahiovik, the 'place to stay all night and fish'. But in Richardson's journal he insists that the party 'encamped on the very spot where the Massacre of the Esquimaux was transacted by Hearne's party' (Houston 1984a, 77) and that 'the ground is still strewed with human skulls' (77). He also endeavoured to cordon off the site of the killings as a kind of memorial space, noting that it was 'overgrown with rank grass, [and] appears to be avoided as a place of encampment'

Figure 9.2   'Bloody Fall' by George Back

Source: Franklin 1823, reproduced with permission from the original held in the W.D. Jordan Special Collections and Music Library, Queen's University at Kingston, Ontario.

(77–78). Richardson and his colleagues were writing for, and shaped by, a culture in which landscapes were believed to reflect and promote emotions (Rugg 1999), and the party invested evocative, emotional detail in everything they encountered on site. Franklin suggested that he could see signs of struggle and violence on the bones themselves, and made a point of re-affirming Hearne's naming of the place as Bloody Fall in his travel narrative (1823). Richardson conjured a seasonal witness to the tragedy; wildflowers mourning a wild grave, flowers black with the blood of the fallen Inuit. Indeed, in spite of the fact that Richardson would have encountered specimens of the black-tipped groundsel throughout his travels in the Coronation Gulf area, he chose 'Bloody Fall' as the type locality for *Senecio lugens*, forever linking the plant with this 'historic' site, and appealing to public knowledge of Hearne's journey. In this way, the naming of *Senecio lugens* can be understood as an effort to augment the sensational and narrative appeal of the expedition report. Conjuring the Bloody Falls massacre in the botanical appendix may well have been a canny appeal to Romantic interest in death, grief, and emotion at the time when Richardson was striving to make a name for himself as both botanist and Arctic explorer.

It seems to me, though, that this mournful flower might actually have very little to do with the victims of the massacre, and might instead reference Richardson's own emotional journey. *Senecio lugens* is just as likely a tribute to the deeply mournful Hearne, who was unable to place his own permanent mark on the landscape, as it is to the mournful Inuit. Reviews of Hearne's travel narrative emphasized Hearne's emotional turmoil at being an impotent witness to such a 'barbarous' scene and all quote Hearne's claim to be unable to reflect back on the event without 'shedding tears' (e.g., Monthly Review 1796, 250). Hearne's 'affecting narrative' was deemed to be 'a proper mixture of indignation at the brutalities, and of compassion for the miseries, of those wretched savages' (250), suggesting that Hearne's emotional reaction to the massacre was a reasonable subject for botanical commemoration.[5] *Senecio lugens* also seems to reference Richardson's own sense of apprehension in what felt to him an incomplete, unsettling landscape. Not only was Bloody Falls lacking in a suitable monument to its past, Richardson regularly commented on the isolation, sterility, and barrenness of the landscape, its absence of depth, boundaries, and human life. The arctic landscape was famously bewildering to the British in this way (Spufford 1996; Wylie 2002), as was the culture of the Inuit who inhabited it. *Senecio lugens*, named over a year after Richardson had left the Coppermine River, seems to capture the great sense of loss and incoherence Richardson experienced on his journey; the failure of the land to make sense, to remember, and the seeming indifference of the land to human survival.

In the weeks following their stay at Bloody Falls, the party began to starve. Winter came early, the herds of caribou upon which the expedition had relied headed south, and the crew was forced to subsist on lichen and the leather from their shoes (Franklin 1823). Nine of the twenty crew members died of starvation that fall as they walked southwards across the open tundra. Surviving members are widely believed to have eaten the fallen. On October 20th, 1821, one voyageur, mad with starvation, shot and killed midshipman Robert Hood, an expedition member. Richardson retaliated by shooting the man in the back of the head (Houston 1984a, 156). Days later, finally reaching Fort Enterprise, where the men anticipated finding stores of meat left for them by Akaitcho, a Yellowknife Indian chief, they found an empty, snowed-in cabin, with no supplies. Franklin describes this moment as one of 'infinite disappointment and grief' (1823, 433), but he also suggests that 'it would be impossible to describe our sensations after entering [the] miserable abode, and discovering how we had been neglected' (433). Franklin may have

---

5   Although some have noted that Richardson was critical of the cartographic errors presented in Hearne's narrative, suggesting that Hearne deliberately misrepresented certain latitude and longitude positions to refute critiques about his having actually journeyed to the Northern Ocean (Richardson 1836), this critique was part of an otherwise congratulatory and admiring account of Hearne's work. Janice Cavell (2007) has recently and persuasively argued that the depiction of Richardson as hostile towards Hearne (as argued, for example, by historian Richard Glover, 1951) was grounded more in Glover's own protectiveness of Hearne's reputation than in historical fact.

affected an inability to express his deeply felt sensations in the pages of his travel narrative, but Richardson seems to have found words in the botanical appendix. Richardson named a plant found in the woods near Fort Enterprise *Geocaulon lividum*. In this case, there is no hiding the very personal, emotional reasons for the name. Although 'lividum' is usually used botanically to reference a bluish-black or lead colour (Hyam 1994, 292), there is nothing black or blue about this plant (Hind 2007). Also known as false toadflax, *geocaulon lividum* is bright green, with leaves that are frequently streaked with yellow, and berries that range from scarlet to orange. 'Lividum' also means 'livid', 'spiteful', or 'envious', and it seems that this second meaning is what Richardson referenced in the naming of the plant – livid at the apparent betrayal of a promise by Akaitcho to provision the Fort, livid with the man who had murdered Hood and his own decision to take a life, a heart black with the state of the decaying crew.

What can we conclude, then, from these personal, emotional traces in Richardson's nomenclature? Certainly, they represent a challenge to contemporary notions of science as emotionless, and they register some otherwise unnamed emotional contours to science and exploration in this era. Each specimen named and conveyed to Britain to be dissected and displayed holds a story, and it appears as though Richardson purposely injected records of his feelings and experiences into his botanical collection, albeit somewhat discreetly. The plants, birds, and fish Richardson named on this journey perform feelings Richardson himself was not permitted to display as an expedition member but he managed to convey as a botanist. As MacLaren (1984) notes, Franklin accepted no doubts, complaints, fears, or vulnerabilities in his crew, and Richardson himself insisted on decorum, tidiness, and good cheer even as the men withered of starvation at Fort Enterprise (Franklin 1823). The names Richardson assigned challenge the detachment and universalizing logic Linnaean taxonomy was supposed to embody; they insist on the naturalist *as* an embodied, moving, feeling creature, and are emblematic of an era in which botanists wrestled with the secularization of their profession. The notion that science is as emotional and embodied a pursuit as any other is not particularly radical, of course, but the inscription of powerful, personal emotions in scientific text, and the conjuring of animate, emotional flora challenges scientific claims to detachment and universality in the very moment meant to exemplify such claims. It would appear as though cultures of natural history in the early nineteenth century accommodated these seemingly contradictory expectations.

But there is also a kind of tyranny in these names, an insistence that the landscape reflect the vagaries of a few men, briefly encamped on lands that hold many more stories than Hearne or Richardson understood. Even if the flower's name was chosen as a sign of respect to the murdered Inuit, it is important to emphasize the already and continually storied landscape known to local inhabitants. Bloody Falls is not just the stopping place of Hearne or Franklin, nor the Arctic tourists who continue to journey to the site. In Inuinnaqtun this place is named Kugluk, 'the rapids', and each fishing, hunting, gathering, and camping site at the rapids has its own specific name. Indeed, the people of Kugluktuk have been working to change the official name of the site back to 'Kugluk' for years, and explicitly reject the memorialization of the massacre event as 'Bloody Falls' (Havioyoak 2008). Archeological evidence suggests that the rapids have hosted human camps for over three thousand years (McGhee 1990), and today, residents of nearby Kugluktuk still travel to the rapids in late summer to fish for arctic char. The 1771 massacre is almost wholly remembered at the falls in English. In Inuit oral accounts, the story is located many miles away, at a hilltop where the surviving Inuit chased their Dene rivals and had their revenge (Adjun 2007; Ahegona 2006; Hikok 2007; McGrath 1993). Theirs is not only a tale of mourning, but also of heroism, nobility, and morality. Similarly, the yellow flowers that bloom along the banks of the Coppermine River have no specific name in Innuinaqtun (Davis 2006; Taipana 2007); they are known simply as 'flower'. In Inuit epistemology, nothing is given a specific name unless it has a specific purpose (Thorpe 2000). The black-tipped groundsel is not edible, medicinal, or useful for navigation. It is simply a yellow flower that lights up the land every July, a part of 'Nunatsiaq', the beautiful land.

Inuit understandings of death and mourning also jar against Richardson's eternally mournful flower. Traditionally, mourning was a brief but intense practice and survivors were encouraged to move on as soon as possible (Jenness 1922; Korhonen and Anawak 2006; Niptanatiak 2007). In his ethnographic writings Jenness contended that the Copper Inuit actively worked to 'banish every unpleasant memory from their minds' (171) and that even the day after a funeral they would 'resume their ordinary occupations and try to forget their loss' (174). Still today Inuit Elders in Kugluktuk advise their community not to linger in their grief: 'our ancestors didn't mourn. Even now, the Elders tell us not to mourn. You have to move on. You can't spend time mourning' (Niptanatiak 2006). Inuit understandings of grief and death merit a much more detailed discussion than this chapter allows, but it would seem that *Senecio lugens* jars against Inuit understandings of mourning, death, and remembrance.

Further questions about this mournful flower emerge from recent research into Hearne's voyage, based largely on Inuit and Dene testimony, suggesting that Hearne did not actually witness the massacre and was in fact two days travel from the site when the killings took place (MacLaren 1991; McGrath 1993). Richardson himself questioned whether Hearne witnessed the massacre in a private letter to his wife (1821), although he never publicly questioned Hearne's testimony and reinforced Hearne's authority in the expedition's report. From its beginnings, then, the link between the rapids, the massacre, and the European gaze seems to have been a kind of performance, an imagined geography of 'being there'. The Franklin expedition was instrumental in confirming Hearne's story and furnishing Bloody Falls with material traces of the past, but if one peers a bit more closely at one of those furnishings, *Senecio lugens*, it becomes clear that this geography of emotion and witness was much more reflective of a series of other 'locals' – the concerns and desires of British botanists, the rise of Romanticism in British culture, and the personal experiences of the Franklin expedition members – than of either Kugluk or the massacre itself. And yet *Senecio lugens* is also suggestive of Richardson's struggle to narrate his journey within particular ideals of travel, science, and exploration, no matter how much the actual voyage strayed from those ideals; to find voice for the deeply emotional geographies of the expedition and to commemorate Hearne's own emotional narrative. It seems, then, that the flower both confirms and unravels Richardson's emotional Arctic geography. Even as he used botanical names to produce a sense of immediacy, divinity, and sentimentality in the Arctic, to produce the Arctic in the image of science, Romanticism, and heroic travel, the links between the specimen of black-tipped groundsel he collected, its Latin name, and the history it references are at the very least tenuous, and recent research suggests they were entirely imagined. The scalar tensions here are not so much between local knowledge and its scientific translation, then, but rather between the 'local' of Richardson's audience and his own localized experiences throughout the expedition, between a story localized at Bloody Falls and a story stretching across the surrounding tundra, and between a single specimen of *Senecio lugens* and a riverbank bright with yellow blooms.

Conclusion

Do glaciers listen? Do flowers mourn? As in Cruikshank's work, the focus of this chapter has been not so much to definitively answer this question, but rather to explore the multiple and shifting ways in which places and things can be known, named, remembered, and felt, and to call attention to the powerful claim of particular ways of knowing and feeling at Bloody Falls. Richardson's own position on this question – do flowers mourn? – is unclear, and I have examined three contexts within which the naming of *Senecio lugens* can be understood. Among teleological botanists adjusting to the secularism of Linnaean naming protocols flowers could in some sense be conceived of as 'mourning', and a name like *Senecio lugens* expressed this sensibility. The poetic image of a flower marking the death of the Inuit also appealed to a more general Romantic British audience, an audience fascinated by imperial discoveries, avidly consuming travel narratives, and preoccupied with death. *Senecio lugens* might also express Richardson's more personal experiences along the trip and index a moment of connection felt between the tragedies of the Franklin expedition and the evocative massacre story passed on by Hearne. Indeed, the diversity of possible interpretations of this flower's mournful name underscores both the richness and the potential of emotional geographic inquiry for understanding histories of science, exploration, and encounter.

And yet it is not enough to recuperate some of the emotionality of this historical moment; indeed, attending to the emotional geographies of exploration and empire opens ambivalent political possibilities. The production of Bloody Falls as a space of death and mourning has been resisted by the people of Kugluktuk for generations. They reject not only the emotional tone of this memorialization but also the details of Euro-Canadian descriptions of the event and the attribution of guilt to the Dene, highlighting instead the imperial designs of explorers like

Hearne. Even while attending to the complexity of feeling and experience indexed by the naming of *Senecio lugens*, then, I have attempted in this chapter to call attention to what is *not* named by this flower. Not only do Inuit understandings of memory, mourning, and death fail to register in this name, but so, too, do the centuries of other stories and feelings associated with Kugluk and with the flowers along the Coppermine River. *Senecio lugens* mourns for particular people in particular ways, but for some it is not mournful at all.

Acknowledgements

I would like to thank Anne Secord for such a close and critical reading of an earlier version of this chapter. Thanks also to Laura Cameron, Joyce Davidson, and Mick Smith for their thoughtful comments and suggestions throughout the editorial process.

References

Adjun, C. (2007), 'Personal Communication' (20 August 2007, Kugluktuk, NU).

Ahegona, A. (2006), 'Personal Communication' (4 July 2006, Kugluktuk, NU).

Allen, D. E. (1976), *The Naturalist in Britain: A Social History* (London: Allen Lane).

Anderson, K. and Smith, S. (2001), 'Editorial: Emotional Geographies', *Transactions of the Institute of British Geographers* 26: 7–10.

Behrisch, E. (2003), '"Far as the Eye Can Reach": Scientific Exploration and Explorers' Poetry in the Arctic, 1832–1852', *Victorian Poetry* 41:1, 73–91.

Bewell, A. (1996), '"On the Banks of the South Sea": botany and sexual controversy in the late eighteenth century' in D. P. Miller and P. H. Reill (eds.), *Visions of Empire: Voyages, Botany, and Representations of Nature* (Cambridge: Cambridge University Press).

Bloom, L. (1993), *Gender on Ice: American Ideologies of Polar Expeditions* (Minneapolis: University of Minnesota Press).

Bravo, M. (1999), 'Ethnographic Navigation and the Geographical Gift', in D. Livingstone (ed.), *Geography and Enlightenment* (Chicago: University of Chicago Press).

Brooke, J. H. (1991), *Science and Religion: Some Historical Perspectives* (Cambridge: Cambridge University Press).

Brooke, J. H. (2001), 'Religious Belief and the Content of the Sciences', *Osiris* 16: 2nd Series, 3–28.

Browne, J. (1996), 'Botany in the boudoir and garden: the Banksian context', in D. P. Miller and P. H. Reill (eds.), *Visions of Empire: Voyages, Botany, and Representations of Nature* (Cambridge: Cambridge University Press).

Cavell, J. (2007), 'The Hidden Crime of Dr. Richardson', *Polar Record* 43: 225, 155–164.

Charters, M. L. (2003–2005), 'California Plant Names: Latin and Greek Meanings and Derivations, A Dictionary of Botanical Etymology'. Sierra Madre, CA. Available online at: <http://www.calflora.net/botanicalnames/index2.html> Last accessed 8 June 2008.

Cruikshank, J. (2005), *Do Glaciers Listen? Local Knowledge, Colonial Encounters, & Social Imagination* (Vancouver: University of British Columbia Press).

Cunningham, A. and Jardine, N. (eds.) (1990), *Romanticism and the Sciences* (Cambridge: Cambridge University Press).

Curl, J. S. (2000), *The Victorian Celebration of Death* (Phoenix Mill, UK: Sutton Publishing).

Darwin, E. (1799), *The Botanic Garden* (London: J. Johnson).

Daston, L. and Galison, P. (2007), *Objectivity* (New York: Zone Books).

Davidson, J., Bondi, L. and Smith, M. (eds.) (2005), *Emotional Geographies* (London: Ashgate).

Davis, J. D. (2006), 'Personal Communication' (6 July 2006, Kugluktuk, NU).

Driver, F. (2004), 'Distance and Disturbance: Travel, Exploration and Knowledge in the Nineteenth Century', *Transactions of the Royal Historical Society* 14: 73–92.

Fossett, R. (2001), *In Order to Live Untroubled: Inuit of the Central Arctic, 1550–1940* (Winnipeg: University of Manitoba Press).

Franklin, J. (1823), *Narrative of a journey to the shores of the Polar Sea, in the years 1819, 20, 21 and 22.. with an appendix on various subjects relating to science and natural history* (London: J. Murray).

Fulford, T., Lee, D. and Kitson, P. J. (2004), *Literature, Science, and Exploration in the Romantic Era* (Cambridge: Cambridge University Press).

Glover, R. (1951), 'A Note on John Richardson's "Digression Concerning Hearne's Route"', *Canadian Historical Review* 32: 3, 252–263.

Grace, S. (1997), 'Gendering Northern Narrative', in J. Moss (ed.), *Echoing Silence: Essays on Arctic Narrative* (Ottawa: University of Ottawa Press).

Grace, S. (2002), *Canada and the Idea of North* (Montreal and Kingston: McGill-Queen's University Press).

Haraway, D. (1988), 'Situated Knowledge: The Science Question in Feminism and the Privilege of Partial Perspective', *Feminist Studies* 14: 575–599.

Havioyoak, D. (2008), 'Personal Communication' (6 March 2008, Cambridge Bay, NU).

Hearne, S. (1795), *A Journey from Prince of Wales's Fort in Hudson's Bay to the Northern Ocean 1769, 1770, 1771, 1772* (London: A. Strachan and T. Cadell).

Heringman, N. (2003), 'Introduction: The Commerce of Literature and Natural History', in N. Heringman (ed.), *Romantic Science: the Literary Forms of Natural History* (Albany: State University of New York Press).

Hikok, N. (2007), 'Personal Communication' (9 August 2007, Kugluktuk, NU).

Hind, D. J. N. (2007), 'Personal Communication' (15 November 2007, Kew Gardens Herbarium, London).

Houston, C. S. (1984a), *Arctic Ordeal: the Journal of John Richardson, Surgeon-Naturalist with John Franklin 1820–22* (Kingston & Montreal: McGill-Queen's University Press).

Houston, C. S. (1984b), 'New Light on Dr. John Richardson', *Canadian Medical Association Journal* 131: 653–660.

Hulan, R. (2002), *Northern Experience and the Myths of Canadian Culture* (Montreal and Kingston: McGill-Queen's University Press).

Hyam, R. (1994), *Plants and Their Names: A Concise Dictionary* (Oxford: Oxford University Press).

Jaeger, E. C. (1955), *A Source-Book of Biological Names and Terms* (Springfield, IL: Thomas).

Jenness, D. (1922), *The Life of the Copper Eskimo. Report of the Canadian Arctic Expedition, 1913–18* (Ottawa: Acland).

Kelley, T. M. (2003), 'Romantic Exemplarity: Botany and "Material" Culture', in N. Heringman (ed.), *Romantic Science: The Literary Forms of Natural History* (Albany: State University of New York Press).

Korhonen, M. and Anawak, C. (2006), 'Suicide Prevention: Inuit Traditional Practices that Encouraged Resilience and Coping' (Ajunnginniq Centre, National Aboriginal Health Organization, Ottawa, ON).

Levere, T. (1993), *Science and the Canadian Arctic: A Century of Exploration 1818–1918* (Cambridge: Cambridge University Press).

MacLaren, I. S. (1984), 'Retaining Captaincy of the Soul: Response to Nature in the First Franklin Expedition', *Essays on Canadian Writing* 28: 57–92.

MacLaren, I. S. (1991), 'Samuel Hearne's Accounts of the Massacre at Bloody Fall, 17 July 1771', *Ariel* 22: 1, 25–51.

Matless, D. (2003), 'Original Theories: Science and the Currency of the Local', *Cultural Geographies* 10: 3, 354–378.

McGhee, R. (1990), *Canadian Arctic Prehistory* (Gatineau, QC: Canadian Museum of Civilization).

McGrath, R. (1993), 'Samuel Hearne and the Inuit Oral Tradition', *Studies in Canadian Literature* 18: 2, 94–109.

McIlraith, J. (1868), *Life of Sir John Richardson, C.R., LL.D., F.R.S. Lond., Hon. F.R.S. Edin., Inspector of Naval Hospitals and Fleets, &c. &c. &c* (London: Longmans, Green, & Co.).

McOuat, G. (2001), 'Cataloguing Power: Delineating "Competent Naturalists" and the Meaning of Species in the British Museum', *The British Journal for the History of Science* 34: 1, 1–28.

Monthly Review (1796), 'Review of Samuel Hearne, A Journey from Prince of Wales's Fort in Hudson's Bay, to the Northern Ocean 1769, 1770, 1771, 1772', *Monthly Review* 20: 246–251.

Niptanatiak, A. (2006), 'Personal Communication' (5 July 2006, Kugluktuk, NU).

Niptanatiak, A. (2007), 'Personal Communication' (20 August 2007, Kugluktuk, NU).

Pálsson, G. (2004), 'Race and the Intimate in Arctic Exploration', *Ethnos* 69: 3, 363–386.

Pálsson, G. (2005), *Travelling Passions: The Hidden Life of Vilhjalmur Stefansson* (Winnipeg: University of Manitoba Press).

Pinch, A. (1996), *Strange Fits of Passion: Epistemologies of Emotion, Hume to Austen* (Stanford: Stanford University Press).

Powell, R. (2005), 'Northern Cultures: Myths, Geographies, and Representational Practices', *Cultural Geographies* 12: 3, 371–378.

Powell, R. (2007), 'Geographies of Science: Histories, Localities, Practices, Futures', *Progress in Human Geography* 31: 3, 309–329.

Rehbock, P. (1983), *The Philosophical Naturalists: Themes in Early Nineteenth Century British Biology* (Madison: University of Wisconsin Press).

Richardson, J. (1819), Letter from J. Richardson to W. Hooker, London, 8 May 1819. *Sir W. Hooker's Letters to J. Richardson 1819–1843* (London: Kew Gardens Library and Archives).

Richardson, J. (1821), 'Letter to Mary Richardson, 18 July 1821 [transcription]', *C. Stuart Houston fonds* (Saskatoon: Special Collections, University of Saskatchewan).

Richardson, J. (1823a), *Franklin's 1st Journey App. 2nd Ed.* (London: Kew Gardens Library and Archives).

Richardson, J. (1823b), Letter from J. Richardson to J. Franklin, 6 August 1823. *Richardson, J. Letters to J. Franklin 1823–1842* (London: Kew Gardens Library and Archives).

Richardson, J. (1823c), Letter from J. Richardson to J. Franklin, 6 July 1823. *Richardson, J. Letters to J. Franklin 1823-1842* (London: Kew Gardens Library and Archives).

Richardson, J. (1823d), Letter from J. Richardson to J. Franklin, 24 June 1823. *Richardson, J. Letters to J. Franklin 1823–1842* (London: Kew Gardens Library and Archives).

Richardson, J. (1823–1842), 'Richardson, J. Letters to J. Franklin 1823–1842' (London: Kew Gardens Library and Archives).

Richardson, J. (1824), Letter from J. Richardson to J. Franklin, Chatham, 22 July 1824. *Richardson, J. Letters to J. Franklin 1823–1842* (London: Kew Gardens Library and Archives).

Richardson, J. (1836), 'Digression Concerning Hearne's Route', in G. Back (ed.), *Narrative of the Arctic land expedition to the mouth of the Great Fish River, and along the shores of the Arctic Ocean, in the years 1833, 1834 and 1835* (Philadelphia: E.L. Carey and A. Hart).

Rollason Driscoll, H. A. (2002), 'The Genesis of "A Journey to the Northern Ocean": A dissertation concerning the transactions and occurrences related to Samuel Hearne's Coppermine River narrative, including information on his letters, journals, draft manuscripts, and published work' (University of Alberta, Canada).

Royal Society (2007), List of Fellows of the Royal Society 1660 – 2007 (London: Royal Society Library and Information Services).

Rugg, J. (1999), 'From Reason to Regulation: 1760–1850', in P. C. Jupp and C. Gittings (eds.), *Death in England: An Illustrated History* (Manchester: Manchester University Press).

Schiebinger, L. (2004), *Plants and Empire* (Cambridge, Mass.: Harvard University Press).

Secord, J. (2004), 'Knowledge in Transit', *Isis* 95: 654–672.

Shtier, A. (1997), 'Gender and "Modern" Botany in Victorian England', *Osiris* 12: 2nd Series, 29–38.

Spufford, F. (1996), *I May be Some Time: Ice and the English Imagination* (London: Faber).

St. Onge, D. (1982), 'The Coppermine River: Art and Reality', *Canadian Geographic* 102: 4, 28–31.

Taipana, L. (2007), 'Personal Communication' (8 August 2007, Kugluktuk, NU).

Thorpe, N. L. (2000), 'Contributions of Inuit Ecological Knowledge to Understanding the Impacts of Climate Change on the Bathurst Caribou Herd in the Kitikmeot Region, Nunavut' (Vancouver, BC: Simon Fraser University).

White, P. (2003), *Thomas Huxley: Making the 'Man of Science'* (Cambridge: Cambridge University Press).

Wylie, J. (2002), 'Becoming Icy: Scott and Amundsen's South Polar Voyages, 1910–1913', *Cultural Geographies* 9: 3, 249–265.

PART 4

Belonging

Chapter 10

Telling Tales: Nostalgia, Collective Identity and an Ex-Mining Village

Katy Bennett

Introduction: On feeling nostalgic

Wheatley Hill's Centenary Heritage Gala Parade started at 10 o'clock sharp on June 30th, 2007 outside the Heritage Centre. The parade was led by a man dressed in Edwardian[1] costume to look like Peter Lee, someone long since dead but symbolically important to the village. Behind this iconic figure the bright silks of lodge banners billowed in the breeze, conveying images and statements important to the identity of this ex-mining village and the former Durham coalfield. Following the banners marched the Peterlee Brass Band[2] and local residents, some of whom were dressed in costumes representing various bygone eras. As the parade progressed through the streets of Wheatley Hill, eyes, for the moment, were drawn away from the shuttered shops and ragged buildings suggestive of a village in economic decline. The parade finished at the Old Scouts' Hut, which had its doors open and stalls prepared for the crowds escaping the start of the rain.

The Gala Parade was one event in a four day period when residents of Wheatley Hill indulged in some nostalgic reverie to celebrate the centenary of what used to be the Chapel of Rest and is now the Heritage Centre (see Figure 10.1). The celebrations embraced a selective heritage of an ex-mining village as particular moments and eras were re/presented and symbols such as brass bands and banners were appropriated to affect a sense of collective identity, belonging and continuity. This chapter looks at the role of nostalgia in the creation of narratives around collective identity, drawing heavily upon Fred Davis' (1979) book *Yearning for Yesterday: A Sociology Of Nostalgia*, still one of the few texts to critically engage

---

1   The Edwardian era in the United Kingdom extends from 1901–1910 during the reign of King Edward VII. It is also sometimes associated with a further four year period to 1914, the start of World War I.

2   Peterlee, named after Peter Lee, was one of 21 New Towns established between 1946 and 1970 following the New Towns Act of 1946 which was designed to improve living and working conditions in post-war Britain. It was instituted in 1948 with the aim of creating an urban centre and alternative employment opportunities in a district dominated by mining. Its design was, like others, influenced by the philosophy and practice of Ebenezer Howard <http://www.englishpartnerships.co.uk/newtowns.htm>.

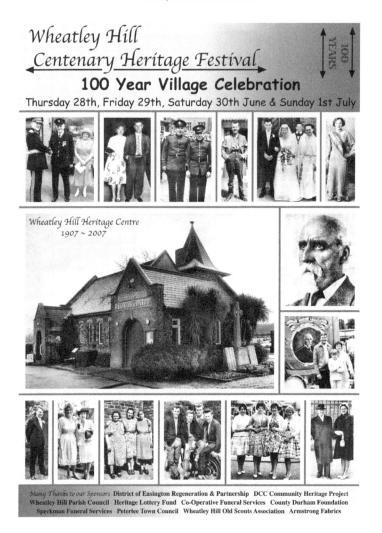

Figure 10.1   Promotional leaflet for the Wheatley Hill Centenary Festival

Source: Author.

with the concept and to have considerable influence on academic research (Legg 2004; Strangleman 1999, 2007).

The focus of this chapter is a community that has dealt with extraordinary change and social upheaval. Wheatley Hill is an ex-mining village in County Durham, part of what was once a significant coal producing area in the north east of England. It is a fragment of a rural landscape punctuated by other villages that similarly expanded dramatically to service the coal-mining industry. At its peak in 1913 the Durham coalfield employed more than 165,000 miners, yet today no pits remain open in the area (Garside 1971; Waddington et al. 2001). In the 1960s, 50,000 jobs were lost in coal mining in Durham, and Wheatley Hill, in 1968, was amongst those pits that closed with 490 miners made redundant (Beynon, Hudson and Sadler 1991). Whilst national economic strategies and international market forces were responsible for many pit closures in the 1960s, the neo-liberal policies of the Conservative government of Margaret Thatcher struck a death knell to the coal mining industry in the 1980s. Despite the Miners' Strike against proposed closures in 1984/5, the speed of pit closures was devastating and in 1993 the last of the collieries in Durham closed. The eradication of the mining industry in Durham exacerbated problems for Wheatley Hill leaving the former coalfield bereft of adequate employment opportunities with services and facilities that the miners once supported struggling to survive.

This chapter explores the way that residents of Wheatley Hill engage in an emotionally mediated process of nostalgic reflection as part of an attempt to cope with on-going changes to a landscape important to their sense of individual and collective identities. Following Davis (1979), I argue that nostalgia helps people to feel a sense of continuity regarding their identity in the face of apparent discontinuity. For residents and especially village elders of Wheatley Hill, this discontinuity is caused by the knock-on effects of the pit closure to a landscape where rhythms and practices were once heavily influenced by the demands of the pit. As people adapt regarding places in transition, they themselves, and their relations with others, become changed in the process. Exploring experiences and expressions of such shifting relations, this chapter aims to contribute to a prominent strand of recent emotional geographies research, investigating diverse forms of affective relational contours between individuals and landscapes. Whereas others provide insights into relations that are, for example, restorative and therapeutic, or shaped by bereavement and loss (Davidson, Bondi and Smith 2005; Hockey, Penhale and Sibley 2005), this chapter explores nostalgic feelings, practices and performances of a particular group of people as they cope with change, and attempt to (re)create a sense of collective identity.

Although much of the focus of this chapter is a group of residents as they celebrate – and watch others celebrate – their heritage, nostalgia is simultaneously a private experience, allowing individuals to preserve a sense of self in the face of change and others' ascriptions and expectations of them. Dwelling on a sense of who they were to understand who they are, people sometimes delve into their past, particularly to especially formative experiences as children or young adults, to find their 'secret selves' (Davis 1979, 39), perhaps incorporating some semblance of their past into their current reality to make it more assimilable. When people divulge their own and identify with each other's expressions of nostalgic experience, a positive sense of partially shared or 'collective' identity can be seen and felt to emerge. However, while Davis (1979) argues that this past has to be individually experienced (that people cannot feel nostalgic for a time that existed before they were born) I maintain that the memories of others (secondary memories) can seep into the terrain of our own memories and sense of nostalgia through, for example, their sharing of stories, photographs and other objects (Lowenthal 1985).

Despite the necessary focus on the past, a central point of Davis' (1979) work is that nostalgia is more revealing about present circumstances than any past reality. What separates nostalgia from other subjective states also oriented to the past, such as remembrance or reminiscence, is its relationship with present moods. Critical to nostalgia is a past given 'special qualities' (Davis 1979, 13) and made especially significant because of how it is juxtaposed with present feeling. Davis wrote:

> to remember the places of our youth is not the same as to feel nostalgic over
> them; nor does even active reminiscence – however happy, benign, or tortured
> its content – necessarily capture the subjective state we associate with nostalgic
> feeling (Davis 1979, 13).

Nostalgia, however, is a complex concept and has only relatively recently been used to describe states of 'feeling'. Tracking some of its use, Davis (1979) showed how nostalgia was historically classed as a 'condition' connected with homesickness:

> *Nostalgia* is from the Greek *nostos*, to return home, and *algia*, a painful condition
> – thus, a painful yearning to return home. Coined by the Swiss physician
> Johannes Hofer in the late seventeenth century, the term was meant to designate
> a familiar, if not especially frequent, condition of extreme homesickness among
> Swiss mercenaries fighting for their native land in the legions of one or another
> European despot (Davis 1979, 1).

Davis (1979) did not offer 'the difficult, extensive, and tedious historical survey of the word's usages' (Davis 1979, 5), but did write about nostalgia's 'unmooring from its pathological base, with its demilitarization and demedicalization' (1979, 4), so that by the twentieth century it was 'much more likely to be classed with such familiar emotions as love, jealousy and fear than with such "conditions" as melancholia, obsessive compulsion or claustrophobia' (Davis 1979, 5). For Davis nostalgia described emotional experience embodying various needs and desires to engage with a past for a sense of continuity and the sake of identity. Attempting to illuminate the experience, he wrote:

> There is perhaps no word that better evokes the odd mix of present discontents,
> of yearning, of joy clouded with sadness and of small paradises lost (Davis
> 1979, 29).

The focus of Davis' book is the 'contents, contours and contexts' (1979, 7) of nostalgia. For Davis nostalgia is experienced through 'yearning', a process requiring a self-awareness that pitches the self in relation to 'paradises lost' and a present that has triggered nostalgic reverie. Davis (1979) considered nostalgia to be a form of consciousness requiring reflexivity. In his interpretative framework, he described three ascending orders of nostalgia, with each 'order' of nostalgia requiring a greater degree of reflexivity than the previous one. The first order he referred to as 'simple nostalgia' (1979, 17) and the warm glow of a subjective state that harbours the belief that things were better in the past. Elsewhere, Davis (1979) adds that this feeling might be tinged with sadness, but a 'nice sort of sadness' (1979, 14) that simultaneously manages to feel good. In the second order, or 'reflexive nostalgia' (1979, 21), rather more questions are asked regarding the accuracy of, and motivation behind, nostalgic claims. In the third order, or 'interpreted nostalgia' (1979, 24), the subjective state is objectified and questions are asked regarding its sources, character, significance and psychological purpose.

The interpretative framework is important to this chapter because it links nostalgia with the never-ending work of (re)constructing an identity. The second and third orders of nostalgia entail individuals being self-conscious about the experience and expression of such feelings. This is illustrated when people question or challenge claims regarding the past on which their nostalgia focuses, and when they explore why they delve into their pasts. The framework also recognizes that nostalgic feelings may escape any representation and analysis. Of course as soon as such words as 'simple nostalgia' are used to capture the 'warm glow' of an experience, such a feeling is pulled into the contours of an experience that is named, objectified and perhaps far removed from the phenomenon itself. Davis (1979) recognized that being too self-conscious regarding nostalgia erodes the ability to experience it.

This chapter first explores the identification of a community through the formal celebration of its heritage.[3] It looks in some detail at the significance of the Gala Parade to a community dealing with change. In this section the work of the social anthropologist Anthony Cohen (1985a, 1985b) is important. In his work on the symbolic construction of community, Cohen was much less concerned with the geographic and social structures (and definitions) of community than with its use in relation to issues of identity (Gilbert 1995). This meant that his focus was on how a group of individuals came to identify itself as a community, and particularly the symbols it used to unite around and differentiate itself from others (Cohen 1985a). Drawing on his argument I show how the symbolic and nostalgic significance of the lodge banner and life of Peter Lee are used to smooth the different experiences of residents of Wheatley Hill into seemingly coherent accounts that define a community and its collective memory (May and Morrison 2005).

---

3   This chapter is based on participant observation and in-depth group work with women aged over 55 who lived in Wheatley Hill. The group met on three occasions in the Autumn/Winter of 2006 in Wheatley Hill Community Centre. This research is part of a broader project exploring identities and regeneration in the former coalfields of East Durham (RES 148-25-0025). The project is one of 25 funded under the Economic and Social Research Council's Identities and Social Action Programme.

The chapter then considers less formal and more intimate, although no less public, performances of nostalgia through the story-telling of a group of women who have lived in the village for many years. Whilst a community's celebration of its heritage (and symbols) separates and excludes those who do not belong from those who do, the women's tales add a further dimension to the issue of exclusion. Sidelined at the Festival were painful times, such as the decades following pit closure, and experiences and representations of women beyond relatively standard and stereotypically 'feminine' portrayals. As a community attempts to tell a smooth, simple story about itself, nostalgia helps people to cope with the present as they make sense of it through a journey into the past that either avoids or erases painful experience, or recasts it in more emotionally manageable terms.

The second part also begins to reveal the more subtle threads of community life which

> are felt, experienced, understood, but almost never explicitly expressed. They provide a subterranean level of meaning which is not readily accessible to the cultural outsider. They are the substance of belonging, and they belie the apparent familiarity to outsiders of the culture's structural forms (Cohen 1985b, 11).

These threads are the everyday, intimate experience of life that encompass such things as the use of language, practices of story-telling, joking and a shared knowledge of genealogy. The women drew upon more than familiar symbols to identify a sense of community, revealing through discussion, joking, telling tales and silence a more intimate knowledge of each other, place and culture than that on public display. Much of this subtlety passes unnoticed by outsiders (like me), but there are perhaps glimmers of it in the second part of the chapter.

Telling Tales 1: Nostalgia and Public Celebration of Collective Identity

> The weather forecast was bad, but the rain was holding off, allowing the banners to billow in the breeze, Peterlee Brass Band to prepare for their performance and some of the villagers to showcase their costumes as they waited outside the Heritage Centre to begin the Gala Parade (Figure 10.2).[4] Inside the small space of the Heritage Centre, the last of the tea was being hastily drunk, biscuits put away and photographs quickly taken. Bodies brushed past old photographs, displays and cabinets as they left the room, seats around the table and big replica oven vacated (Figure 10.3). Women wearing white pinnies[5] and caps finished the washing up and left the building to join the parade (Figure 10.4). Together

---

4   This is based on my research diary which includes writing on my participant observation work, notes from talks given in the Heritage Centre and information collected and displayed by residents in the Heritage Centre.

5   Abbreviation of pinafore.

Figure 10.2   Residents with the Wheatley Hill Lodge banner

Source: Author's photo.

Figure 10.3   Dressed up for the centenary celebrations

Source: Author's photo.

Figure 10.4   Kitchen antics during the centenary celebrations
Source: Author's photo.

the costumes represented different eras of the past century as women, men
and children in Edwardian costume mixed with a Second World War army re-
enactment group. In the middle of it all a woman in a 1950s skirt, pop socks and
plimsolls chatted animatedly to a friend. As the parade was about to begin the
men carrying the Wheatley Hill Lodge banner moved to the front, ready to lead
the procession.

The Lodge banner was central to the nostalgia of the Gala Parade, connecting
residents of Wheatley Hill to a coalmining heritage as they processed through the
streets of the village. Rich in symbolic meaning, the banner represents a sense of
continuity for those residents, who, every year, parade it through the village before
taking it to join people with banners from other (ex) mining villages at the Durham
Gala. It connects people with previous generations who lived in the village and
links Wheatley Hill to other places that shaped the Durham coalfield.

Lodge banners comprise art work and silk attached to a bar, hooked to end poles
which are held and carried by men. On the front of Wheatley Hill Lodge banner
is an image of Peter Lee, who will be introduced in more detail later, set against
a background of red and purple silk. The current banner was recently produced
in 2000 to replace an older version. Like other Lodge banners, several versions
of the Wheatley Hill Lodge banner have been created over the years as they have
aged, been damaged or lost. The earliest version of the Wheatley Hill Lodge
banner, created in 1873, showed Capital and Labour – 'Let us reason together'
– the reverse featuring a checkweighman at work accompanied by the statement
'A just balance is our delight'. Later versions of the banner, going back to at least

1947, feature Peter Lee on the front (letter from Billy Middleton, Wheatley Hill History Club, undated).

The Lodge banner is symbolically important, drawing people together and affecting a sense of collective identity, even though it probably means different things to individuals. Two of its meanings are important here. Firstly, as already indicated, it draws Wheatley Hill into a broader sense of collective identity, connecting it to a heritage that it shares with other communities comprising the former Durham coalfield. This is because each banner is one piece of a jigsaw representing key figures (men important to local, county, national or international political life), sentiments, beliefs and attitudes of the Durham Miners' Association (DMA) and a coalmining region. Together they portray the:

> physical manifestations of a deeply rooted occupational identity. The entire history, social, political and economic life of the Durham coalfield can be seen through the imagery of the banners. Whilst each banner is unique, collectively they are a physical representation of occupation, of working class politics and aspiration, of collectivism and of community welfarism (Stephenson and Wray 2005).

Rather like the Wheatley Hill Gala Parade, but on a much larger scale, the Lodge banners of the Durham coalfield process together once a year through the streets of Durham city at the Durham Gala.[6] Behind each banner traditionally marched Lodge officials, the colliery brass band and community members in a ritual that began in 1871 to celebrate social solidarity and the mining communities that comprise the Durham coalfield (Beynon and Austrin 1994). Historically, the Durham Gala was a day out in July for mining families as they travelled into the city by foot, bus or train armed with their picnics and ready for a day of socializing and meeting up with mining families from other villages and towns across the Durham coalfield. It was the 'Big Meeting' for men who shared an occupational identity that placed them '*within* but not *part* of society' (Beynon and Austrin 1994, 36), a day for mining families to be centre stage as they confronted what Beynon and Austrin (1994) refer to as the 'the Durham system' and a reminder of the political clout, if not economic power, of mining communities. The Durham Gala was also a time to remember those killed in pit accidents with the banners of Lodges struck by accidents and death in the previous year draped in black crepe.

Secondly, the lodge banner is meaningful because it represents the distinctiveness of Wheatley Hill, conveying a sense of its particular sentiments, beliefs and heritage.

---

6   Although the last pit in the county closed in 1993, the Durham Miners' Gala continues, with a broader remit that embraces representation from other unions and organizations. Much of the day though is a nostalgic celebration of mining communities, reproducing and reimagining a sense of identity for older generations whilst socialising younger generations into mining traditions and heritage.

Accompanying the Wheatley Hill Lodge banner, at the front of the procession, was a tall, white haired, bearded man in Edwardian dress. This was a local man dressed to look like Peter Lee, the legendary figure with a nearby New Town named after him and particularly significant to the history of Wheatley Hill, figuring on the front of its Lodge banner. As the gala parade progressed along Front Street faces appeared at windows, neighbours chatted to one another outside their homes, sporadically breaking off from their conversations to wave at familiar faces as they passed by (Figure 10.5). A woman, her hair in curlers underneath a pink scarf, cheered and laughed as she spotted people she knew. Numbers began to swell as individuals joined the parade, passing by closed down shops that once thrived in this ex-mining village. As the parade neared the finish the first spots of rain began to fall leaving only a small amount of time for photographs to be taken in front of the banners, next to the man dressed as Peter Lee, before they were quickly and carefully packed away. Chairs for the band to sit and play remained empty and the WW2 re-enactment group gingerly sheltered under trees alongside army memorabilia laid out on the ground. A young man with a baby in a pushchair, his face pale, his cap shielding his eyes studied some of the things before moving away unsteadily and lighting a cigarette. Everyone else crammed inside the Old Scouts' Hut, browsing the various stalls selling everything from marmalade to local history books. In the bar area women in white bonnets sat making clippy mats,[7] watched by bystanders (Figure 10.6).

Figure 10.5   Parading through Wheatley Hill

Source: Author's photo.

---

7   Alternatively called 'proddy mats', these are made by pulling small ribbons of material from cut up clothes through a stiff woven base to create a muliticoloured mat.

**Figure 10.6   Making clippy mats in the Old Scouts' hut during the Centenary
Heritage Festival**

Source: Author's photo.

Peter Lee is symbolically important to the collective identity of Wheatley Hill embodying values and attitudes important to its sense of community. Born on July 20th, 1864 in Trimdon Grange, County Durham, Peter Lee's working life began young, when he left school aged 10, and was punctuated by travel which led him to work in mills and mines in England, and abroad in North America and South Africa. It was in 1902 when he moved with his wife and family to Wheatley Hill to be checkweighman, a position appointed and paid for by miners to check the coal hewers' weight and to ensure a fair deal, that his association with the village began. Through his work on the Parish (and District) Council Peter Lee was the driving force behind the construction of roads, a sewage system, street lighting, piped water into houses, cemetery and the Chapel of Rest (now the Heritage Centre) in Wheatley Hill. Driven by his belief and uptake of ideas of Methodist Christianity, which he translated from a social religion into municipal socialism, his desire for social and welfare reform deepened and broadened as he played an increasingly important role in the DMA and Local Government (Beynon and Austrin, 1994). When Lee died on June 16th, 1935 his association with Wheatley Hill was cemented as his final wish was to be buried in its cemetery, despite having lived many years in Durham City.

Although his lifetime stretched across periods of immense hardship for mining families when mine owners inflicted reduced wages on their employees, there were lock outs and the General Strike of 1926, very few residents overtly recall or refer to the pain of such times in their conversations about the past. What they do 'remember' through nostalgic representations of Peter Lee is a man who overcame hardship to become an iconic figure. In the Heritage Centre during the Festival, the man dressed as Peter Lee talked about Lee's life. Two particular issues regarding the theme of overcoming problems and disadvantage were emphasized. Firstly, as a young man Lee was known for his drinking and fighting, but he changed his fortunes through travel, working hard and educating himself. Lee therefore provides a line of connection between troubled young men in Wheatley Hill today, such as the young man with the child in a push chair, and its past. Lee is perhaps a symbol of continuity as well as hope.

Secondly, as already discussed, Peter Lee was instrumental to the implementation of new facilities and services in a place that lacked these. Today, with the knock-on effects of pit closures, low income households[8] and a population that has declined from more than 5,000 when the pit was open to just over 3,000[9] in 2001, Wheatley Hill has witnessed the loss of many of its facilities and services. This is particularly devastating to a place with an ageing population in which more than a third of its residents are living with limiting long-term illnesses. The present problems of Wheatley Hill are obviously different to those of its past, but through the efforts of Lee and others, positive change happened then suggesting that it might happen again. Already, like the Chapel of Rest which has a new use, the Miners' Welfare Club has been turned into a successful Community Centre through the hard work of some local residents. Through nostalgic recollections of a man, who symbolises change and overcoming hardship, 'the present seems less frightening and more assimilable than it would otherwise appear' (Davis 1979, 36).

Telling Tales 2: Exclusive Nostalgia

Whilst the previous section embraced public, formal performances of nostalgia through the gala parade and a community's celebration of its heritage, this section engages with nostalgia at a more intimate level through the tales of six women, all over the age of fifty-five. Iris, Maggie, Ann, Joan, Dawn and Mary[10] evoked a sense of community that reached beyond the more obvious symbols and structures used to identify it as they joked, disagreed, prompted one another and fell into silence.

---

8   Forty-seven percent of people aged 16 to 74 in Wheatley Hill were classed as economically inactive in 2001(http://www.neighbourhood.statistics.gov.uk/dissemination/).

9   The population of Wheatley Hill CP Parish was 3,181 in 2001 (http://www. neighbourhood.statistics.gov.uk/dissemination/).

10   All names have been changed.

The women knew not only about each other's lives, but also the local practices of behaving appropriately regarding such knowledge. There are two parts to this section. The first explores why particular eras and decades, to the exclusion of others, were on show at the Heritage Festival and significant to the nostalgia of the women. The second part highlights the exclusive reliance on gender stereotypes of the Heritage Festival as the telling tales of women expand upon their identification in the story-telling of a community.

Exclusive Times

Whilst particular eras and decades were poignantly celebrated during Wheatley Hill's Centenary Heritage Festival, other decades were notable for their absence. The two periods that dominated the festival and the tales of the women were the Edwardian Era and a period of time stretching from the 1940s to the 1960s. What appears to connect these epochs is that people are somehow able to erase any pain from memories or navigate around a potentially painful experience, a necessity for Davis (1979) who wrote that 'nostalgia is memory with the pain removed' (1979, 37). As discussed at the start of this chapter, feeling nostalgic might mean sadness, but 'a nice sort of sadness', that does not distract from the sense of its warm glow. When too much pain creeps into the experience, then it usually ceases to be what might be described as nostalgia.

The Edwardian era was significant at the Heritage Festival because of its association with Peter Lee. It is a safe territory for nostalgic celebrations of a culture because, as already discussed, few people who participated in the festival would have actually experienced the hardship and difficulties that people faced then. Furthermore, the repercussions of that era are known, having been played out and etched into the history of the village. People often talked as if they could remember such a time, probably because of the story-telling and memories of others that mediated their own. Woven through the women's narratives of childhood were (what would later be revealed as secondary) memories such as Dawn's account of how to preserve cheese without a refrigerator (wrap it in wet greaseproof paper), Joan's recollection of washing clothes in a poss tub[11] and Mary's tale of the greengrocer who used to shout 'oranges and lemonges' as he travelled around the streets.

The 1940s to 1960s were another important time celebrated at the Festival because this was when most of those involved in its organization were children and young adults. For Ann, Maggie, Joan, Iris and Dawn the 1950s and 1960s accommodated childhood and teenage years before marriage. This was the particular period of the women's lives when they could identify with each others' stories as they recounted tales of playing in the streets with their friends and performing childish, harmless pranks on local people with nicknames such as 'Hawkeye'.

11   Wooden cylinder with a wooden baton ('poss stick') used to wash clothes.

Summer holidays were spent in nearby Crimdon where horses and carts carried sideboards, beds and chairs down to a tent or chalet occupied by mining families holidaying together on the coast. The group also bonded over their teenage years as they remembered, prompted by Joan, the fashion and clothes from an era when petticoats were made from sixty yards of net and starched with sugar and water that used to attract the wasps, skirts were reversible and belts synched in waists. Through accounts of the clothes that they wore, the women slid into memories of balls and all night dances and the social life of their community. Underpinning such nostalgia was a village with a pit that was still open, a buoyant population and a wealth of services and facilities, including a dance hall and cinemas.

The decades following the pit closure in 1968 barely figured in the nostalgic reverie of the Heritage Festival or the women's conversation, perhaps because they marked the start of the slow decline of the village and a time in the women's lives when pain could be less easily erased from memories. The colliery was at the hub of village life, employing men and dictating the routine of households. Its closure meant the loosening of ties between households as new jobs were found elsewhere, early retirement was taken and families moved out of the village. Mary evoked this sense of community fragmentation caused by pit closure when talking about neighbours:

> we don't have many now. Not like we used to have. We used to help each other. If anyone was ill. You were never on your own. If anybody was ill the neighbours used to go in and see you.

Whilst Mary still has neighbours, she has lost a sense of emotional connection to them. The newness of neighbours is not the issue: pit villages have always been familiar with the flow of people in and out to service the demands of the colliery. Perhaps what is more significant is the lost sense of identification with others who no longer take part in the rituals and practices that revolved around the pit. Pit closure in Wheatley Hill meant not only job loss, but also an end to the practices important to its identification. The demise of some routines was instantaneous, such as the flow of men heading towards the pit at particular times of day and night, others were more gradual as they faded from the landscape of village life. The Miners' Welfare Club, a place the women remembered attending for wedding receptions and all-night parties, fell into disrepair and for a while became a place to be avoided. Places such as this had provided a venue for socialising and learning those practices that strengthen relations between individuals and make 'real' neighbours.

Furthermore, for Ann, Joan, Mary, Iris, Dawn and Maggie, life after pit closure was also shaped by the trials and tribulations of adulthood as they set up home and were responsible for their own households. After marriage, the women's lives were more evidently marked by difference rather than a bank of common experience that could be mined to shape a sense of collective identity. With adulthood came responsibility, children and balancing work at home with employment at local factories, shops, schools, hospitals and care homes. Any sense of nostalgia was less pronounced as the women were confronted with sometimes painful memories of living with men who were dealing with redundancy or who worked away from home for long periods of time; men who cherished them or failed to do so. For one woman marriage had ended in divorce, for another a child had been left disabled from a horrific accident, for all concerned, the welfare of family and household members had generated daily worries.

Past episodes or moments of lives that become painful to dwell upon are not the fuel of nostalgia. Strands of life haunted by feelings of isolation despite (or because of) people nearby and painful experiences do not figure in nostalgic narratives around collective identity as people cope with present circumstances. This meant that entire decades (following pit closure) did not figure in Wheatley Hill's Heritage Festival, perhaps showing how long it takes a community to mourn the loss of something so central to its identity.

Exclusive Gender Stereotypes

Throughout the Festival and gala parade women were closely aligned to private worlds encompassing unpaid work and narratives of domesticity as they paraded through the streets in pinnies and bonnets, bent over clippy mats in the Old Scouts' Huts and served tea to visitors in the Church and Heritage Centre. The full and broad experience of women was often over looked as nostalgic celebrations created a simple narrative of community that located the lives and work of women within the four walls of their homes. But the story-telling of the six women frequently dwelt on experiences of paid work and representations of women at work in the public space of the village. Although the nature of their work often meant that the women were still performing domestic chores, the stage was public rather than private space and the spotlight was on senses of selves rather different to those identified in the story-telling of a community at the Festival.

As adults, all of the women worked from leaving school at fourteen (in the case of Mary) or sixteen (for the others). Most of them remained in paid work for much of their working lives, punctuated only by childbirth and childcare when children were very young. Some of the women changed jobs regularly, shifting from one factory to another for better pay. Maggie was remarkable for the sheer number of different jobs that she had had over the years, working in numerous factories, for hospitals and 'on the post' for Royal Mail. Others were remarkable because they had worked in the same job for many years, like Joan who had been the dinner lady at a local school for thirty-two years.

The nostalgic tales of work provide powerful insights about how the women portrayed themselves. They dwelt on the camaraderie of the workplace and friendships formed over sneaked 'fag (cigarette) breaks' in the ladies washrooms, backsides resting against sinks whilst talking to each other via the mirrors on the walls. They talked of developing friendships with colleagues whilst outsmarting bosses and managers and outwitting the next (night) shift of (usually male) workers through the 'surprises' they left, such as empty bobbins for those who had left their machines inadequately prepared for the women's shift in a wool factory. Fun-loving and humorous, perhaps cheeky and naughty, the women's nostalgic tales of themselves at work contrasted with representations of women as white pinnie and bonnet wearing figures of stereotypically feminine domesticity seen at the Heritage Festival. A story that perhaps best illustrates contrasting representations of women is a tale that Maggie was asked to re-tell about an event which happened whilst she was working in a hospital ('Tell them about that man Maggie', 'you'll love this', 'Go on and tell them, hee hee'…). The story involved a male patient, Maggie and her superior, a Sister, a 'spinster who'd looked after her parents all her life'. One day the Sister asked Maggie to help her get Tommy, a patient to the toilet:

> I says 'Where's your sticks Tommy?' 'Divvent need sticks when I've got two bonny lasses like yee to tek us'. So when you got him to the door, the wall was here so I had to run round, through the day room and come in the other way to fetch him through. So we get him through, gets him onto the toilet and I did away. I says 'Just give us a shout when you need us'. All of a sudden I heard this unearthly squeal and the sister, oh my god. 'Oh! Oh!'. I run down and I says 'Whatever's the matter? Oh it's Thomas!' So I looked at him and he's giggling, sitting on the toilet. I says 'What's thaa done now?' I says 'What have you done this time?' He says 'Well it's hor', he says, 'I just lifted it out like this and said does thaa fancy a bit of this?' She was mortified. I says 'Put it away Tommy', I says 'Look at her, she's having a seizure'. Honest, we've had some laughs in there'.

On the one hand this story smacks of the sexual intimidation of a woman trying to do her job. On the other, and meant by this group, it is a story that juxtaposes one version of a woman, the sort not dissimilar to the one pouring tea and making clippy mats at the Heritage Festival, with the representation of a rather different kind of woman – one at work in a public place and having a laugh.

A desire to add nuance to stereotypical representations of women was also conveyed through stories of women played out in the public space of community life. Looming large in their nostalgic accounts of a community past were tremendous women such as the aforementioned 'Hawkeye' who used to chase them (and men) with a big bamboo cane through the streets of the village when they were playing pranks or being naughty. Similarly:

> Vera: There was an old woman lived over the road to me Aunt Mary. I used to stop at my Aunt Mary's a lot and she (the old woman) used to stand outside her house and she smoked a pipe and she would fight a man.
>
> Dawn: uh hum, she would take any man on.

Tales of their childhood were littered with powerful women who were central to community life and the life cycle of the village, attending and helping other women during childbirth through to washing and laying out dead bodies, placing pennies on their eyes in preparation for burial. The women, especially Iris, portrayed themselves like their predecessors as strong and determined, through, for example, their work regarding the conversion of the Miners' Welfare Club into a Community Centre. They have battled for funding and sought local help and free goods to refurbish and transform the place. The women remembered when Iris collected some plants donated to the Community Centre, carrying them through a busy shopping centre to her car:

> Iris: Hey I did, that's true. We went and got some lovely plants didn't we and I was like Tarzan in the jungle, one in this hand and one carrying two trees. What we've done for this community centre is unbelievable.

Conclusion

The aim of this chapter has been to show how a focus on emotion opens up space for thought and reflection for a geographer dealing with issues of community and identity in a post-industrial landscape. With a central focus on nostalgia, I have explored how residents of Wheatley Hill, an ex-mining village, cope with the effects of pit closure and attempt to create a sense of collective identity and continuity in the face of apparent discontinuity. A central part of its focus was the nostalgic appropriation of symbols, such as the Lodge banner, at Wheatley Hill's Centenary Heritage Festival, that linked residents to particular decades and practices important to creating a sense of community and belonging. The power of such symbols is that although they mean different things to people, they work to unite individuals, separating them from outsiders unable to relate to them. They also represent a sense of continuity and hope, connecting people to a past peppered with (sometimes similar) problems that the village survived.

Nostalgic performances at the Heritage Festival wove a relatively smooth, coherent story about a community, excluding tales and experiences that deviated from its narrative of collective identity. Cast to the shadows of the festival were experiences and representations of women beyond their stereotypically feminine portrayal, and other 'subjects' such as the evidently painful decades following pit closure were similarly marginalized. Clearly, nostalgia links the present to a *particular version* of the past, from which painful experience is often avoided or somehow erased to positively enhance present emotional experience and sense of community. Amidst the festival, people were brought together, involved in an event, dressing up, playing a role, standing close to others, exchanging words, sharing snippets of nostalgic reverie, watching, laughing, silent. Part of an event that might fuel tomorrow's nostalgia.

Acknowledgements

Thanks to the residents of Wheatley Hill, the six women who took part in the in-depth group work, Emily Nolan for helping me to organize and run the groups and the ESRC for funding the research. Finally, thanks to Joyce Davidson, Liz Bondi, Mick Smith and Giles Mohan for their comments on this chapter.

References

Beynon, H. and Austrin, T. (1994), *Masters and servants: class and patronage in the making of a Labour Organisation* (London: Rivers Oram Press).

Beynon, H., Hudson, R and Sadler, D. (1991), *A tale of two industries: The contraction of coal and steel in the North East of England* (Buckingham: Open University Press).

Cohen, A. (1985a), *The symbolic construction of community* (London: Tavistock Publications Ltd).

— (1985b), 'Belonging: the experience of culture', in A. Cohen (ed.), *Belonging: Identity and social organisation in British rural cultures* (Manchester: Manchester University Press), 1–17.

Cowie, J. and Heathcott, J. (eds.) (2005), *Beyond the ruins: The meanings of deindustrialisation* (New York: Cornell University Press).

Davidson, J. Bondi, L. and Smith, M. (2005), 'Introduction: Geography's "Emotional Turn"', in J. Davidson, L. Bondi, and M. Smith (eds.), *Emotional geographies* (Aldershot: Ashgate).

Davidson, J. Bondi, L. and Smith, M. (eds.) (2005), *Emotional geographies* (Aldershot: Ashgate).

Davis, F. (1979), *Yearning for yesterday, a sociology of nostalgia* (New York: The Free Press).

Garside, W.R. (1971), *The Durham Miners 1919–1960* (Allen and Unwin).

Gilbert, D. (1995), 'Imagined Communities and Mining Communities', *Labour History Review* 60:2, 47–55.

Hockey, J., Penhale, B and Sibley, D. (2005), 'Environments of memory: Home space, Later Life and Grief', in J. Davidson, L. Bondi, and M. Smith (eds.), *Emotional geographies* (Aldershot: Ashgate).

Legg, S. (2004), 'Memory and nostalgia', *Cultural Geographies* 11, 99–107.

Lowenthall, D. (1985), *The past is a foreign country* (Cambridge: Cambridge University Press).

May, S. and Morrison, L. (2005), 'Making sense of restructuring: Narratives of accommodation among downsized workers', in J. Cowie and J. Heathcott (eds.), *Beyond the ruins: The meanings of deindustrialisation* (New York: Cornell University Press), 259–283.

Stephenson, S. and Wray, D. (2005), 'Emotional regeneration through community action in post-industrial mining communities: The New Herrington Miners' Banner Partnership', *Capital and Class* 87, 175–199.

Strangleman, T. (1999), 'The nostalgia of organisations and the organisation of nostalgia: Past and present in the contemporary railway industry', *Sociology* 33:4, 725–746.

— (2007), 'The nostalgia for permanence at work? The end of work and its commentators', *Sociological Review* 55:1, 81–103.

Waddington, D., Critcher, C., Dicks, B., and D. Parry (2001), *Out of the Ashes? The social impact of industrial contraction and regeneration on Britain's mining communities* (Norwich: The Stationery Office).

Chapter 11

Death and Bingo? The Royal Canadian Legion's Unexpected Spaces of Emotion

Deborah Thien

'I think the legion plays a very important place in people's lives… In a way, it becomes a way of life'

(Josephine Selkirk 2006)

Introduction

After 83 years of service, the material manifestations, activities, and symbolic postures of The Royal Canadian Legion are seemingly ubiquitous within the Canadian landscape, particularly in the small communities that make up rural, remote and northern Canada. Legion branches are housed in familiar square buildings on main streets marked by the easily recognized blue and gold Legion crest (Figure 11.1). Inside, plaques, maps, photos, scrapbooks, history books, and other historical materials honour fallen soldiers and celebrate the local heroes of Canada's war and peace-time actions. The Legion's high-visibility Poppy Campaign raises funds for its best known public event: the November 11, Remembrance Day ceremonies. Remembrance Day events, held 'lest we forget' the sacrifices of war veterans, reliably draw big crowds and garner national news coverage. As a national veteran's organization, the Legion represents (a version of) the Canadian nation, shaped by history and tradition. Remembrance is the *raison d'être* of the Legion; however, the Legion halls also host all manner of local events: weddings, funerals, Christmas parties, Elvis impersonators, social dances and Bingo games. Consequently, respect for this iconic institution, its ongoing homage to Canadian military heroism and a now elderly veteran population sits alongside an equally popular characterization of the Legion as an unsavoury 'beer and bingo' hall, a place understood to be literally in its death throes as the numbers of World War I, II and Korean War veterans dwindle.

In this chapter, I argue that the Legion is not the static entity these stock characterizations suggest. Upon closer examination, the Legion is an intriguing locus for social, spatial, and emotional relations, differently produced in specific places, and out of which sometimes unexpected feeling subjects take shape in contrast to the usual suspects (old men telling war stories, the 'beer and bingo' crowd) popularly attributed to the Legion landscape. In this examination of the Legion's emotional geographies, I suggest that the familiar representations of the Legion belie the complexities of Legion spaces and their co-implication in always emergent subjectivities. I argue that this national veterans' organization in offering a special place for remembrance, for honouring, for celebration, also engages in exclusionary practices, that are nonetheless subject to revision; that for many people the Legion is a 'way of life' (Selkirk 2006), even as it exists to commemorate death; that Legion branches evoke an 'environment of memory' (Hockey et al 2005) within which deep emotional responses may be sanctioned, while tacitly or overtly foreclosing other(s) emotion. When the doors are opened, the Legion is revealed as a deeply evocative and dynamic site.

In the following sections, I first set the context for my arguments by detailing the theoretical and methodological framework for this research. Then, I provide an overview of the Legion's history and culture. Next, I consider how Legion branches emerge within distinct place-based contexts. I argue that the Legion is not so easily read as a singular ubiquitous entity; rather, it is a locus for unstable intersections of militarism, masculinity, 'race', and emotion, contributing to the production of and constituted by myriad 'feeling subjects'.

Setting the Context

Despite its longstanding presence in the Canadian landscape, very little independent research exists on the Royal Canadian Legion and its compelling effects (but see Walton-Roberts 1998; Morton and Wright 1987). Legion-sponsored publications (e.g. Walpole 2005; Hale 1995; Bowering 1960) focus on detailed, historical, and somewhat idealized accounts of a stalwart veterans' organization. This study offers a critical engagement with the Legion and its membership as represented through Legion publications and limited existing research; in newspaper and magazine articles (including a Canadian national paper, *The Globe and Mail*; local papers such as Sayward's *Compass* and Prince George's *Citizen*; and the magazine, *Canadian Geographic*); the Legion website; views of Legion members and users from primary and secondary sources; and fieldwork observations. Original qualitative data were gathered from formal (i.e. semi-structured and recorded) interviews, informal interviews, participant observation, and photographs in the northern city of Prince George, BC, as well as in seven Northern Vancouver Island, BC communities, which range in population from 30,000 to less than 1000: Port Hardy, Port McNeill, Port Alice, Alert Bay, Sayward, Campbell River and Quadra Island.

My research objective is to consider the Royal Canadian Legion as a locus for geographies of emotion. A number of arguments about places, people and emotion frame this analysis. First, places are not neutral backdrops for human activities, nor are they fixed in time and space; rather places are constituted in relational ways, dependent on 'that throwntogetherness, the unavoidable challenge of negotiating a here-and-now' (Massey 2005, 14). Legion branches, though they reference the same umbrella institution, are not perfectly replicated in each community; rather, they are uniquely, if relatedly configured by ongoing negotiations in their place-based contexts. Second, people are not self-evident, readily circumscribed containers of experiences, emotions, or anything else; rather, individual identities, or subjectivities, can be understood as 'temporary determination[s]' generated via 'contingent social relations' (Natter and Jones III 1997, 149; see also Butler 1990, 1997). In the case of Legion members, although specific guidelines exist for Legion membership, these guidelines have changed over time, and have been treated more and less flexibly, formally respected or informally maintained, in differing branch locations. So, the possibilities of 'a' Legion identity have varied and continue to vary, for example, along gendered and racialized lines. These contingencies of identity do not preclude attention to individuals' relations to or with the Legion; indeed, I suggest this precisely invites attention to how individuals have engaged with, emerged from, and represented the Legion's socio-cultural landscapes.

Conceptually, emotion is defined in part by its place amongst a set of long-established and well-documented dichotomies which have set the mind in opposition to the body; femininity as masculinity's shadow other; and rationality against emotionality (Rose 1993). An effect of these histories is an overdetermined link between men, superior mental capacities, and reason, and between women, the messiness of embodiment, and emotion (Lupton 1998). The Legion, a predominantly male and militaristic institution is thus not immediately recognizable as a (typical) space of or for emotion. Emotional geographies research to date has encouraged a more nuanced consideration of the ways in which emotions and spaces are relationally constituted (e.g. Bondi 1999; Anderson and Smith 2001; Davidson and Bondi 2004; Thrift 2004; Davidson et al. 2005; Thien 2005a; Hockey et al. 2005; Bondi 2005). In seeking to understand how emotions shape spatial phenomena *and* how spatial phenomena shape emotions, this relational perspective moves 'us' away from surveying the world as if 'we' were unaffected by it, and also relinquishes the possibility that emotions are enclosed within (in brains, hearts, souls, psyches), never leaking out to affect the world around us. Instead, emotions may be experienced as 'individually lived' but are also 'socially derived through relationships with others within specific structural conditions' (Philipose 2006, 63). The emphasis on this interaction, this 'back-and-forth', is a key element of this conceptual framework. Such a framework positions emotion as dynamic, actively constituting those psychic, social, and spatial relations, and acknowledging the important role of emotion in subjectivity. In this case of the Legion, its militaristic and masculinist, material and symbolic spaces unexpectedly offer an important space for feelings, and thus allow non-normative feeling subjects to take (at least temporary) shape.

In considering 'feeling subjects', I am positioning subjects as 'structure[s] in formation' (Butler 1997, 10) discursively, reiteratively and continuously shaped. To paraphrase Judith Butler, the terms by which we render ourselves as subjects, intelligible to self and others, are the terms of social normativity, a relational context that is only ever partially recoverable by any one person, because we can never be fully self-knowing (Butler 2005, 20–21). These terms, or 'relations of dependency' (Butler 2005, 20), allow us to recognize what is normative, even as we may fail to meet such norms, or indeed disrupt or call them into question. Deborah Lupton (1998, 172), for example, has argued that to be a 'civilized' emotional self, is to be 'cognizant of when it is appropriate to repress the expression of one's feelings and when it is appropriate to reveal them, and to act accordingly'; that is, to recognize socially normative emotionality. Further, this cognizance is intertwined with normative practices of gender, which broadly locate femininity as emotional and masculinity as unfeeling. Yet, as the subject is always in flux, the 'emotional self' is 'responsive to changes in sociocultural meaning and representation and in individuals' own biographies of interactions with others as well as with inanimate things (such as places and objects)' (Lupton 1998, 168). The feeling subject, then, is ever forming, but not in a time-space vacuum; rather, this subject 'in formation' takes shape, becomes intelligible, in concert with places and objects[1] encountered. As a consequence, distinct if mobile constellations – of individuals, their feelings and the places of or for feeling – can be assessed as collectively (if not always collaboratively, or knowingly) constituting the dynamic processes of feeling subjects.

The Royal Canadian Legion: A Brief History

The Canadian soldiers who survived 'the Great War' returned home to a complicated new reality. Following World War I, there was little in the way of an organized process to assist veterans, their families, or their communities to cope with the economic, social or emotional fall-out of the war.[2] Consequently, several veterans' organizations were formed in Canada between 1917 and 1925 (Hale 1995; Bowering 1960). Of these, the Great War Veterans Association (GWVA), formed in 1917, was the largest and became the foundation for the Royal Canadian Legion. The GWVA was to set some significant precedents for the Legion. The GWVA actively advocated for veterans on matters of healthcare. Additionally, the organization addressed the concerns of mostly non-professional Canadian soldiers who resented war-time preferential treatment on the basis of rank by proposing all veterans meet equally as 'comrades' in their organization (Hale 1995). Despite espousing equality within its organization, the GWVA, a founding member of the British Empire Service League formed in Capetown, South Africa in 1921 (Bowering 1960), simultaneously was deeply entrenched in a largely inequitable

---

1   In the psychoanalytic theory of the object relations tradition, the 'object' may be many things, including another person, an item, or a place towards which a person's psychic interest is directed (Bondi 1999; Chodorow 1978).

2   Where state or donor-funded care was not available, the burden of caregiving undoubtedly fell to family members and communities. The issue of such informal care for veterans with emotional or psychological scars in the context of Canadian deinstitutionalization of mental health services is the subject of another paper.

colonial framework: 'empire unity and empire settlement', typical of a prevalent discourse of Anglo-Canadian superiority in these interwar years (Pickles 2000, 82). This early history generated uneven ground upon which eventual Legion members would be differently racially placed.

The Royal Canadian Legion was formed in November 1925, in Winnipeg, Manitoba, in a move to unite the disparate Canadian veterans' associations. The Legion's major commitments, following the example of the GWVA, were to advocate for care for veterans, promote national unity, and remember the sacrifices of soldiers (Bowering 1960, 15). And, as with the GWVA, 'in the Legion, you checked your rank at the clubhouse door' (Hale 1995, 100), symbolically leaving military and class status at the entrance as people did their coats at the coat-check. All ex-service personnel held the same title: 'Ordinary Member'. This definitive flattening of military hierarchies was an important moment of recognition for the working-class men. The club setting also offered a comfortable collective space for a predominantly male clientele to find companionship and shared understanding of their wartime experiences. The consumption of alcohol was and remains a major element of the social club aspect of the Legion. A former Legion president, Robert Kohaly, noted of the early and persistent association of the Legion with beer-drinking veterans: 'You can't blame them, because they drank a hell of a lot of muddy water out of a helmet to pay for it' (Hale 1995, 181). Josephine Selkirk, a venerated ordinary member from 1947, who has collected nearly all the honours that the Legion dispenses (Peebles 2005), expresses her sense, that if 'the boys went a bit wild', it was only to be expected; the companionship and drinking offered a way to 'cope' with the aftermath of terrible hardships in both World Wars without actively discussing difficult events (Selkirk 2006).

By the time the Legion's commitments were tested by World War II, the average legionnaire was 49 years old, and the Legion was an organized and experienced veterans' body (Hale 1995). Schooled by the experiences of World War I veterans, concern for the soldiers' wellbeing during this Second World War was paramount (Walpole 2005, 19). To address the eventual rehabilitation of soldiers, the Legion's War Services included education, personal services (including counselling from Legion personnel and war padres), entertainment, recreation and travel services to overseas on-leave personnel (Hale 1995). The Legion's concerns for their forces stemmed in part from the fear that returning soldiers would be dependent upon public welfare, a fear realized by many World War I veterans. In addition, battle shock cost the war effort, so soldiers' fear and anxiety needed to be carefully managed.[3] In 1941, the Legion fought for and won 'free treatment of all [disability] conditions for one year during the cessation of hostilities' (Hale 1995, 74).[4] During Canada's military intervention in the Korean War, the Legion again provided

---

3 See Deborah Cowen's (2008b) extended discussion of the political links between welfare and warfare.

4 Canada did not have free universal medicare until the 1960s. Until then, care for the disabled was piecemeal, provided by charities in some cases (Morton and Wright 1987).

'support and comfort' to the troops (The Royal Canadian Legion 2007). Service, not just for veterans, but for local communities, youth, and sports, became an important aspect of Legion affairs following World War II. Through these outreach efforts, the Legion became 'ingrained in the social life of every community where a branch was located' (Hale 1995, 167).

The Changing Spaces and Feeling Places of the Royal Canadian Legion

Historically, support for the Legion has been strong, at both local and national levels. In particular, the Legion's organization of Remembrance Day, which commemorates the signing of the Armistice of World War I in 1918, continues to command a large audience.[5] This event's ritual elements – the walk to the graveyard, the laying of the wreaths, the shared silence, tears, saluting of gravestones and cenotaphs, the presence of elderly veterans standing proudly, tea and coffee at the hall – collectively offer a significant space for socially sanctioned expressions of loss and the public practices associated with death. These familiar memorial services, the poppy campaign, and Legion-sponsored annual pilgrimages of remembrance to overseas war sites continue to connect Canadian legionnaires to an international, highly emotive network. In the post-war years, however, the Legion has had to work harder to garner community support as the 'beer and bingo' image has prevailed over the gravitas of the historical symbolism of the institution and as the composition of its traditional (white, male) membership has been challenged. In part, these shifts in perception are a reflection of the Legion's necessarily changing membership. In 2007, World War II veterans were an average age of 82, and the Korean War veterans were not far behind, in their late seventies (Roy-Sole 2007). As the Legion's ordinary membership has dwindled, the Legion has had little choice, though not without heated debate, to open wide its doors.

Today's Legion offers four possible routes to membership within its Canadian Branches: Ordinary Member (military service past or present); Associate Member (family of ordinary member, some civil servants); Affiliate Voting Member (neither ordinary nor associate, but supportive of Legion's aims and objectives); and Affiliate Non-Voting Member (non-Canadian citizen, non-Commonwealth subject from an Allied nation who support the aims and objectives) (The Royal Canadian Legion 2007). Membership in an organization which views itself as having 'evolved from the voice of social reform, to [Canada's] moral conscience, to the leading proponent for the maintenance of traditional values' (Hale 1995, 2) comes with certain expectations. Legion members are expected to share an attachment to Legion values, its history and traditions, demonstrated in and through attendance at the Legion branches, and through representation as a national membership body.

---

5   The *Globe and Mail* reports that Canada is the only remaining Commonwealth nation to mark Remembrance Day on November 11 and argues this is a demonstration of the Legion's still powerful social clout (Sullivan 2005).

The application for membership requires potential members to certify that they do not have a dishonourable discharge, desertion or evasion of service on their record; that they are not affiliated with any group whose interests would conflict with the Legion, and that they are not a communist, fascist or anarchist. Those unsupportive of the Legion's aims and objectives are not welcomed to join. In this process of establishing membership – who is in and who is out – normative feeling subjects take (temporary) shape in ongoing and dynamic ways, constituted by and constituting an integral part of the Legion's emotional geographies.

Undoubtedly, the Legion's branches share many material and symbolic commonalities, which contribute to engaging and shaping members in normative, recognizable ways, as the membership regulations above suggest; however, at a closer glance, Legion branches have always been dependent on the specificities of location. Legion branches are differentially affected by particular intersections of local interests, levels of participation, the involved personalities, and in many of these fishing and/or forestry communities, the existing or severed ties with industry and economy. Legion branches are not, therefore, straightforward replications of one another, but are differently articulated 'intersubjectively experienced' places (Milligan et al. 2005, 51) according to these dynamic and relational factors. Indeed, while Legion buildings look very much the same (Figures 11.1, 11.2 and 11.3), in fact, an examination of the material histories of British Columbia's branches reveals a great diversity. Legion branches were, for example, former hospitals, feed stores, army buildings, navy buildings, Quonset huts, and Panabodes,[6] many renovated extensively over the years, while others were purpose-built structures constructed locally by volunteer labour once enough funds and a sufficient membership had gathered together (Walpole 2005). As another example, the Legion branches, while offering a public place for members of a local community to gather, still retain aspects of a private club, such as the process of 'signing in'. John Rowell, currently a Legion secretary, likes this long-standing practice of the Legion, which he feels enforces regulation of and responsibility from the membership. This set-up, he suggests, offers some control over who comes in, allows for a process of approval, and initially limits the influence which can be exercised by a new member (new associate members cannot vote for one year) (Rowell 2007). As discussed further below, this practice has been used to shape or restrict membership in a locally-mediated fashion, sometimes in contravention of Legion charter. In visiting different Legion branches, I noted this practice was sometimes observed by those entering, and sometimes not. As a non-member, I was signed in by members sometimes and at others, signed myself in as a guest or did not sign in at all. The Legion's iconic status as an organization lends itself to a vision of recognizable repetition; yet, the Legion branches demonstrate a unique character in relation to their particular relational contexts.

---

6 The Panabode is a Canadian pre-fabricated log cabin.

Figure 11.1   Port Alice Legion
Source: Photo by author, 2007.

Figure 11.2   Port Hardy Legion
Source: Photo by author, 2007.

Figure 11.3   Port McNeill Legion
Source: Photo by author, 2007.

Soldiering on: Militarism, Masculinity and Emotion

> Our roots, whatever happens, will always be in the military (Mary Ann Burdett, Dominion Command president, quoted in Deacon 2004).

> People think of the Legion as bunch of old men sitting around drinking and telling war stories (Legion member, Port Hardy Legion 2007).

Each Legion branch is simultaneously the same and also different, a locus for normative, if unstable, social discourses, for example, of militarism, masculinity, and whiteness, which contribute to the production of and are constituted by 'feeling subjects'. In some important ways, the Legion is both a military and a masculine space, and as such the feeling subjects that emerge in ongoing ways in relation to the Legion are also both military and masculine subjects. In both its discursive and material 'shaping of civilian space and social relations', the Legion has been governed by 'military objectives, rationales and structures' (Woodward 2005, 719-721) in general, and by the imperial desires of Anglo-Canada in particular. As such, the Legion is a military and a masculine space; the Legion branches, uniquely constituted, are a locus for normative, if unstable, social discourses of militarism and masculinity and these have shaped the emotional geographies therein. For example, Legion membership used to be strictly limited to those who had served in the Canadian Armed Forces: the predominantly male 'ordinary member'. This traditional or ordinary member of the Legion is described by members as drawing from a shared set of feelings and experiences as shared only by other soldiers: 'the sense of service and comradeship which all ex-service personnel seek to express in one way or another' (John Anderson quoted in Hale 1995, 115). As time has passed, determining the ins and outs of membership in the Royal Canadian Legion has proven a contentious issue. Woodward (2005, 729) has noted: 'Armed Forces, and defence institutions, take great care in producing and promoting specific portrayals of themselves and their activities in order to legitimize and justify their activities in places, spaces, environments and landscapes'. It is not surprising, then, that the boundaries of belonging have been closely managed. Veterans are not a coherent or cohesive group; indeed not all veterans belong to the Legion,[7] but the ordinary members who do are not only understood to subscribe explicitly to the Legion's militaristic agenda (for example, through the process of application, but also demonstrated in everyday ways – in parking lots of the Legions visited for this study, vehicles displaying commemorative veteran plates and 'support our troops' bumper stickers featured prominently), but also to embody the aims of the Legion with dignity and fellow feeling.

As a military institution, the Legion is largely a masculine institution; the social and legal practices of militarism have reinforced this gendered terrain resulting in the predominance of men as soldiers, and thus as ordinary members. While stereotypically gendered notions of emotionality are rarely directly mapped onto male or female experience, idealized expressions of masculinity and femininity contain strongly prescriptive notions of appropriate emotionality, developed through social, spatial and political means. The associations of men with the authority of order, rationality, and self-control, in contrast to women's emotional disorder have a long history (see Lupton 1998, 105–136 for a more detailed tracing). Thus, while 'hegemonic masculinity' may not be identical with the lives of men as soldiers, a dominant cultural association 'shapes a society-wide sense of masculine reality, and, therefore, operates in the cultural domain as on-hand material to be actualized, altered or challenged through practice' (Connell and Messerschmidt 2006, 849). This idealized expression of masculinity unfolds in and through its social constitution; for example, 'there is a circulation of models of admired masculine

---

7 Non-Legion members are not the subject of this study, but of course also hold opinions about the legion. Many veterans have chosen other Veterans Associations (e.g. First Nations Veterans Associations) or found fellowship in their officers' clubs or church. During the period of the Legion's early evolution, a strong pacifist counter-discourse, emerging from the Anabaptists, Quakers, Mennonite, and Hutterite settlements of 18–19th century Canada, as well as labour and farm and women's organizations. But, historian Mark Moss argues, this pacifist movement was too radical, too marginal to challenge the might and militarism of British imperialism (see Moss 2001, 140–146). Additionally, Legion members do not necessarily patronize the branches. As one woman explained, she likes to 'pay her dues' but has no time to visit, and she noted, it's 'more of a men's place' (Quadra Island 2007).

conduct, which may be exalted by churches, narrated by mass media, or celebrated by the state' (Connell and Messerschmidt 2006, 838). In this light, militarism is seemingly inseparable from the ideals of masculinity (Enloe 2000).

If the hegemonic masculinity of the modern period, specifically in western Anglophone and north European cultures (Lupton 1998, 106), is practiced through demonstrations of strength and courage, through control over self, space, and feminized emotion, war is the 'ultimate forum' for this exercise (Moss 2001, 15) and the soldier is positioned as the ultimate (un)feeling subject. Military conduct hinges on the mastering, containment or repression of all things feminine, including emotion, despite the literal presence of, and with potentially dangerous consequences for, enlisted women (Cowen 2008a). The coding of the masculine as 'unemotional' has established men as appropriately minded for soldiering (Lupton 1998, 106). Importantly, this desired unfeeling state does not denote the absence of emotion, nor the lack of feeling, but rather, its (supposed) control. The normatively masculinist soldier is called upon to have self-control; to have the strength to deny pain, fear or rage; to assert independence of spirit; to be ennobled by adversity, even in the case of the severe physical disabilities that can result from soldiering (Cohen 2001). Emotional content is clearly not absent – not in the fields of war, nor in the literal or symbolic spaces of the Royal Canadian Legion – but emotional expression tends towards idealized descriptions of bravery, fortitude, valour. Tales of battle in letters, poems, and historical narratives demonstrate this, from the stoicism of the soldier writing home from the front, to the noble sentiments of loss and lost love famously expressed in Canadian John McCrae's poem of 1915, *In Flanders Fields*:

We are the Dead. Short days ago

We lived, felt dawn, saw sunset glow,

Loved and were loved, and now we lie

In Flanders fields (The Royal Canadian Legion, 2008)

For many veterans, there is a great deal of pride associated with Legion service, as it foregrounds past military service (Fairbairn 2000). The Legion makes space for such an ongoing reiteration of national service as heroic, brave, and in latter times, peace-keeping. The dignity thus afforded by association with the Legion is considerable, to the extent that members report imposters posing as veterans. These poseurs, one respondent suggested, are seeking valuable recognition in a culture of patriotism, fellowship and good will. This type of recognition is indeed important; for aging ordinary members, such collective remembrance is an opportunity for the validation of elders' individual memories (Hockey et al 2005). In this regard, the Legions offer a network of peer support for elderly veteran members. Members hold positions of importance; they are recognized as part of a living historical record. As keepers of such memories, these usually elderly members are regarded with a respect they may not command in other social spheres.

While masculinity is undoubtedly privileged as a productive, normative discourse of militarism, the complex nuances of gendered relations within militaristic domains (Enloe 2000) are evident in the Legion's geographies. In formal and informal research encounters in this study of the Legion, strong emotional responses not culturally associated with militarism or masculinity, such as tears, and the willingness to recount emotionally difficult memories took place as individuals described their relationships to and within the Legion. Thus the normatively militaristic and masculine Legion spaces popularly associated with stereotypically 'manly' attitudes, paradoxically offer a significant space for male feeling in a manner normatively accorded to a feminized repertoire of emotionality (that of community, sharing of emotions, emotionality through community).

The Legion has made distinct, if arguably limited, spaces for women. The Ladies Auxiliary (LA) was once most women's sole avenue to Legion membership and remains predominantly a women's institution within the broader organization. 'To the Ladies!', the final chapter of Bowering's Legion history (1960, 229) lauds the LA as 'a tower of strength' and argues that 'no history of the Legion would be complete without at least some small tribute to their contribution.' Bowering celebrates the LA members for taking on 'monotonous' chores, providing a 'cheery welcome' and caring for veterans (1960, 230). The business of caring, especially the strength to take care of others, is, of course, profoundly gendered emotional territory (see Hochschild 2003; Robinson 1999). And yet, this prescribed women's space within the Legion has also been susceptible to revision. As recently as 2007, the LA in a small British Columbian community proposed a radical change of Legion business. Galvanized by a group of women who did not want to see their group dissolve due to low numbers, this auxiliary group opened their LA membership to men. This move was initially met with great resistance at other local Legion branches in British Columbia. As Bonnie West, a current Zone Commander noted, the general consensus on making such a change was 'hell, no!' (West 2007). However, the new arrangement was validated at the annual Legion convention when the potential legal ramifications of the equality legislation contained within the Canadian Charter of Human Rights and Freedoms[8] were raised (West 2007). In another example of the bounds of the Legion's spaces being more flexible than they might initially appear, the online Magazine, *Lesbian Quarterly* (2008)

---

8   The 'Equality Rights' of the Charter (Section 15, 1–2) state: 15. (1) Every individual is equal before and under the law and has the right to the equal protection and equal benefit of the law without discrimination and, in particular, without discrimination based on race, national or ethnic origin, colour, religion, sex, age or mental or physical disability. (2) Subsection (1) does not preclude any law, program or activity that has as its object the amelioration of conditions of disadvantaged individuals or groups including those that are disadvantaged because of race, national or ethnic origin, colour, religion, sex, age or mental or physical disability (Canadian Charter of Rights and Freedoms 1982).

advertises the rental of a local Legion hall near Vancouver, British Columbia, for monthly women's dances, despite a seemingly strongly heterosexist culture. Thus, while militarism and heteronormative masculinity are deeply intertwined, they are not unassailably so (Enloe 2000, 235).

'Race' and the Legion: Managing Emotion through Racializing Place

> The modern colonial view suggests that emotionality reflects weakness in racialized and gendered ways, whereas control over emotions marks masculinity and 'whiteness' (Philipose 2006, 61).

Deborah Cowen notes in her informative genealogy of the soldier-citizen in Canada, 'Canada was never homogenously – only hegemonically – white; ethno-racial diversity has long been an essential part of the "building of Canada"' (Cowen 2005, 670). The Legion's hegemonic whiteness is demonstrated in both material and symbolic ways. The visual record (e.g. photos and films) of the Legion tends to foreground Anglophone Canadians of European ancestry, such as the famed Victoria Cross holder, Billy Bishop. The Legion is a predominantly English-language organization. These representations of Legion veterans contribute to nationalist narratives of Anglo-Canadian identity as representative of veteran identity. Yet, the Legion is simultaneously and contradictorily home to a heterogeneous mix of 'raced' and national identities, representing the diverse identities of Canadian troops. The legion's history is not as 'white' as much of the visual record might suggest. For example, in the Legion's early years, Branch No. 9 (British Columbia) was formed by Japanese-Canadian veterans (Morton and Wright 1987). However, the achievement of 'whiteness', Kobayashi and Peake suggest (2000, 393), manifests not in particular bodies, but in the normative control of daily practices, and especially through the occupation of (social, cultural, political) space. Branch No. 9's charter was revoked in 1942, as Japanese-Canadians became suspect citizens and were sent to internment camps.

The place of indigenous Canadian veterans with the Legion is similarly illustrative. A *Legion Magazine* feature describes the extraordinary contribution of 'Canada's indigenous peoples', who despite indignities and loss of human rights, visited upon them by the Canadian state, 'answered [the calls to war] in droves' (Salat 2006). According to Veterans Affairs, some 7,000 First Nations people served in World War I, World War II and the Korean War, as well as unknown numbers of Inuit and Métis (Salat 2006); however, First Nations' groups suggest the number is likely much higher (First Nations University of Canada 2007). Yet, few Legions have an overt First Nations presence, a likely consequence of a long history of racial exclusion:

> Aboriginal veterans seldom had access to Royal Canadian Legion branches and newsletters. These were very helpful to most other veterans, informing them

about the benefits available and helping them find out how to obtain them. In addition, they provided a useful means for discussing and comparing experiences on the subject. However, status Indians were usually barred from participation in the Legion, because Legions served liquor, and Aboriginal men subject to the *Indian Act* could not attend functions where liquor was served. Exclusion of Indian veterans from Legions was extremely discriminatory, considering they had fought, been wounded and died alongside their non-Aboriginal comrades (Royal Commission on Aboriginal Affairs 1996).[9]

As the Royal Commission describes, not only were First Nations Veterans excluded from the Legion branches, but also, they were excluded from opportunities afforded by such spaces, including the potential to share and/or compare their experiences of and feelings about war. That this was a deeply felt exclusion was made clear in the Alert Bay Legion, where a First Nations member offered his commentary on the First Nations-Legion relationship. He noted with deadpan humour that there was 'some history' with the Legion: 'some good, some not so good' (Legion member, Alert Bay 2007). He went on to note that 'they [the Legion] didn't use to let us in here, right up to the '80s' (Alert Bay 2007). He further emphasized that his past experiences of exclusion made him all the more determined to 'take his seat' in today's Legion. In contemporary times, this Legion branch displays a strong First Nations historical and contemporary presence. Inside the branch, a prominent display is dedicated to a First Nations local war hero. A non-First Nations member commented of his fellow First Nations legionnaires, 'this Legion couldn't exist without them' (Alert Bay 2007). While times, and laws, have changed, the emotional undercurrents of this historical exclusion linger on in the feelings and practices of some indigenous veterans,[10] indicating the different positioning of other(s) feelings within the Legion, but signalling also the shifting landscapes of this organization.

The Legion's particular intersections of militarism, 'race', identity, and place are further demonstrated by responses to the wearing of Sikh turbans to a Remembrance Day celebration at the Newton Legion branch in Surrey, British Columbia (Hale 1995; Walton-Roberts 1998). The Remembrance Day ceremonies are a profoundly charged emotional event for veterans and offer the most

---

9 'Status Indian' is a term of legal status which designates a person registered as an Indian under the Indian Act of Canada (Indian and Northern Affairs Canada 2003). This much amended and often contested federal legislation outlines Canada's federal obligations as regards 'Indians', and regulates the management of lands, monies and other resources. While the term 'First Nation' is not legally defined, it is commonly employed as a social and political alternative to 'Indian' to refer to indigenous people with or without the legal status as defined by the Indian Act (Fiske and Browne 2006).

10 As suggested earlier, some indigenous veterans have chosen to belong to First Nations Veterans Associations. This choice in a small community such as Alert Bay, however, would mean the loss of the Legion branch as a significant place for socializing.

spectacular and public venue for the Legion's agenda of remembrance. Preceding the 1993 Remembrance Day celebrations, the Newton branch issued an invitation to minority groups to join them. Local World War II Sikh veterans attended the ceremonies but were then prevented from entering the Legion building because they were unwilling to remove their turbans in accordance with Legion regulations (Walton-Roberts 1998). Hale, a frequent chronicler of the Legion's activities writes: 'The Legion tradition of removing your hat in clubrooms as a mark of respect for fallen comrades is well known and widespread' (1995, 248). In the ensuing furious debates – against racial and religious prejudice, for the preservation of 'tradition' – the turban as a symbol of (other) identity and (another) place became the focus of broader arguments about tradition and control over space (Walton-Roberts 1998, 320–322). Walton-Roberts takes a post-colonial perspective, remarking that the conflict revealed how 'the legacy of *European* settlement and tradition' (1998, 321), is privileged by the Legion. The achievement of whiteness is in part reflected by control over emotions, including the power to determine where and when emotions are out of place (Philipose 2006). In this framing, the Legion's 'tradition' specifically racializes its Sikh members by (continuing to) master their spatial and emotional access. The painful wrangle over turbans is demonstrative of the emotional intensities which swirl around the practices of the Royal Canadian Legion, and also indicative of the ways in which some feelings matter more in the Legion's normative spaces. Yet, this spatial conflict over emotional territory also demonstrates the dynamic and intertwined processes of place-making and the dynamism of feeling subjects who are emergent in uneven and sometimes unexpected ways.

Conclusions

Tracing the geographies of the Legion across time and in place is suggestive of the ways in which a seemingly ubiquitous, militaristic, traditionally masculine organization has in fact many ongoing iterations. While recognizable notions of the Legion and of the Legion membership exist, the dynamic interface of places, people, and their relations destabilize the fixity of these representations. For example, the membership criteria have not been as stable or as fixed over time, nor have the bounds of membership been uncontested; therefore, the normative feeling subject has shifted and changed in and through always emergent 'relations of dependency' (Butler 2005, 20). In both explicit and subtle ways, in keeping with the organization's historical, militaristic, colonial roots, the Legion's feeling subjects have been subjected not only by gender, but also by racialization, a process of whitening the social landscape (Kobayashi and Peake 2000). The Legion is thus, in some important ways, framed by the modern 'colonial present' (Gregory 2004) within which whiteness and masculinity prevail. Yet, as the diversity of people, places and emotions flow back-and-forth, in the 'throwntogetherness' of Legion branches, dynamic feeling subjects continue to evolve in multiple ways.

As the above sections detail, differently feeling subjects are continuously in flux, constituted by complex reiterations of hegemonically militaristic and masculinist practices, racialized interactions, and disruptions to the socio-cultural spaces and in the particular places of the Royal Canadian Legion.

Legion branches, then, are not simply faded backdrops for predictable activity; rather they are fashioned in unexpected and often highly emotive moments, vis-à-vis this ongoing back-and-forth of socio-spatial negotiation. These branches have played an important role in making a space for at least some feelings; validating emotional expression in a seemingly constrained masculinist and militaristic field, unexpectedly offering sanctioned space for coping with emotions. As such, the Legions undoubtedly provide an important space for some to care about – and for – self and others, channelling feelings, for example, through extensive, detailed and ritualized memorial processes within which (legitimate) participants take part, as well as through everyday practices, like taking one's seat at the bar. In this fashion, the Legion is perhaps unintentionally contributing to challenging normative spaces of and for emotion. In the Legion branches, men's emotions are meaningful; caring by and for a mainly male veteran population has mattered significantly in the history of this institution. Arguably, such acknowledgement and demand for attention to male experiences of distress, loss, the traumas of war, have led to expanding the available emotional range usually attributed to men.

In making space for some feeling, the Legion's landscapes have arguably limited others by virtue of promoting normative masculinity and whiteness. And yet, as other(ed) subjects have engaged with Legion spaces, its normative landscapes have been inevitably subjected to ongoing revision. Attention to these processes of emotion, to feeling subjects, and to emotional geographies, is crucial for acknowledging the spatial politics of emotion. For a seemingly mundane and ubiquitous presence, often misrepresented as place solely for beer, bingo and old men, the Legion has a deeply emotional terrain to explore. An investigation into the Legion as locus for emotional geographies thus offers insights into a rich and complex set of social, spatial, and emotional relations which have implications for understanding the spatialized politics of emotion.

Acknowledgements

My deepest thanks to the Legion members who generously shared their stories, memories and experiences. I thank Greg Halseth for supporting this project in its earliest form, and Regine Halseth and Sarah Cady for their research assistance. My thanks to Joyce Davidson and Laura Cameron for organizing the second Emotional Geographies conference, at which an earlier version of this paper was presented and to Joyce Davidson and Liz Bondi for their considerable editorial expertise. My gratitude also for the receptive colloquia audiences at University of Arizona and University of Groningen who asked valuable questions of this material. I gratefully acknowledge a SSHRC postdoctoral award (756-2006-0406), the University of

Northern British Columbia (UNBC), and California State University, Long Beach for funding this work. All errors of interpretation within are my own.

References

Anderson, K. and Smith, S. J. (2001) 'Editorial: Emotional Geographies', *Transactions of the Institute of British Geographers* 26, 7–10.

Bondi, L. (1999) 'Object Relations Theory', in McDowell, L. and Sharp, J. P. (Eds.) *A Feminist Glossary of Human Geography* (London: Arnold), 188–190.

Bondi, L. (2005) 'The Place of Emotions in Research: From Partitioning Emotion and Reason to the Emotional Dynamics of Research Relationships', in J. Davidson,L. Bondi, and M. Smith (eds.), *Emotional Geographies* (Aldershot: Ashgate).

Bondi, L., Davidson, J. and Smith, M. (2005) 'Introduction: Geography's Emotional Turn', in J. Davidson, L. Bondi, and M. Smith (eds.), *Emotional Geographies* (Aldershot: Ashgate).

Bowering, C. H. (1960) *Service: The Story of the Canadian Legion* (Legion House, Ottawa: Dominion Command, Canadian Legion).

Butler, J. (1990) *Gender Trouble: Feminism and the Subversion of Identity* (New York: Routledge).

Butler, J. (1997) *The Psychic Life of Power: Theories in Subjection* (Stanford, CA: Stanford UP).

Butler, J. (2005) *Giving an Account of Oneself* (New York, Fordham UP).

Canadian War Museum (2001) Canadian War Museum: Canada's national museum of military history <http://www.warmuseum.ca> Last accessed 18 July, 2007.

Chodorow, N. J. (1978) *The Reproduction of Mothering: Psychoanalysis and the Sociology of Gender* (Berkeley; London: University of California Press).

Cohen, D. (2000) *The War Come Home: Disabled Veterans in Britain and Germany, 1914–1939* (Berkeley; Los Angeles; London: University of California Press).

Connell, R. W. and Messerschmidt, J. W. (2005) 'Hegemonic Masculinity: Rethinking the Concept', *Gender Society* 19:6, 829–859.

Cowen, D. (2005) 'Welfare Warriors: Towards a Genealogy of the Soldier Citizen in Canada', *Antipode* 37:4, 654–678.

Cowen, D. (2008a) 'Reorienting Recruitment: Towards a 'Different' Military?', in *Military Workforce: The Soldier and Social Citizenship in Canada* (Toronto: University of Toronto Press).

Cowen, D. (2008b) 'The Military Labour of Social Citizenship', in *Military Workforce: The Soldier and Social Citizenship in Canada* (Toronto: University of Toronto Press).

Davidson, J. and Bondi, L. (2004) 'Spatialising Affect; Affecting Space: An Introduction', *Gender Place and Culture* 11:3, 373–374.

Deacon, J. (2004) 'They Soldier On', *Maclean's* 117: 46, 118.

Enloe, C. H. (2000) *Maneuvers: The International Politics of Militarizing Women's Lives* (Berkeley, California: University of California Press).

Fairbairn, B. (2000) 'We Asked and You Told Us', *Legion Magazine.*

First Nations University of Canada (2007) Saskatchewan First Nations Veterans Memorial <http://www.firstnationsveterans.ca> Last accessed 27 August 2007.

Fiske, J.-A. and Browne, A., J. (2006) 'Aboriginal Citizen, Discredited Medical Subject: Paradoxical Constructions of Aboriginal Women's Subjectivity in Canadian Health Care Policies', *Policy Sciences* 39:1, 91.

Gregory, D. (2004) *The Colonial Present: Afghanistan, Palestine, Iraq* (Malden, MA: Blackwell).

Hale, J. (1995) *Branching Out: The Story of the Royal Canadian Legion* (Ottawa, The Royal Canadian Legion).

Hochschild, A. R. (2003) *The Managed Heart: Commercialization of Human Feeling* (Berkeley, Calif.; London: University of California Press).

Hockey, J., Penhale, B. and Sibley, D. (2005) 'Environments of Memory: Home Space, Later Life and Grief', in J. Davidson, L. Bondi, and M. Smith (eds.), *Emotional Geographies* (Aldershot: Ashgate).

Indian and Northern Affairs Canada (2003) Terminology <http://www.ainc-inac.gc.ca/pr/info/tln_e.html>, February 11, 2008.

Kobayashi, A. and Peake, L. (2000) 'Racism Out of Place: Thoughts on Whiteness and an Antiracist Geography in the New Millennium', *Annals of the Association of American Geographers* 90:2, 392.

Lesbian Quarterly (2008) Lesbian Agenda Calendar http://www.sophiakelly.ca/lq/, March 29.

Lupton, D. (1998) *The Emotional Self: A Sociocultural Exploration* (London: Sage).

Milligan, C., Bingley, A. and Gattrell, A. (2005) 'Healing and Feeling: The Place of Emotions in Later Life', in J. Davidson, L. Bondi, and M. Smith (eds.), *Emotional Geographies* (Aldershot: Ashgate).

Morton, D. and G. Wright (1987) *Winning the Second Battle: Canadian Veterans and the Return to Civilian Life, 1915–1930* (Toronto: University of Toronto Press).

Moss, M. (2001) *Manliness and Militarism: Educating Young Boys in Ontario for War* (Don Mills, Ontario: Oxford University Press).

Natter, W. and Jones III, J. P. (1997) 'Identity, Space, and other Uncertainties', in G. Benko, and U. Strohmayer (eds.) *Space and Social Theory: Interpreting Modernity and Postmodernity* (Oxford; Cambridge, Mass., Blackwell).

Peebles, F. (2005) Legion Honours its Own. *The Prince George Citizen* (Prince George, BC).

Philipose, L. (2006) 'The Politics of Pain and the End of Empire', *International Feminist Journal of Politics* 9:1, 60–81.

Pickles, K. (2000) 'Exhibiting Canada: Empire, Migration and the 1928 English Schoolgirl Tour', *Gender, Place and Culture: A Journal of Feminist Geography* 7:1, 81–96.

Robinson, F. (1999) *Globalizing Care: Ethics, Feminist Theory, and International Relations* (Boulder, CO., Westview Press).

Rose, G. (1993) 'The Geographical Imagination' *Feminist and Geography: The Limits of Geographical Knowledge* (Minneapolis, University of Minnesota Press).

Rowell, J. (2007) Interview. Alert Bay, B.C. July 13.

Roy-Sole, M. (2007) 'Meet me at the Legion', *Canadian Geographic*, Nov/Dec, 66–74.

Royal Commission on Aboriginal Peoples (1996) 'Chapter Twelve: Veterans'. Report of the Royal Commission on Aboriginal Peoples'. Indian and Northern Affairs Canada (ed.), Ottawa: The Commission. Vol 1, Part II, Chap 12.

Salat, N. (2006) A Spiritual Homecoming. *Legion Magazine.* Jan/Feb, <http://www.legionmagazine.com/features/memoirspilgrimages/06-01.asp#1>, August 1.

Selkirk, J. (2006). Interview. Prince George, BC. April 20.

Sullivan, P. (2005) 'Legionnaires Moving National Headquarters; Sagging Membership, Serious Cash Crunch Taking Their Toll on Royal Canadian Legion', *The Globe and Mail*, Monday, May 16.

The Royal Canadian Legion (2007) The Royal Canadian Legion <http://legion.ca/asp/docs/home/home_e.asp>, August 13.

The Royal Canadian Legion (2008) Poppy and Remembrance <http://www.legion.ca/asp/docs/rempoppy/allabout_e.asp> February 21.

Thien, D. (2005) 'After or Beyond Feeling?: A Consideration of Emotion and Affect in Geography', *Area- Institute of British Geographers* 37:4, 450–456.

Thrift, N. (2004) 'Intensities of Feeling: Towards a Spatial Politics of Affect', *Geografiska Annaler Series B* 86:1, 57–78.

Walpole, S. (2005) The Royal Canadian Legion: Past, Present and Future. *The Royal Canadian Legion, BC/Yukon Branches: 2005 Commemorative History.* Halifax, Nova Scotia, Fenety Marketing Services.

Walton-Roberts, M. (1998) 'Three Readings of the Turban: Sikh Identity in Greater Vancouver', *Urban Geography* 19:4, 311–331.

West, B. (2007). Interview. Port McNeill, BC. July 12.

Woodward, R. (1998) '"It's a Man's Life!": Soldiers, Masculinity and the Countryside', *Gender, Place and Culture* 5:3, 277–300.

Woodward, R. (2005) 'From Military Geography to Militarism's Geographies: Disciplinary Engagements with the Geographies of Militarism and Military Activities', *Progress in Human Geography* 29, 718–740.

Chapter 12

'I Love the Goddamn River':
Masculinity, Emotion and Ethics of Place

Cheryl Lousley

This chapter discusses the rural novels of two contemporary Canadian writers: David Adams Richards and the late Matt Cohen. Both novelists focus on rural men marginalized by changing socio-economic forces which disrupt and reconfigure relationships to the land. In his Salem novels, especially *The Disinherited* (1974) and *The Sweet Second Summer of Kitty Malone* (1979), Cohen traces the post-war decline of the precarious family farms carved out of the thin soil of the Canadian Shield during the colonial settlement of eastern Ontario. The process culminates with rural gentrification in his final novel, *Elizabeth and After* (1999), where the landscape has been transformed from fields and bush to strip malls, car lots, and suburban homes. Richards's novels present a bleak portrait of the Miramichi River region of New Brunswick, where pulp mills and forestry operations decimate the woods and the salmon streams. The two writers address different trends in the post-industrial transformation of rural areas: Cohen describes a region shifting from a production to consumption-based economy; Richards portrays a region declining into an ecological sacrifice zone. But both focus on how the transformation of the rural environment transforms emotional and social relationships of place.

I take as my title a statement by eighty-two-year-old illiterate woodsman Simon Terry, or 'Old Simon', a central character in Richards's novel *Lives of Short Duration* (1981). Simon Terry's gruff expression of his sense of place – 'I love the goddamn river' (118) – is striking because it is one of the only times in Richards's *oeuvre* when a character speaks about the emotional connection to place. Nevertheless, the moral compass in Richards's and Cohen's rural fiction is usually provided by either a male character like Old Simon, who has lived his life on the land in deep knowledge and appreciation of a particular place, or a son who seeks his place in the world after being cut off from the land. The unusualness of Old Simon's outburst – and the fact that it goes unheard – underscores how the ethical subjectivity of these male characters rests on their very inarticulateness. Their masculinity is constructed through an inarticulate, pre-conscious, emotional-bodily response to the living world that emerges as the epitome of human subjectivity and morality. In environmental and geographical thought, it is the relationships between femininity and the natural that are most commonly discussed. It may therefore seem surprising that ethical subjectivity is constructed in these novels through the emotional embodiment of male characters. But there exists a literary and cultural celebration of masculine vitality that extends back to the nineteenth century. This chapter explores how a contemporary version of masculine vitality is being harnessed for environmental ethics.

Emotional Natures and Cultures of Masculinity

Although gender is a significant category of analysis in environmental thought and literary criticism, masculinity has received little attention. Two sets of discussion dominate the scholarship on gender and environment: first, cultural associations of women and nature (and the extent to which women and nature are debased and devalued due to this association) and, second, sociological relationships between women and environments due to gendered divisions of labour, space, and risk. Both sets of debate have become increasingly nuanced in appreciation that, rather than a singular or monolithic construction of hegemonic femininity, femininity is historically specific and mobilized through shifting ideologies of class, race, nation, ability, and nature. Stacy Alaimo's (2000) landmark ecocritical text *Undomesticated Ground* shows how feminist writers have differently imagined possible intersections of femininity and nature to articulate a range of feminist and environmental politics in different periods. A similar appreciation of the range of articulations of masculinity has not yet developed, as eco-feminist critics have tended to focus on how masculinity functions hegemonically as a seemingly neutral or unmarked subject position (see Plumwood 1993; Legler 1998). In *Eco-Man*, Mark Allister (2004) gathers together the first collection of essays on masculinity and environmentalism, but the collection focuses primarily on providing positive images of men in nature, rather than examining constructions of masculinity. As one reviewer has noted, the result is a preference for 'white men in rural nature' (Webler 2006: 120), a tendency that feminist and environmental justice scholars have long decried in ecocriticism for presuming that non-urban and pastoral environments are the 'natural' site for environmental ethics and politics, and that white men are the unmarked subjects who know nature best (see Murphy 1995; Stein 1997; Legler 1998).

The ecocritical association of 'white men' and 'rural nature' is not accidental, but can be traced to the nineteenth century rise of 'passionate manhood' or 'primitive manhood' in American and British culture (Rotundo 1993: 222, 227).[1] Culminating at the turn of the century in U.S. President Theodore Roosevelt's 'doctrine of the strenuous life' through outdoor work, adventure, and big-game hunting (Rotundo 1993, 226), primitivist masculinity involved celebrating the primal and vigorous man as an antidote to mass industrial society and consumer culture. Men were encouraged to cultivate physical strength and bodily prowess,

---

1   See Campbell and Bell (2000) for an introduction to rurality and masculinity; see Mangan and Walvin (1987) for a transatlantic perspective on the rise of Victorian manliness.

and movements such as the Boy Scouts aimed to 'introduce coddled boys to the wilderness' to develop manly skills (Rotundo 1993: 258). Non-urban settings were central to primitivist masculinity as the space where men could be self-reliant and physically active, but also functioned symbolically as the authentic source of masculinity. John Mackenzie (1998: 174–5) describes how Baden-Powell's Boy Scout movement found a model in the British imperial 'big-game hunters' who 'required all the most virile attributes of the imperial male; courage, endurance, individualism, sportsmanship (combining the moral etiquette of the sportsman with both horsemanship and marksmanship), resourcefulness, a mastery of environmental signs and a knowledge of natural history'. In the United States, the popular figure of the cowboy, roaming freely on the western frontier, epitomized this natural masculinity. Once seen as an uncouth 'herder' (Kimmel 1995: 127), the cowboy emerges as a romantic figure by the late nineteenth century, described in a 1902 popular novel as 'a natural nobleman, formed not by civilization and its institutions but the spontaneous influence of the land working on an innate goodness' (see Kimmel 1995: 129).

Michael Kimmel (1995: 116) describes the phenomenon of primitivist masculinity as a retreat from feminization:

> By feminization I refer both to real women, whose feminizing clutches – as teachers, mothers, and Sunday School teachers – were seen as threatening to turn robust [boys] into emasculated pipsqueaks, and also to an increasingly urban and industrial culture, a culture that increasingly denied men the opportunities for manly adventure and a sense of connectedness with their work.

In *American Manhood*, E. Anthony Rotundo (1993: 253) concurs that the masculine passions for wilderness, sport, and adventure are 'inversions of "feminized" Victorian civilization' and relates them to the growing presence and influence of women within public life. Rotundo emphasizes how the ideology of primitive manhood evolved from an increased distinction between genders, and to some extent a gender reversal in conceptions of ethics (see also, Chapman and Hendler 1999). The Victorian era saw a 'growing tendency to look at men as creatures of impulse and passion, even as animals or savages' (Rotundo 1993: 227); women concomitantly come to be seen as the guardians of virtue, civilization, and domesticity, tasked with civilizing men and society. In response, passionate or virile masculinity celebrates the animalized man and disparages 'civilization' as artificial, sentimental, and effeminate. A later figure like Ernest Hemingway, another 'big-game hunter,' is often taken to exemplify virile masculinity in his literary and lived romance of the bullfight and aesthetics of emotional restraint.

Mark Seltzer (1992) points to the importance of changing labour conditions, especially the replacement of male labour by machines, in the development of anxiety about masculine agency and virility. He argues that masculine identities must be read not only in relation to femininity, but also in relation to consumption and the machine. Wilderness experience and outdoor work represent an assertion of the vitality of the human body and emotional self against the cold determinism of the machine, as much as a robust manhood positioned against feminized consumer culture.[2] However, Seltzer cautions that the machine is only ambivalently opposed to the natural male body. It also functions as an exhilarating prosthesis, as augmentation and symbol of male power (160). The contributors to *Eco-Man* present their positive images of men in nature over and against that primitivist or frontier masculinity associated with the mastery and conquest of a feminized nature (see Marx 1964; Kolodny 1975), since this seems to exemplify masculinity as machismo and to equate masculinity with violence and supremacy. But they still share with the primitivist turn the rejection of a deadening machine and consumer culture, especially how 'Many men and boys fantasize about controlling big machines in the service of whacking around and altering the natural world' (Allister 2004, 3; see also Handelman 2004). Scott Slovic (2004), Patrick Murphy (2004), and other contributors to *Eco-Man* therefore emphasize men's caring and nurturing relationships with nature and other people; men are not merely aggressive and destructive, these essays assert, but also emotionally involved and concerned.

Following Nina Beym's (1985) classic essay 'Melodramas of Beset Manhood', recent scholarship on gender in literature tends to emphasize how texts are marked by ambiguity and tension; we should not categorize texts as either libratory or oppressive, but rather seek to appreciate their historically specific articulations of gender, emotion, and ethics (see Howard 1999; Chapman and Hendler 1999).[3] For example, Stephen Clifford (1998) and Thomas Strychacz (2003) re-read gender in the writing of Hemingway and D.H. Lawrence, the two modernist writers most associated with a masculinity based on a vigorous life in nature, to suggest that their constructions of masculinity are less stable and heroic than has been implied. Mary Allen's (1983) analysis of Hemingway's bullfighting texts also shows the complexity of his work for ecocriticism. She suggests that Hemingway's description of the bullfight is articulated in terms of love for the animal: the man shows his respect and admiration by confronting the physical animal.

What recent scholarship on gender and emotion in literature especially emphasizes is that masculinities should no longer be discussed in terms of a retreat from or restraint in emotion, but rather as particular articulations of emotion. As Milette Shamir and Jennifer Travis (2002: 3) relate in *Boys Don't Cry?*, a common-sense equation is that 'emotion is sentiment and sentiment is female'. Neither is accurate. The naturalized connection between women and emotion is akin to (and another version of) the naturalized connection between women and nature that feminists have sought so long to disrupt and historicize. Most significantly, the rise of primitivist masculinity coincides with a shift in notions of emotion. Michael

---

2   Seltzer notes that the standardized production of men through nature experience, ironically, takes its cue from industrial production.

3   Shamir and Travis (2002) similarly criticise Kimmel for presenting too hegemonic an image of masculinity.

Bell (2000: 11) argues that understandings of emotion, and its relationship to truth and ethics, have changed significantly from the eighteenth century 'cult of sentiment' to the present day, where sentimentalism has a derogatory and feminized connotation. A prime example is the derisive reduction of love of nature to the epithet 'tree-hugger' – a label that implies one is naïvely self-indulgent and excessively emotional. Bell (2000: 170) argues that it is important to recognize the attack on sentiment not as a shift away from emotion, but as a shift to a new conception of feeling. He describes the difference between eighteenth and twentieth century notions of feeling or emotion in terms of sincerity versus authenticity, or 'truth of feeling' versus 'truth to feeling' respectively.

The eighteenth century culture of sentiment involved cultivating a mode of personal feeling that coincided with professed principles (Bell 2000: 166): one needed to make feelings true to already existing moral precepts and social conventions. In the shift to truth to feeling, feelings become *a priori*: 'the revisionary reaction of the nineteenth century effected a downward transposition of feeling into the unconscious. Self-consciousness was seen as almost intrinsically insincere ... since there can be no access to unconscious feeling' (Bell 2000: 166). Feelings became 'true' because located prior to thought and reason. They became defined as the passions and appetites that exist beneath or before conscious reflection. To be *true* to feeling means to acknowledge and pursue those inner desires, not to repress them under a cover of social nicety and false kindness (Bell 2000: 189). Virtue therefore lies in authenticity rather than appropriate behaviour.

Bell (2000) explains that because the twentieth century notion that feelings, *qua* preconscious and authentic responses to the world, are genuine is so widely accepted, the notion of sentiment, or crafting feeling, is repugnant and seemingly hypocritical. In her essay 'What is Sentimentality?' June Howard (1999: 65) presents a similar assessment, noting how 'according to the common sense of the modern world, feelings well up naturally inside individuals – tropes of interiority and self-expression are difficult to resist'. Howard (1999) argues that the modern association of emotion with authenticity, and concomitant antipathy towards sentimentality, has been a key barrier to theorizing emotion: emotion is commonly conceived as that which cannot be theorized; it is the opposite of reason (whether deemed an irrational scourge, or seen as an antidote to modernity). By rejecting the dichotomous choice between authentic and inauthentic emotions – or the (female) presence and (male) absence of emotion – we can begin to appreciate the historically specific modes through which emotion is articulated and enacted (Howard 1999: 69), including, I will add, for environmental ethics.

In contemporary Canadian fiction, such as Cohen's *Elizabeth and After* and Richards's *Lives of Short Duration*, the 'big game' hunter becomes a figure not of authenticity but ridicule; he is an example of a man anxious about his masculinity, hanging on to an outdated and misplaced notion of masculine power. Nevertheless, as I will show, this version of masculinity is ridiculed by associating it with social scripts and expectations; in other words, it fails because deemed inauthentic. In their rural novels, Cohen and Richards depict and validate the emotional and ethical relationships some rural men have with animals and place, but their environmental ethics remain based on a celebration of the vitality of the masculine body against prescribed, feminized convention. These literary constructions of masculinity are important for ecocriticism and environmental thought because they highlight the role of subjectivity, not just nature, in ethical paradigms.

Performing Masculinity, Being Animal

The rise of 'passionate manhood' historically coincides with the descent of emotion into the unconscious described by Bell (2000). As Rotundo (1993: 6) describes, over the course of the nineteenth century, 'the body itself became a vital component of manhood: strength, appearance, and athletic skill mattered more than in previous centuries'. In tandem, the reigning conception of morality shifts from externally imposed rules to internally felt truth. Both discourses share the notion of a core or genuine self buried beneath layers of social convention. One effect of this idea is that feelings appear most authentic or true when expressed as actions of the body, rather than expressions in language. As Paul Sheehan (2002) explains, language and narrative are often taken as emblematic of individual expression, but they also represent social conventions and rules. Narrative, especially, represents the self-conscious ordering or plotting of motives and emotions. The subject therefore appears authentic not only when associated with bodily action, but when opposed to scripted performance or narrative. Rotundo (1993: 224) describes how this opposition between body and script was central to the relationship between gender and morality in the late nineteenth century: 'middle-class men saw action – even unthinking action – as manly and viewed 'the etcetera of creed' as a sign of effeminacy'. A doctrinal approach to morality suggested subservience. However, an opposition between authentic subjectivity and narrative creates a challenge for literary texts, where emotion is necessarily expressed in language and narrative.

   Although Hemingway's modernist aesthetic of emotional restraint may offer one response, Sheehan (2002) points to the figure of the animal as playing a significant role in textual negotiations of subjectivity. Animals can represent either the bodily dimension of human life or a subjectivity prior to or outside language and consciousness. On one hand, animals are invoked in modern thought and literature as a contrast to the human. Animals are portrayed and imagined as bodies driven by instinct against which humans appear free-willed, spontaneous, and emotional. On the other hand, animals are imagined as purer forms of the emotional human, not alienated from their bodies and free from social expectations and the constraints of mechanical regularity in work and behaviour. For example, companion animals, such as cats or dogs, are often portrayed as more spontaneous, authentic, and consistent than humans in showing affection. As Sheehan (2002) points out, these contrasting conceptions of the animal rely on an opposition between narrative and subjectivity: when the embodied animal is overdetermined by instinct, the self-conscious human appears free from the constraints of narrative; when the embodied animal appears free from social norms and laws, the self-conscious human is trapped within narrative. Sheehan (2002) suggests that this set of oppositions helps illuminate how some modern fiction relies on the animalization of the human as a means of textually portraying an authentic or spontaneous human subject. A turn to the figure of the animal, and the animalization of the human, can therefore be a productive site from which to examine constructions of masculinity and its role in affective relationships to place, especially given the long-standing association of virile masculinity with wild animals.

Cohen's novel *Elizabeth and After* locates the possibility for an affective relationship to place in the figure of Carl McKelvey. Carl emerges as an ethical subject through a series of oppositions between different performances of masculinity. His father, William, represents the rural patriarch who resorts to a flamboyant display of machismo once he no longer has a role in the rural economy. Sneaking out of the seniors' residence, he steals a white Cadillac from the car lot of the real estate magnate who conned him out of his farm, and he drives it into the lake.

> Five years ago…he couldn't have come back…this way, practically a tourist, and stood looking at what had once been his and before that his father's without feeling loss twisting through his gut like a knife. The only way he could have returned was the way he had: in Luke Richardson's white bomb, with his foot on the floor (Cohen 1999: 358).

William is 'practically a tourist' because he has lost social status and authority with the sale of his property and the end of his working life. The rural man whose livelihood and identity is premised on work-in-nature has little role to play in the leisure economy. William's last act of bravado, with its reliance on the machine, is thus poignantly symbolic of both his powerlessness and his attempt to hang on to the social privileges of rural patriarchy.

Luke Richardson, the real estate dealer, also resorts to displays of machismo; in his case, in order to legitimize his economic domination of the town, which is based on capital not work. Luke's masculinity is performed through mastery of the animal. He considers Carl, his tenant and employee, an animal that he can unleash on his enemies: '"What I've got on him," Luke said, "is that if he steps out of line I'll let you at him. You can bury him alive. Let the birds eat out his eyes. Whatever you want. But not until I say so"' (288). Luke's animal discourse seems to reduce Carl to the level of a vicious dog, foaming at the mouth for a chance to rip apart his prey. But his excessive language reverses their roles. Luke's brutality – his symbolic animalness, or inhumanity – is demonstrated by the way he treats and conceives of animals, whether the birds or Carl, as instruments of violence.

In contrast, Carl's *humanity* is demonstrated when his actions show identification with non-human animals as fellow living creatures: 'When he flushed a pair of grouse from beneath some junipers he didn't even think of shooting. As their wings beat against the ground his heart hammered along' (329). The parallel

Cohen draws between Carl's beating heart and the bird's wings implies they share a common bodily essence as living beings. It is Carl's pre-conscious, bodily recognition of this commonality that differentiates him, on ethical grounds, from Luke. Both men hunt. But Carl '*didn't even think* of shooting' (329, my emphasis). He *spontaneously* demonstrates compassion in response to animals. By contrast, Luke's hunting practices make the animal, like the landscape he packages for the real-estate market, an object to visually consume (282):

> 'There I am with it.' Richardson pointed to a picture of himself standing beside the fallen moose. He was holding his rifle by the barrel, its butt on the ground, and in his fringed leather jacket and wide hat he seemed to think he was some kind of old-time buffalo-hunting pioneer instead of a small-town business man who'd hired a booze-soaked farmer to lead him through some second growth to an animal all past and no future.

The double-barrelled adjectives parody the machismo Luke anxiously tries to assert by mastering the animal. The ironic gap between Luke's pose and his historically specific conditions shows him to be a pale imitation of imperialist hunting safaris (see MacKenzie 1987; Dunaway 2000); the moose is doomed by habitat loss, providing little challenge and sportsmanship in hunting it down. Luke relies on the photo, his 'trophy' moose heads and antlers (282), and boastful stories to obtain the masculinity 'naturally' available to Carl: Luke follows a narrative in order to appear manly; Carl unselfconsciously accepts his own physicality.

This same dynamic is also at work in Richards's novel *Lives of Short Duration*, which, as its title implies, presents a place mired in social, moral, and ecological decay and debasement. The region's forests are logged, the river polluted, and woods knowledge abandoned for the rewards of modernity: fast food, cars, and consumer goods. The post-war generations import and imitate the changing fashions from elsewhere, making no connection between the capitalist system that produces these goods and their own cultural decline and political disenfranchisement. Old Simon Terri, who, barely able to walk, nevertheless walks out of the hospital and goes to die out in the woods, symbolizes the life once lived in careful attunement to the land. Earning money in his later years as a hunting guide for American sportsmen, Simon is witness to a scene of bloody cruelty that echoes Luke Richardson's hollow pose with his trophy moose. Wealthy son of the town's mill owner, Randolf Alewood hires Simon and Daniel Ward, a Micmac guide, to take him moose hunting. After running away on his first encounter with a moose, Alewood returns the next day and belly-shoots a spring moose calf:

> He puts eight bullets into it. And with his hands covered with blood he began laughing – and as Daniel Ward began to cut the calf, with its snout that was ugly enough to be human, and flies in the dry air, he began to laugh too, and Alewood couldn't stay away from it and couldn't stop putting his hands in the calf's blood.

'See the shot I made – that first shot – '

'That was a good shot,' Daniel Ward said. And Simon hated them both and turned away (Richards 1981: 137).

Simon's disgust is not with hunting, but rather with the disrespect the poor etiquette shows the animal. Daniel Ward, as Simon realizes, feeds Alewood's sense of manly mastery for the sake of the alcohol he will provide, but both woodsmen are traumatized by the incident and their participation in it. When sober, Daniel Ward refuses to talk about it; Simon recalls the events as he walks out to Cold Stream, a special place in the woods where he and Daniel once spotted a rare and elusive panther, and where Simon's body is found the next spring. Juxtaposing the calf's death with the treasured memory of the panther affirms Simon's appreciation for animals – the animal more special than the 'ugly…human' (137).

The place of animals as a moral testing ground of authentic masculinity is also evident in Richards's novel *Evening Snow Will Bring Such Peace* (1990). At the close of the novel, the violent and maligned Ivan Basterache sacrifices his life in his efforts to save a horse left trapped in a forest fire by his negligent father, Antony. The act is the culmination of the moral contrast developed through the course of the novel between the abusive and hypocritical father and the compassionate, faithful son. While both fail and hurt the people in their lives, Antony's callous hypocrisy extends to his treatment of animals, whereas Ivan is consistently careful, especially in his hunting and trapping. In his simple altruism, this lowly character stands as the epitome of the masculine subject, as indicated by his epitaph: 'Ivan Basterache / A Man / 1957–1979' (226). At once exceptional yet ordinary, Ivan represents the 'common man' whose humanity is revealed by his kindness to the 'common animal'.

In his book *Corporal Compassion* (2006) Ralph Acampora builds on the philosophy of truth to feeling to argue that the intuitive inclusion of non-human animals within the sphere of moral value is based on identification and recognition of shared *bodiment*[4] rather than the possibility of a shared *subjectivity* that most animal rights debates focus on (and which hinge on the question of self-consciousness). Indeed, he suggests that the ethical recognition of humans is but a narrow subset of this broader sense of appreciation for life, or compassion for other bodied beings. Sheehan (2002) emphasizes that the animal is not only an object of human compassion, but is also embedded within conceptions of the human as a compassionate being. The figure of the animal plays a central role in constructing the pre-discursive subjectivity of the human. Animals 'mediate between the human and the natural realms – between the culture-oriented subject world and the insensate object world, precariously placed between pure instinct and a rudimentary form of consciousness' (Sheehan 2002:33). The animal can therefore appear *both*

---

4 Acampora (2006: xiv–xv; 135–6 n.9) uses 'bodiment' to avoid the distinction between inner self and outer body implicit in the term '*em*bodiment'.

as a lesser being, because seemingly trapped in the mechanical determinism of biological instinct, *and* as a purer form of the human, an authentic being outside the mechanical rigidity of social scripts and expectations. The human as animal – acting spontaneously, from the body – and the human identified with animals are therefore taken to be purer or more moral *humans*. In parallel, the human who degrades or mistreats animals – symbols of innocence because uncontaminated by social or political prejudices – appears as a *brute*, as *in*human.

Men Without Words

Carl and Ivan are examples of what Sheehan calls the 'denarrativized' human subject (2002: 27). Spontaneous and non-rational, their compassion is 'animalized,' in part, because not *scripted* or *articulated*. As a form of truth to feeling, their compassion is set prior to language, even consciousness. Sheehan uses the phrase '*de*-narrativized' rather than 'true' or 'natural' to highlight how it is a form of subjectivity constructed in resistance to the ordering functions of language, narrative, and time. Joe in Richards's novel *Nights Below Station Street* (1988) is another denarrativized male figure. He saves the life of the urbane Vye at the end of the novel despite a debilitating back injury that keeps him unemployed and at home. Joe's comfort and familiarity in the woods places him in a situation to act heroically without doing so self-consciously, as the narrator explicitly emphasizes:

> Then he crossed the brook with Vye on his back, and made his way up through clear cut and slag and moved toward his truck. His back pained only slightly but he did not feel it so much – *not knowing the process of how this had all happened*, only understanding that it was now irrevocable because it had (1988: 225, my emphasis).

Joe's inability to perceive or articulate the code of conduct he follows keeps its essence prior to narration. Moreover, his profound experiential knowledge is contrasted with the effeminate banter of Vye, who talks incessantly and condescendingly, while lacking the sense to dress or drive properly in winter, much less navigate through the woods.

Even when expressed in words, such as when Old Simon proclaims his love for the river, the experience of place is placed outside the realm of narration:

> 'I love the goddamn river', he said.

> 'What's that Papa? – You kids shut up, me and Papa are havin a conversation'....

> 'No matter, no matter', Simon said ... (Richards 1981: 118).

Simon's brief interjection in the gibberish talk of his granddaughter Lois and her children confirm what the narrator has presented about the inarticulateness of his woods experience (94):

> And what could you tell them? That you made 74¢ a day and had to walk 40 miles on snowshoes, and had built camps from cedar and skids with the bow ribs made from roots and had stayed up two months in the woods alone and could smell fourteen different kinds of snow?

The rhetorical question shows the depth of Simon's experiential knowledge – too substantial to be easily relayed and explained – while ultimately demonstrating its tragic irrelevance in the globalized consumer culture that now dominates the river.

Joe and Old Simon have a sense of place for which Kevin Dulse in *The Coming of Winter* (1974) yearns but cannot attain because the woods are no longer intact. The conditions that produce this virile and virtuous masculinity have been lost. Kevin believes the woods provide a space for authentic action where he is free from the mundane repetitiveness of his mill job: 'it was Saturday; he did not wish to think of lifting crates, nor did he wish to think of Sunday when there was never anything to do but wait for Monday's shift' (6). But his alienation from the land is demonstrated in the novel's opening scene when he carelessly shoots a cow while hunting in a small gully, a forest remnant in the midst of agricultural lands. The incident shows Kevin lacks judgment, while establishing him as a sympathetic character because of the bodily compassion he has for the injured animal. The injured cow calls Kevin to account by her repeated 'bellowing and whining' (7). His emotional response to her cries is equally physical; and the emphasis on the shared physicality of Kevin and the cow positions them in an ethical relationship. When he is confronted by a farmer, by contrast, the limited dialogue suggests words entrap rather than reveal the self: '"I told you – ", Kevin began again, his voice sounding like someone else's' (10). Only when conceived as body and action – as emotional utterance – do words appear true to Kevin: 'The man, why hadn't the man yelled?' (12). The woods represent a lost way of being that Kevin cannot articulate, especially to his domineering girlfriend Pamela, whose incessant questions and talk represent the oppressive, civilizing force of 'winter', which comes at the close of the novel in the form of their marriage. The most Kevin can do is retain his dignity by staying outside narrative through silence: 'he said nothing and did not look at her, as if the pride in him refrained from all comment' (152). Like Old Simon's return to silence after Lois fails to hear him, Kevin, in saying nothing, preserves the dignity of authentic masculine action outside the feminine realm of words.

Gender(ed) Narratives

An alignment of masculinity and the body and femininity and language may appear to reverse eco-feminist arguments about the reduction of women to the bodily and natural in hegemonic versions of western history and culture (Merchant 1980; Plumwood 1993). But, in parallel with the contradictory position of the animal, women represent *both* a gateway to biological origins, by association with wildness, animality, sexuality, and in land-as-woman iconography, *and* the civilizing force or narrative that quells the natural male self. Cohen presents the authentic or natural man as a possibility not attained by patriarch Richard Thomas in *The Disinherited* by positioning him between two contrasting versions of femininity: the wild, desirable whore who cannot be domesticated and the civilized, frigid, wife who is responsible for maintaining cultural standards. Katherine Malone is the wild exception to all the wife figures in the novel. Her free-spirited spring play is explicitly contrasted with the ordered home, yard and appearance of Richard's wife, Miranda:

> Katherine coming up to meet him, barefoot and her skirt covered with mud and hay: so much less ordered than Miranda, good-natured body that let itself be pulled out of shape by circumstance. And stands facing her, she is almost as tall as he is, her hand inside his arm, carrying herself unprotected from things, the opposite of Miranda who will go into the garden when it is warmer with her bright kerchief wrapped around her hair and her tennis shoes, lotion on to keep away the bugs and cotton gloves to save her hands from blisters – Miranda knowing which side is which and gradually taking the city to the country with lilac bushes in the front yard and sneaking flowers in among the vegetables (Cohen 1993a: 124).

Miranda's concern with the appearance of her body is replicated in her primarily aesthetic approach to the land, which leaves her always an urban outsider. Katherine, on the other hand, is 'almost as tall as he is' (124); she is Richard's almost-equal since she remains independent of all the urban, cultural pressures that hide and contort women's bodies, like the planted flowers that suppress the unpredictable wildness of the land. Katherine's sensual body stands in for the land itself and Richard's affair with her is the closest he comes to an intimate, non-alienated relationship with nature. She functions as the never fully possessed gateway to the natural and uninhibited for the rational, civilized man, albeit only so long as Richard meets her outside her own domestic duties and place.

The actual attainment of an intimate, unselfconscious relationship between men and land appears in Cohen's novel *The Sweet Second Summer of Kitty Malone*. Having chosen to live in the brief 'sweetness' of the present instead of working the land for some future prosperity, the Frank brothers temporarily achieve a union of body and land from years of familiarity (212):

> Where on the road they had been stumbling and bumping into one another, they now moved quietly, each step balanced and absorbed in the soft earth, every dip or twist in the path, every branch anticipated and known, built into the flow of their motion as if this path, these trees and bushes and the ground they grew out of were part of their own bodies.

Half-drunk, neither brother is aware of this fluid connection between body and land: it merely exists, beneath their conscious awareness. Carl McKelvey seeks a similar assurance of identity in his primal return to the woods at the close of *Elizabeth and After*:

> Everyone else was part of the Empire; whether they knew it or not they were conscript soldiers, life-crushers, mindfucked robots blindly tramping over other life-crushed, mindfucked lives. Not McKelvey…He had forgotten about Fred, Luke, anything that wasn't air, water, earth (Cohen 1999, 333).

These men connect with the land not only by forgetting other men, but also in finding freedom from women. Carl achieves his independence by weaning himself from the memories of his ex-wife and dead mother (the Elizabeth whose haunting importance is evident in the novel's title). The Frank brothers are bachelors until the end of *The Sweet Second Summer of Kitty Malone* when marriage appears to risk emasculation and alienation from land and place. When Kitty admonishes Pat, 'Don't act like a hick,' at a city restaurant, Pat remains silent but inside is 'angry at this woman for trying to lead him away from himself, angry at himself for being so desperate to escape' (Cohen 1993b: 225). Through their speech, the women represent a civilizing force that threatens to make self-conscious the inner self, or 'natural man'.

Marriage in these novels is an overdetermined narrative – a mechanical structure that can imprison men with its expectations and routines. Sheehan (2002) argues that the vitality of the human is not only asserted against the regimented predictability of the machine (as Seltzer [1992] describes), but also the mechanical attributes of narrative; it is not machines per se, but *mechanization* that is taken as a threat to masculinity – and to life. The essentially 'human' – and therefore moral – becomes that which is fallible, spontaneous, and emotional, in opposition to the plotted, predictable, and mass operation of the amoral machine or narrative. Strikingly, Cohen uses organic rather than mechanical imagery to describe the relationship between men, machinery, and the land in *The Sweet Second Summer of Kitty Malone*. Mark Frank's junkyard appears as a garden: 'the bits of chrome and mirrors shone through the air more brightly than any flowers' (Cohen 1993b: 42). The vehicles are humanized, the technology naturalized: 'The warm yellow light turned the welding pit into a comfortable cave. And it made the belly of Randy Blair's truck seem like the insides of a person. Long tubes and shadows implied a vast network of iron arteries and nerves' (45). The junkyard is depicted as a less exploitative form of farming because it can be worked on a 'denarrativized' scale and rhythm (39):

> he was glad to have escaped the backbreaking and endless hours of tending and ploughing. Instead of being faced by animals and manure, looking out his own front door he was able to gaze on a cornucopia of ancient and rusting cars and trucks. They filled the barnyard better than pigs and cows ever had.

The point is made explicit through a contrast with the regulated hours Mark's brother Pat must keep in his job as a mechanic in town: 'They were expecting him at eight o'clock; by noon he would be dead on his feet and counting the hours until closing … his time measured out on the greasy white card they kept for him in the office' (221).

Richards's depiction of the mill in *The Coming of Winter* similarly critiques the deadening effects of mechanical time. Decontamination of the effluent depends on a mechanical regularity impossible under Kevin's working conditions:

> It was his responsibility to pour lime into the yellow water, making it clearer when he did. Four hundred bags a day he was to pour though many days he only managed 250 bags. There were three shifts and he was never found at fault, and also many times he would be called away in the middle of his pouring to do some other task so that always it was at best a very uneven job (Richards 1974: 73).

The passage opposes the 'mechanical time' on which the mill processes are rationalized and the 'biological time' of living beings, human and nonhuman. Kevin is burned by the lime he shovels; the salmon are killed off by the poisonous effluent: 'As long as there's mills on this river, there ain't going to be salmon – no matter how long they close the season' (150). Life is affirmed – *and defined* – in opposition to ordered time and imposed regularity, which are variously associated with women, mechanical operations, and narrative.

Conclusions

Richards's and Cohen's novels show emotional bodiment to be central to constructions of masculine identity, and also to imagining moral lives in rural places. They juxtapose spontaneous responses of the (male) body with mechanical and/or inauthentic scripts of (female) society to affirm the vitality of being alive in a living world. The populist image of virile masculinity thereby becomes the groundwork for an ethical appreciation of place. But, in turn, it is compassion for animals and bodily connection to place that enables these male subjects to garner sympathy and stand as moral exemplars. Outcast, silent, and misunderstood within their own social milieu, these men receive the sympathy of readers *through narratives of the animal*. We can know and care about these men only through the animalization made possible by narrative: the demonstration of animal-to-animal corporeal compassion. The men are the figures through which the reader can appreciate place; the animals the figures through which the reader can find sympathy with the men. Simon's love for the river is exemplary precisely because it is pre-discursive, or what Sheehan more accurately describes as 'denarrativized:' by remaining outside narrative, his love remains pure, spontaneous, and authentic – like an animal.

Narrating ethics through the animal and other categories associated with the natural (such as place, body, and emotion) may appear to be a slippery slope to irrationality and ideological obfuscation. Locating ethics in the emotional responses of the body, especially when they appear to be prior to thought and outside social influence, tends to make the decidedly Christian values of compassion, sacrifice, and duty appear natural or innate, and to similarly 'naturalize' the association of strength, silence, and steadfastness with masculinity. The problem with 'naturalization' of the social is that ideas that privilege groups in power can appear true or inevitable, rather than historically specific patterns that can be changed through political action. Cohen and Richards do situate masculine identities within particular historical and geographical junctures: their plots hinge on how transformations of rural lands and livelihoods disrupt and challenge cultural values and identities. But they juxtapose history against the emotional self, as the novel's orientation to individual experience is bound to do. This tension sits at the heart of the problem of ethics in the modern world: where does one look for moral direction, if not to life itself? Too often the political critique of 'naturalization' results in discounting or evading the role of emotion, embodiment, animals, and places in our moral imaginations. Instead, we might strive to better understand the narrative categories and strategies through which environmental ethics are articulated *and felt* – in lived experience and in reading. To do so, as I have shown here, we need to attend not so much to representations of nature, but to constructions of subjectivity. Subjectivity is an important avenue of analysis not because it is the seemingly natural home or locus of emotion and ethics, but because realist writers such as Cohen and Richards present an indifferent, non-teleological natural world. The ethical sensibility that underwrites their narratives, in other words, is not located in nature, but rather in particular human qualities and relations (including with nonhumans).

References

Acampora, R. (2006), *Corporal Compassion: Animal Ethics and Philosophy of Body* (Pittsburgh: University of Pittsburgh Press).

Alaimo, S. (2000), *Undomesticated Ground: Recasting Nature as Feminist Space* (Ithaca: Cornell University Press).

Allen, M. (1983), *Animals in American Literature* (Urbana: University of Illinois Press).

Allister, M., (ed.) (2004), *Eco-Man: New Perspectives on Masculinity and Nature* (Charlottesville: University of Virginia Press).

— (2004). 'Introduction', in M. Allister (ed.), *Eco-Man: New Perspectives on Masculinity and Nature* (Charlottesville: University of Virginia Press), 1–13.

Baym, N. (1985), 'Melodramas of Beset Manhood', in E. Showalter, *The New Feminist Criticism: Essays on Women, Literature, and Theory* (New York: Pantheon), 63–80.

Bell, M. (2000), *Sentimentalism, Ethics and the Culture of Feeling* (New York: Palgrave).

Chapman, M. and G. Hendler (1999), *Sentimental Men: Masculinity and the Politics of Affect in American Culture* (Berkeley; London: University of California Press).

Clifford, S. P. (1998), *Beyond the heroic 'I': Reading Lawrence, Hemingway, and 'Masculinity'* (Lewisburg, Pa.: Bucknell University Press).

Cohen, M. (1993a/1974), *The Disinherited* (Kingston: Quarry Press).

— (1993b/1979), *The Sweet Second Summer of Kitty Malone* (Toronto: Random House).

— (1999), *Elizabeth and After* (Toronto: Random House).

Dunaway, F. (2000), 'Hunting with the Camera: Nature Photography, Manliness, and Modern Memory, 1890–1930', *Journal of American Studies* 34:2, 207–230.

Handelman, A. (2004), 'Anecdote of the Car: The Diminished Thing', in M. Allister (ed.), *Eco-Man: New Perspectives on Masculinity and Nature* (Charlottesville: University of Virginia Press), 83–97.

Howard, J. (1999), 'What Is Sentimentality?' *American Literary History* 11:1, 63–81.

Kimmel, M. S. (1995), 'Born to Run': Nineteenth-Century Fantasies of Masculine Retreat and Re-creation (or The Historical Rust on Iron John)', in M. S. Kimmel, *The Politics of Manhood: Profeminist Men Respond to the Mythopoetic Men's Movement (and the Mythopoetic Leaders Answer)* (Philadelphia: Temple University Press), 115–150.

Kolodny, A. (1975), *The Lay of the Land: Metaphor as Experience and History in American Life and Letters* (Chapel Hill: North Carolina Press).

Legler, G. (1998), 'Body Politics in American Nature Writing: "Who May Contest for What the Body of Nature Will be?"', in R. Kerridge and N. Sammells, *Writing the Environment: Ecocriticism and Literature* (London: Zed Books), 71–87.

MacKenzie, J. M. (1987), 'The Imperial Pioneer and Hunter and the British Masculine Stereotype', in J. A. Mangan and J. Walvin (eds.), *Manliness and Morality: Middle-class Masculinity in Britain and America, 1800–1940* (Manchester: Manchester University Press), 176–198.

Mangan, J. A. and J. Walvin, Eds. (1987), *Manliness and Morality: Middle-class Masculinity in Britain and America, 1800–1940* (Manchester: Manchester University Press).

Marx, L. (1964), *The Machine in the Garden: Technology and the Pastoral Ideal in America* (New York: Oxford University Press).

Merchant, C. (1980), *The Death of Nature: Women, Ecology and the Scientific Revolution* (San Francisco: Harper Collins).

Murphy, P. (2004), 'Nature Nurturing Fathers in a World Beyond Our Control', in. M. Allister (ed.), *Eco-Man: New Perspectives on Masculinity and Nature* (Charlottesville: University of Virginia Press), 196–210.

—— (1995). *Literature, Nature, and Other: Ecofeminist Critiques* (Albany: State University of New York Press).

Plumwood, V. (1993), *Feminism and the Mastery of Nature* (London: Routledge).

Richards, D. A. (1974), *The Coming of Winter* (Toronto: McClelland and Stewart).

—— (1981), *Lives of Short Duration* (Toronto: McClelland and Stewart).

—— (1988), *Nights Below Station Street* (Toronto: McClelland and Stewart).

—— (1990), *Evening Snow Will Bring Such Peace* (Toronto: McClelland and Stewart).

Rotundo, E. A. (1993), *American Manhood: Transformations in Masculinity from the Revolution to the Modern Era* (New York: HarperCollins).

Shamir, M. and J. Travis, Eds. (2002), *Boys Don't Cry? Rethinking Narratives of Masculinity and Emotion in the U.S.* (New York: Columbia University Press).

Sheehan, P. (2002), *Modernism, Narrative, and Humanism* (Cambridge: Cambridge University Press).

Slovic, S. (2004), 'Taking Care: Toward an Ecomasculinist Literary Criticism?', in M. Allister (ed.), *Eco-Man: New Perspectives on Masculinity and Nature* (Charlottesville: University of Virginia Press), 66–80.

Stein, R. (1997), *Shifting the Ground: American Women Writers' Revisions of Nature, Gender, and Race* (Charlottesville: University Press of Virginia).

Strychacz, T. F. (2003), *Hemingway's Theaters of Masculinity* (Baton Rouge, La.: Louisiana State University Press).

Webler, T. (2006), 'Review of *Eco-Man: New Perspectives on Masculinity and Nature*', *Men and Masculinities* 9:1, 118–120.

PART 5

Enchanting

Chapter 13

Enchanting Data: Body, Voice and Tone in Affective Computing

Frances Dyson

In her essay *A 'Feeling for the Cyborg'* Kathleen Woodward asks the crucial question: 'What is the key to believing that a digital life form (a 'bot', for example) possesses subjectivity?' (2004, 191).[1] As Woodward herself reflects, there is no easy answer here, because the attribution of emotions to computers, robots, and cyborgs, is more fantasy than fact, and its fantastic qualities 'serve as a bridge, an intangible but very real prosthesis, one that helps us connect ourselves to the world we have been inventing' (Ibid.). The key to believing that a 'bot' has emotions is intermingled with the same kind of 'suspension of disbelief' that science fiction films demand of the spectator, and for Woodward this suspension is powerful enough to abandon the perspective of critical theory, and address science fiction as a form of 'future fact'.[2] The co-mingling of fiction and fact, present and future, brought together in the oxymoron 'future fact' calls attention to the kinds of reconciliations that have to occur in order for the concept of emotional cyborgs to resonate with existing, historically formed understandings of machine-human interaction. Other such cultural/epistemic shifts have already broken the ground for the idea of a feeling 'bot': we think of the attribution of being (in Artificial Life), of space (in cyberspace) and of reality (in 'virtual' or 'mixed' reality). However, these sweeping redefinitions of cultural and metaphysical categories lack the frisson that the concept of 'affectionate machines' often provokes. Looking to popular culture, we see that the difference between having and simulating emotions, the

---

1   As Woodward reflects: 'The rhetoric of the attribution to and instantiation of emotions in the lifeworld of computers, replicants, and cyborgs, bots and robots, a lifeworld that extends to ours – indeed is ours – serves as ... a coupling device. ... thus the emotions as thematized in science fiction ... and the emotions as they are experienced in our technological habitat ... serve as a bridge, an intangible but very real prosthesis, one that helps us connect ourselves to the world we have been inventing. ... This rhetoric allows us to behave 'as if their emotions are real, as our science fiction insists that they are' (2004, 192–193).

2   Citing MIT researcher Rodney Brooks' belief in the eventuality of 'emotion based intelligent systems', Woodward concludes that the relationships humans form with their machines 'where the process of technocultural feedback loops generates emotional connections' is an example not simply of emotional parity between human and machine, but of the more fundamental 'principal and process of emergence' (Ibid., 191).

difference, in essence, between human and machine, is established, over and over again, within narratives about technology. The film *Blade Runner* epitomizes this cultural obsession with defining and re-defining what being human is through an appeal to the presence, or absence, of 'real' emotions. We think here of the scene where Rachael, a replicant destined to be 'retired' before her emotions can develop, takes the Voight Kampf test – during which her pupils are scanned for signs of a physical response to the emotionally charged questions she is asked. Or Spielberg's *Artificial Intelligence*, where the obsessive life quest of the robot child is to make his adoptive mother love him. There are numerous examples of humans outwitting machines because of their superior emotional integrity, and this quality has acted as a barrier against some of the more technophobic trends in modernity – especially the fear of becoming automatons, or of becoming redundant through automation. But looking outside the sphere of science fiction, to the insights and fantasies of art, and the field of Human Computer Interaction (HCI) that focuses on the computation of affect, I want to re-pose Woodward's question in broader terms, to ask not if a 'bot' can feel, but what it would feel, and what kind of empathetic relationship would be possible between it and us? Specifically, I will read Catherine Richards' *Method and Apparatus for Finding Love*, and Ben Rubin and Mark Hanson's *Listening Post* – both pieces that employ 'bots' to establish relationships between technological systems and humans, and both instances of art creatively unsettling some of the basic assumptions of science.

For anyone unfamiliar with the term, affective computing is generally defined as the identification and measurement of a person's emotional state by tracking bodily indicators such as heart rate, galvanic skin response (sweat), breathing rate, gait, facial expressions, etc. The operating assumption here is that changes in emotional states – what principle researcher Rosalind Picard refers to as 'sentic modulations' – are reflected by certain physiological patterns (for instance, skin conductivity tends to climb when a piece of music is energizing and falls when it is calming), that allow pattern recognition technologies to detect, for instance, user frustration or stress (1997, 23–25). Embedded in specially designed devices, or common accessories like the PDA, these systems provide a material conduit between the present realities of HCI, and future fantasies of feeling machines, thus enabling the cyborgian futurology of media culture to descend from the heights of fiction to the practicalities of everyday life. Wireless, wearable and spatially deployed, able to identify and assess the users emotional state by monitoring their bodily effusions, such devices are seen, by their very attachment, to become 'embodied', making their responses far more complex than say, the rising and falling blips of a heart rate monitor. Based not simply on somatic data, but on a complex, and potentially oneristic, feedback loops, we might wonder what kind of human-machine interactions these feeling gadgets might produce. Such questions are sardonically raised in Richard's *Method and Apparatus for Finding Love*, a purely conceptual piece consisting entirely of a patent proposal for a set of devices that are designed to initiate contact and develop a romantic relationship amongst users.[3] The devices communicate the whereabouts, habits, desires and psycho-physical state of their wearers, exchanging the kind of information that was once undertaken by family and social networks but now exists in the limbo of nomadic, fragmented, postmillennial culture. In collaboration with Martin Snelgrove, Richards designed and wrote the patent, citing its value as 'an apparatus … which, carried by or embedded in a lonely or socially inept individual, communicates with like device in such a way as to divine the likelihood of attraction due to relative sexual, social, intellectual or spiritual interests of the bearers' (Richards and Snelgrove, 2002, 1; hereafter cited as R&S). With the dryness of a patent and the profit criteria of classic economic analysis, the piece markets itself on the basis of its usefulness and efficiency. For instance, the amount of time and emotion devoted to a potential relationship is weighed against the likelihood of that relationship succeeding – creating a risk factor that can be used to calculate the amount of time and energy that should be invested. Such calibrations transform emotional relationships into the equivalent of a stock, subject to brokerage, speculation, and exchange, that fits well within the 'influence of modern marketing' and the decreased possibility for random interaction due to 'behaviors such as living alone, watching television, surfing the net, playing videogames, telebanking, single-parenting, driving alone, telecommuting and eating at drive-through restaurants' (R&S, 17). Further, in a subtle commentary on the dissimulation inherent in technoculture, the piece acknowledges the prevalence of deceit in romantic negotiations: 'Deceptive behaviors occur because progress to courting behavior may be valuable for reasons other than the pursuit of love, such as short-term sexual gratification, power, status, revenge, winning a bet, access to jewelry, art, income streams or other valuable property' (R&S, 18). Even if no property and little time are lost, they cause 'false positives' early in courtship, which waste emotional investment (R&S, 18).

The minor contradictions and odd juxtapositions involved in the economics of romantic love, the simple line drawings that represent the devices and the quaint, at times brutally frank discussions of the characteristics of the target audience, create an environment that accommodates wildly divergent cultural tropes and discursive styles, to some extent reflecting the variance within new media discourse itself. By introducing Greek mythology at the beginning of the patent (Figure 13.1),[4] Richards invokes both a sense of absurdity, and a subtle reference to the ways of knowing – the epistemology – of ancient times, where love was a result of Cupid's arrow, and the arrow was always double edged:

---

3   This piece is a component of the three part exhibition *Excitable Tissues*, exhibited at The Ottawa Arts Gallery, 2004.

4   Cited as Figure 1 in the Patent Draft for Method and Apparatus for Finding Love.

Figure 13.1   The composite of many characters

Source: Catherine Richards © CARCC 2008, with kind permission.

Note: In the embodiment of Figure 1 subject 84a Eros caresses subject 84b Aphrodite includeing direct mechanical activation of a simple pressure switch in nipple device 97, which communicates over wireless link 102a to embedded device 99 in subject 84c Jealousy. The effect is amplified by a further interaction in which wireless link 102b from device 92a held by subject 84dFolly to peach device 92b held by said subject 84b Aphrodite causes it to secrete a sweet-smelling liquid, distracting said subject 84a Eros from the operation of arrow device 92c held by said subject 84b Aphrodite which can be violently but partially embedded in the body of either said subject 84a Eros or said subject 84c Jealousy in order that its actuator 124a may inject an aphrodisiac such as Viagra, hallucinogen such as lysergic acid, muscle relaxant such as curare or other chemical or plural chemicals such as Rohypnol known in the art of seduction. The choice of chemical may be modified by interactions with external parties, such as subject 84e Deceit who carries a handheld device 92d connected by wired link 103 to headband device 92e which is more conveniently placed to make wireless link 102c operable to said arrow device 92c. By means of sensor 123a operable to detect pheromones said handheld device 92d selects which type or combination of chemicals can best be injected to make the behaviour of injected subject 84a Eros or injected subject 84c Jealousy more desirable to both subject 84e Deceit and subject 84b Aphrodite. Other nearby subjects, such as subject 84f Time may have only indirect interactions with the principal emotional drama engulfing subjects 84a through e.

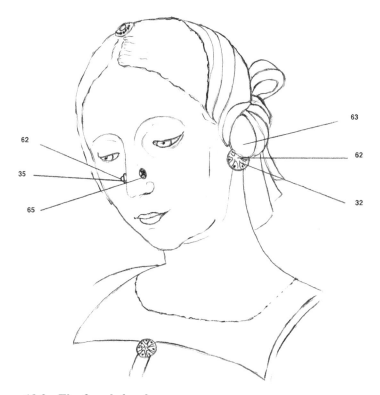

Figure 13.2   The female head

Source: Catherine Richards © CARCC 2008, with kind permission.

Note: Figure 13.2 shows an embodiment of the device in an earring, with sensor operable to detect heart rate and actuator operable to detect heart rate and actuator to whisper message; together with an embodiment of the device in a nose stud, with sensor operable to detect breathing rate and actuator operable to dispense pheromones in order to subliminally excite the subject.

In Figure 1 subject 23a Eros caresses subject 23b Aphrodite including direct mechanical activation of a simple pressure switch in nipple device 36, which communicates over wireless link to 41a to embedded device 38 in subject 23c jealousy. In order that its actuator 63a may inject an aphrodisiac such as Viagra, hallucinogen such as lysergic acid, muscle relaxant such as curare or other chemical or plural chemicals such as Rohyponol known in the art of seduction (R&S, 21).

The evacuation of affect from the language of *Method* evident in the above quote is not simply to satisfy the protocols of the patent office. For Richards, devices like Lovegetty,[5] and services like www.match.com indicate a rise in 'the shopping model of romantic love, contradict(ing) the deeply held romantic model that the culture has which is – "gasp, I walked in the room and I was just struck"'.[6] While the patent eschews the shopping model in favour of the active creation of desire,[7] it articulates the deep cultural disconnect between love and shopping, romance and the bottom line, soul mates and gadgets, by creating a hermeneutic that uses elements of Greek mythology, the iconography present in Renaissance paintings, and references to icons of modernist art such as Duchamp's Large Glass. Noting that, in classical iconography, objects are like 'devices that you have to know how to read',[8] *Method* moves from the symbols of art to the object signs of consumer culture, to the manufacture of affect, using 'the language and the imagery of the patent [to] evacuate what it [*Method*] is pointing at – the interrelationships between people'.[9] This occurs at the level of objects and devices, which, as they become computerized – as they collect more details about their users, also increase their symbolic capital, in ways similar to Renaissance portraits. But where figure drawings unite objects and people in a cosmology that places romantic love within a rich cultural tapestry of fate, desire and sociality, *Method* creates transfers between devices and bodies that are the hallmark of popular fiction, represented by the figure of the cyborg. *Method*'s epistemology then, is based not on the quixotic temperaments of the gods, but on the atomized workings of circuits, transmissions, triggers, monitors, information flows and chemical/neurological and physical reactions. These create an epistemic system

---

5 Lovegetty is a portable wireless device that transmits signals to other devices within a certain range. By making owners aware of other Lovegetty owners, within their vicinity the device aims to establish friendships.

6 Author's discussion with Richards, Daniel Langlois Foundation, Montreal, 2004.

7 Richards points out that previous inventions such as Lovegetty are 'too involved in self-deception, too open to manipulation, to ready to purchase the "front" that will not lead to "lasting" love' (Ibid.).

8 Referring to Figure 1, which illustrates the interrelationships between people, objects, passions and gods, Richard comments that 'you have to know what represents deceit and jealousy, and that the apple is both sweet and dangerous' (Ibid.).

9 Ibid.

considered superior to the often 'incorrect' signals of the body – to which the individual either no longer has access, or requires an instrument that will interpret such communications, transmissions, or what was once quaintly called 'messages from the heart', correctly. Operating on a method-ology of distrust, the patent also emphasizes a need for devices independent of the conscious biases of their wearer. As a result, the devices use physiological signs to measure and signify emotional states, on the assumption that the user has less control of these signifiers – in contrast to appearance, or habits etc., which can be manufactured: 'This apparatus or software method must be subvert (in the sense of hiding knowledge of the underlying patterns of desire from all participants)' (R&S, 19). In order for it to be successful in 'finding love', it must also use means that 'detect patterns of behaviour inductively and modify them' (R&S, 20). Thus the figures provided show 'embodiments' of the device in earrings, nose studs, chest belts, nipple clips, and extend – to absurdity – some of the ideas and technologies within affective computing already discussed.

Figures showing body piercing – often related to the idea of the posthuman body penetrated by foreign objects – also detail the introduction of various behaviour-altering, love-inducing chemicals (such as pharmaceuticals, pheromones, artificial hormones, psychoactive chemicals) as the device interprets physiological signals and, after certain criteria have been met, responds accordingly. 'Actuators' connected to the 'embodiments' will, for instance, 'dispense pheromones in order to subliminally excite the subject' (R&S, 20). Thus the apparatus shifts from a technology that creates, rather than simply records or monitors, the romantic environment. In so doing, it moves from being inert and passive – a wireless, multisystem surveillance system, to a technology that manifest some of the qualities of artificial life. For instance, an 'applet management system' distributes RAM resources 'according to their consumption and success at prediction. … (Evolutionary) algorithms rank applets and distribute resources or cuts in resources' or 'download applets from new acquaintances so that successful ones can propagate' (R&S, 23). But in contrast to the generally quaint and benign activities of a-life creatures or agents, the devices in *Method*, together with the ambiguity of the figures, suggest a far more sinister operation. The famous Michelangelo, appearing at first to be in a state of ecstasy, is fitted with a 'breast device' (Figure 13.2),[10] 'wherein the sensor 62 is operable to detect heart rate and breathing rate and depth, and the actuator 63 is operable to constrict the chest in patterns that simulate rapid breathing or that temporarily restrict breathing altogether' (R&S, 20). Similarly, a nipple device not only detects 'changes in state of erectile tissue' [but will also] 'apply a mild electric shock' (R&S, 20). Electric shocks and suffocation – often associated with instruments of torture, suggest that the fine line between assistance and coercion may be breached, as the devices do whatever they can to establish not necessarily a romantic relationship, but the kind of data profiles necessary to simulate success and thereby ensure their 'viral' propagation.

---

10   Cited as Figure 11 in the Patent Draft for Method and Apparatus for Finding Love.

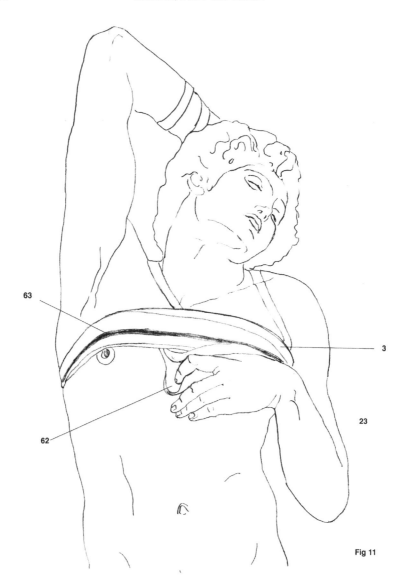

Figure 13.3   Three-quarter view of the man with arms overhead

Source: Catherine Richards © CARCC 2008, with kind permission.

Note: shows an embodiment of the device in a chest belt, with sensors operable to detect heart rate and breathing rate and depth, and actuator operable to constrict chest.

Dissimulation

The problem of deceit and dissimulation on the part of the user identified in *Method*, is also present in the technologies themselves. So-called 'embodied' technologies rely on what are essentially artificial perceptual systems – remote sensing, speech analysis, physiological pattern recognition etc., to collect data, and artificial intelligence systems, to analyze and represent it. At both ends of the perception/knowledge loop then, there are the mediating influences of computing, with its quantitative interpretative methods and modelling limitations, based as they are on the belief in the computation of mind and matter. This kind of solipsism operates on a number of levels: deployed on the body, in the public environment, via the Internet etc., dataveillance systems and 'reality mining' can only glean relevant information about individuals using collection and analytic methods that are already modelled on what researchers believe might be significant indicators of, for instance, emotional states. As Herbert Dreyfus has pointed out, the amount of information collected by wearable devices, pale in comparison to that stored already in the body; therefore, 'all the zillions of facts that are relevant to the body and that the body knows – or rather the subject who has a body knows, can't be accessed' (2001, 18). These systems lack not just what the body knows, but also access to the repositories of affective expression that are constantly being repressed, or manufactured in social situations – a problem that MIT researchers working in the field of affective computing and 'e-rationality' recognize. Picard notes the difficulty in finding the 'real emotions'. Not only do emotions vacillate: joy and anger 'can have different interpretations across individuals within the same culture...' (2001, 1179), but finding the true emotion is almost impossible when the subject knows he or she is being monitored or is part of an experiment.[11] With a new iteration Heisenberg's uncertainty principle, Picard repeats the dictum 'measurement of ground truth disturbs the state of that truth' (2001, 1178).

Acquired Autism

If the quest for true emotion risks encouraging the simulation of emotions rather than their direct experience, affective computing itself, as a field of research, risks interposing the same ambivalent dynamics of belief, riddled with negativity, fantasy and desire, that the attribution of emotions to machines elicits. For if it is

---

11 As Picard notes: 'The most natural setup for gathering genuine emotions is opportunistic: The subject's emotion occurs as a consequence of personally significant circumstances (event-elicited); it occurs while they are in some natural location for them (real-world); the subject feels the emotion internally (feeling); subject behaviour, including expression, is not influenced by knowledge of being in an experiment or being recorded (hidden-recording, other purpose). Such data sets are usually impossible to get because of privacy and ethics concerns' (2001, 1179).

enough that computers mimic rather than experience emotions, perhaps this holds true for humans as well? What would be the consequences of such widespread simulation? What kind of subjectivity would develop as users – that is, anyone who travels, banks, phones, uses the Internet, or occupies public spaces – become increasingly cognizant of their electronic Doppelganger? Will we, for instance, assume the posture of the not guilty as we pass through airport security, training our bodies to fit a normative profile that we know responds adversely to, say, 'over the limit gesticulations' or an 'overly' loud voice? Could the knowledge that every emotional indicator is being monitored induce a kind of social and biological autism? Or would the feedback loops generated between the device and its wearer instead create a sophisticated repertoire of postures designed solely for the camera, while de-skilling, through under-use, spontaneous emotional expression when the camera isn't there. Picard and Klein raise the issue of autism in relation to on-line communication:

> [I]n a sense, everyone is autistic online: today's systems limit your ability to see facial expressions, hear tone of voice, and sense those non-verbal gestures and behaviours that might otherwise help you disambiguate a hastily sent non-angry message from a genuinely angry one. To the extent that people spend more time communicating with each other without sufficient affect channels, they may actually be reducing some of their emotional skills – a kind of 'use it or lose it' opportunity cost (2002, 14).

Autism figures within the literature as both a (technological) impediment to full communication, and a (human) disability that technology might aid. Thus researchers at MIT's Affective Computing Group are developing ESP, The Emotional-Social Intelligence Prosthesis, which is a wearable system that augments the wearer's 'emotional-social intelligence' by sensing, in real time, the wearer's non verbal cues (via the computational modelling of head and facial displays) and then communicating these inferences to the wearer via visual, sound, and tactile feedback. Other projects (for instance, Conductive Chat) use skin conductivity (aka galvanic skin response) to register emotions in, for instance, text-based interpersonal communication such as instant messaging, providing an emotional 'body' track, if you like, along with the writer's text.[12] In many of these

---

12   The researchers recognize that 'Without the benefit of eye contact and other non-verbal "back-channel feedback", text-based chat users frequently resort to typing "emoticons" and extraneous punctuation in an attempt to incorporate contextual affect information in the text communication. [By integrating] users' changing skin conductivity levels (used as a measure of emotional arousal, and high levels are correlated with cognitive states such as high stress, excitement, and attentiveness), EmoteMail does much the same thing, registering the writer's facial expressions and typing speed during the composition of email to indicate emotional states and help the recipient "decode the tone of the email"' MIT Media Lab: Affective Computing Group, <http://affect.media.mit.edu/projects.php>.

projects the underlying premise is that all communications technologies and some people (e.g. those diagnosed with autism) lack what the researches refer to as 'mind-reading' functions, that is, the ability to read the mind through the facial expressions and vocal tone that accompany most human-to-human exchange. In this respect, humans and machines meet on the common ground of a disability that can be addressed technologically only by disassembling the circuit between emotion, expression, representation, and language. Alex Pentland describes the working methodology as 'disambiguating the affect of the speaker without knowledge of the textual component of the linguistic message' implying that meaning interferes with the transmission of emotion, and somehow needs to be put aside, or ignored.[13] Enhancing affective bandwidth through various devices is seen as a way to mitigate what Picard and Klien describe as a general societal lack of 'solid, effective, non-judgmental active listening skills' (2002, 15) and has spawned projects such as Pentland and Nathan Eagle's (2003, 1) *The Relationship Barometer: Mobile Phone Therapy for Couples*, a project that would extract salient features – such as speaking rate, volume, duration, transitions, interruptions etc. from couples conversations, to determine which party is dominating the conversation, which party is doing the listening – and intervene accordingly.[14]

These kinds of interventions and enhancements all involve various degrees of what I am calling 'acquired autism' – a condition that is both generated and remedied within a technologically driven psycho-social environment. Autism – seen as a deficiency, is treated in a way that is similar to deafness, by amplifying, like a hearing aid, the signals that one's emotional intelligence would normally pick up. Yet unlike a hearing aid, these 'emotional aids' are developed to address both a human and technological dis-ability, a technology for technology if you like. The linking of biological and machinic autism is extremely relevant because looking back to the history of telecommunications we find the peculiarly telephonic protocol of asking 'are you there?' or 'who is speaking?' resurfacing, as the question of who, or rather what, is listening, becomes a central concern. The irony here is that telephony has already produced a mediated voice, what could be called the 'telephonic' voice, which has learned to accommodate the effects of transmission. Initially because the static of transmission made every telephone call something of an aural adventure, and currently because the vagaries of network cover place every

---

13   See MIT Media Lab: Affective Computing Group, <http://affect.media.mit.edu/projects.php>

14   Collected over a period of time, the researchers believe that the results would not only yield information about the health of a couple's relationship but potentially intervene in the conversation by 'strengthening the positive features while dampening aspects that have a high probability of propagating a negative interaction. ...[For instance] using such features as volume and pitch, automated mediation techniques [would] help transfer the floor to a participant who is being dominated in a conversation' Eagle and Pentland (2003, 1). See also Celeste Biever's article 'Cellphones turn into smart personal assistants', *New Scientist*, 27 November 2004.

call on the edge of failure, users have learned to filter out hesitation, long pauses, low voices, and multi-syllabic answers. The enormous weight of the transmitted voice, its recuperation of presence against all odds, has balanced these two main eras of telepresence, reconciling the old (telephony) and the new (wireless) in that brief, urgent and now iconic question 'can you hear me now?'. But what was once a quest for clear transmission is now overshadowed by the possibilities that a new kind of 'listening' presents: a listening not to sound, nor to speech, but rather to, or for, an elusive emotional content. Registered through acoustic parameters (such as volume and rhythm) or conversational patterns (such as the dominance of one speaker, the frequency of interruptions etc.) affective content becomes an object of therapeutic interrogation and intervention. The tone of the voice, that particular vocality which represents and carries affect, enters this problematic carrying with the fore- knowledge that it may not be speaking to another human, and that its speech may be monitored by machines that cannot hear.

While the presence and quality of tone might be highly charged, indeed it might carry an enormous amount of meaning, it is also almost impossible to characterize just what exactly this tone is. This presents a substantial challenge for the Affective Computing Group, which is attempting to identify, measure and use tone as a tool in, for instance, predicting the outcomes of social interactions or the dynamics of work place relations in the virtual office.[15] MIT's ongoing research project oriented towards the voice involves building 'models for automatic detection of affect in speech'[16] based on an extensive database of natural speech showing a range of affective variability (Madan, Caneel and Pentland 2004, npn). Instead of using human actors, which the group believes will often produce unnatural 'performed' expression, the raw data – the 'natural speech' – is collected from volunteers who navigate an interface guided by 'embodied conversational agents', defined by Bickmore and Cassells as 'animated anthropomorphic interface agents that are able to engage a user in real-time, multimodal dialogue, using speech, gesture, gaze, posture, intonation, and other verbal and non-verbal behaviors to emulate the experience of face-to-face interaction' (Cassell cited in Bickmore and Cassell 2000, 2). These agents prompt the user to speak about an emotional experience, and by so doing, contribute to the database. Once built, researchers then examine acoustic parameters related to voice quality, intonation, loudness and rhythm, for particular patterns that target the affect in speech rather than the meaning of the words spoken. Although, as Pentland claims, it is possible to construct measures for different types of 'vocal social signalling' that can be computed in real-time using only a PDA, and can successfully predict a speaker's level of engagement with, and even attraction to, another speaker, it would seem that the unique

---

15   According to wearable computing pioneer Pentland: '… [t]one of voice and prosodic style are among the most powerful of these social signals even though (and perhaps because) people are usually unaware of them' (cited in Madan, Caneel and Pentland 2004).

16   See 'Highlighted Projects' online at <http://affect.media.mit.edu/projects.php> Last accessed 8 June 2008.

intersection of sound, speech and music that all cohere in vocal 'tone' resists such measurement in all but the most regulated conversational environments, such as the speed dating sessions that Pentland has used in his experiments (2004, 1).

Dissimulation reappears in this context as an even more tightly bound contradiction. Installed in the workplace, for instance, modelling affect in speech may produce a 'performed' autism that is itself a reaction, or adaptation, to the various prostheses devised to counter the technological autism inherent in machines that have no senses. If any of these technologies are to have a general rather than individual application, the database must be enormous, yet the garnering of affective instances through constant monitoring means that the dissimulation provoked by surveillance will itself block any access to the truth of the subjects text, speech or what ultimately becomes their performance. For Picard a central concern is the voluntary engagement of the user, who should be aware that their speech is being monitored. Yet this is more than simply ethical issue, for it begs the question of how aware we are, and how much control we have over, our vocal tone. We should remember that just as a voice might express affect, it can equally be 'affected', marking the voice as an instrument. Indeed, if tone is registered by acoustic parameters that are independent of, or rather, run as a lyrical or musical correlate to what is spoken, as such they cease to be strictly acoustic, but are better understood through the musical meaning of tone and the performative meaning of intonation. Tone is both more than the sound of speech and less than the meaning of the words spoken. Like physical gesture, it shares with music and sound an ambiguous status, an ill-defined meaning. Tone can slip into the sound of speech, just as the sound of speech can disassemble into babble or accelerate into fury, rage or an almost silent, speechless hiss. Even prior to technologies of recording (the tape recorder), transmission (such as the telephone), or analysis, (voice recognition software), the voice acted as a chimerical transducer. As Milan Dolar describes: '[the voice] makes the utterance possible, but it disappears in it, it goes up in smoke in the meaning being produced' (2006, 15). To this vanishing act, or becoming invisible, must be added a second: sound itself, which disappears in the necessary distinction between voice and sound, the distinction that the listener must make in order to hear 'a voice'.

The voice – perhaps the most duplicitous of bodily signifying systems, always seems to offer a truth that it cannot deliver. As Derrida and others have demonstrated, (and does not need to be repeated here) the voice is not absolute, originary, nor even authentic, but is itself an instrument, and instrumental in establishing the self-presence metaphysics relies on. There is no true, recoverable, a-signifying voice – but more importantly from the era of telephony we have a voice that is already highly modulated, as Derrida remarks '(the telephone) is not an external element of the context; it affects the inside of meaning in the most elementary sense' (1992, 267, my emphasis).[17] In the era of affective computing, there are

---

17 This inside, is also, as Dolar points out, a constitutive illusion: 'If metaphysics ... is carried by the propensity to disavow the part of alterity, the trace of the other, to hold onto

now no ears that could hear this 'true' voice, for the technologies that filter, tag, analyze, and codify the voice, retain it as data, but do not hear it as voice or sound. One has to ask what happens when the voice is treated like data and mined for affect? Doesn't the issue of hearing become mute, because to some extent, it is no longer the voice that speaks nor the ear that hears, but an intermediary taking the form of a device, occupying a database, and 'listening' through systems that filter an already enigmatic and ineffable tone from an already technologized, instrumentalized speech. Within such a configuration, one wonders how affect would arise, because the voice, both physically and metaphysically, assumes an ear – both the ear of the speaker and the ear of the other. The reciprocity, or dual movement between voice and ear, speaker and listener, is both individually and socially enacted, simultaneously, when one speaks. What happens then, when one half of the equation is missing?

Or, returning to MIT's Recognizing Affect in Speech project, when the other half is a computer, how can once ever be sure that the 'affect' registered and represented by a computer (or 'embodied conversational agent') will be any closer to the 'true' or 'natural' expression of that emotion that the affect dictated by an actor or performed by an imposter. Again, the history of telecommunications and the chimerical quality of the voice return to haunt attempts at clear and untroubled transmission, as Bickmore and Cassell's 'REA' shows only too well. REA is an 'an embodied conversational agent whose verbal and nonverbal behaviours are designed in terms of conversational functions' (Bickmore et al 1999, 1).[18] REA has a female figure, and is capable of speech with intonation, and facial and bodily gestures. Her image, projected onto a screen, interacts with the user, responding to their non-verbal cues (sensed via cameras track the head and hand positions in space) and to their voice and manner of speaking (input via microphones attached to the user's clothing). REA can be interrupted, knows how to take turns in a conversation, and moves from social chat ('phatic communication') to task oriented dialogue when the users non verbal cues reach a certain level of 'depth' and 'closeness'. As the researchers note, the most important use of non verbal behaviour in social dialogue is the display of interpersonal attitude (positive or negative) through what they refer to as 'immediacy behaviours' (such as having a

---

some ultimate signified against the disruptive play of differences, to maintain the purity of the origin against supplementarity, then it can do so only by clinging to the privilege of the voice as a source of an originary self presence. The divide between the interior and the exterior, the model of all other metaphysical divides, derives from here. ... This illusion—the illusion par excellence—is thus constitutive of interiority and ultimately consciousness, the self and autonomy' (2006, 15).

18   Bickmore and Cassell describe ECA's as having 'the capability of producing and recognizing non verbal clues in simulations of face-to-face interactions'. The simulations involve users in a real-time, multi-modal conversation with a graphical animation (or 'animated anthropomorphic interface') 'using speech, gesture, gaze, posture, intonation, and other verbal and non-verbal behaviours to emulate the experience of face-to-face interaction' (Cassell cited in Bickmore and Cassell 2005, 2, 5).

close proximity to the conversant, a direct gaze, a smiling face, etc) and 'vocalic behaviours such as more variation in pitch, amplitude, duration and tempo, reinforcing interjections such as "uh-huh" and "mm-hmmm", greater fluency, warmth, pleasantness, expressiveness, and clarity and smoother turn-taking' (Bickmore and Cassell 2000, 9).

REA, and the experiment that Bickmore and Cassells undertook, is mainly designed to assess the usefulness of 'embodied' interactions – where 'embodied' is understood in terms of having a graphical display, and creating an animation that responds primarily to gesture, facial animation, turns of the head, etc. The findings however, have been mixed; in fact, Bickmore and Cassells concluded that sometimes, and with some people, talking on the phone to an agent elicited greater trust than interacting with REA (2000, 21). This supports other findings that speech to a simulated computer system tended to be 'telegraphic and formal, approximating a command language' that people speak more carefully and less naturally, literally measuring their words (Ibid., 11). Without going into the details of this research or its results, it is interesting to note that REA introduces chat, small talk – the thing that establishes trust – by contextualizing the exchange within a technological frame, initially, and most immediately, by referring to the simulation of her own voice and the amplification of the user's. REA begins the conversation with 'that microphone is terrible, I hate using those things'. This opening comment is followed by, 'Sorry about my voice, this is some engineer's idea of natural sounding', and then 'Are you one of our sponsors?'. If the user answers 'yes', REA asks if they were at the last sponsors' meeting and then refers to her voice once again, 'I got so exhausted at the last sponsor meeting I think I was starting to lose my voice by the end'. Presumably having now established 'closeness' with the user, REA moves onto business, 'where would you like to live?'.

A number of themes are apparent here: note that the first means of eliciting affect (warmth) is by referencing the mediation of the user's voice. The discomfort that many people experience at having their voice amplified is recognized by REA, who also hates having her voice amplified, as if her voice, and the user's voice, are the same (they are either both human or both synthetic). REA then apologizes for her voice, indicating to the user that she realizes she is a machine, blaming the 'engineer' – the classic unfeeling, generally male, geek who has an 'idea' of natural sounding voice, but in typical engineer style, that idea has no relation to the actual 'natural' sound that the computer system, REA, understands. So human machine bonding is established via the figure of the engineer, the microphone, and the absence of a 'true' voice. But as REA moves between co-conspirator, confessor and computer 'she' also dis-integrates any direct relationship between the user and the affective expression. With an engineer lurking in the background, this Embodied Conversational Agent (ECA) disavows its agency, its 'voice' as a simulation, entering into the dynamics of dissimulation that *Method* elaborates. Framed already by the technologies of the voice – both old (the microphone) and new (the voice simulator), we discover again multiple layers of adaptations, modifications, representations and simulations create an absurdly implosive, unstable techno-epistemic system – as unearthed as the sonorous and phonic ground it presupposes.

Listening Post

If the truth of the emotions is the very quality that has defined, culturally at least, what being human is, in this respect the search for the 'true' voice – the voice that delivers, unmediated and unperformed, the 'affect' of the subject – verges on the uncanny. Because as we approach or inhabit the posthuman, the affect carried by tone signals more than anything else that a human rather than computerized voice is speaking. In this respect, it is interesting that Ben Rubin and Mark Hansen's *Listening Post* uses a synthesized, computerized voice to 'humanize' the somewhat alien medium of text-based conversations on the Internet. Here, instead of transforming the sounding voice into data, data (the text typed during online exchanges on the internet) is transformed into voice. Instead of computing actual vocal tone as a means of gauging a speakers emotional state, tone is synthetically produced as a way of igniting the cybernetic imagination through an aestheticization of Internet 'chat' and a reframing of information systems that track internet chatter. Both renderings of the voice and its tone – the computational and the aesthetic – impinge upon and renegotiate one of the central questions of posthumanism, one that we see constantly worked over in science, art and culture, namely whether, and how, there could ever be machine subjectivity?

Rubin and Hansen introduce their piece in their article *Listening Post: Giving Voice to Online Communication* with the following:

> The advent of online communication has created a vast landscape of new spaces for public discourse: chat rooms, bulletin boards, and scores of other public on-line forums. While these spaces are public and social in their essence, the experience of 'being in' such a space is silent and solitary. A participant in a chat room has limited sensory access to the collective 'buzz' of that chat room or others nearby – the murmur of human contact that we hear naturally in a park, a plaza or a coffee shop is absent from the online experience (2002, 1).

This introduction also underscores the way that the piece configures sound, as a medium that can, almost magically, become part of the 'being' of the Internet, the sensorium of online infrastructures, and the experience of engaging in real time screen based communications. In a move that has almost come to epitomize the rhetoric of new media, sound – a phenomenon requiring physical space, material objects, movement and atmosphere – is joined with a concept of 'space' that appends, without qualification, the architectural metaphors, the 'chat rooms' of the Internet. This is a move that sound, materially and culturally, enables. In the same way that the voice passes through and between language, music, and noise, as I mentioned with regard to tone, sound moves between material and message, representation and thing. *Listening Post* exemplifies this subliminal movement, as the spatiality of sound, mixes with the fictional space of the Internet, giving the latter a materiality, a medium, through which text can be transformed into voice. The musical elements in the soundtrack invoke music's association with the sublime, the ethereal, the unrepresentable, linking the melodic and tonal qualities of the voice to the conceit of a universal language. As the sonic architectures shift from noise to music, from babble to speech, from chatter to chant, the three types of sound used in the piece: 'click, tone, voice' create a triptych that situates Internet communication, within three familiar and related cultural frames, the first being visual, the second religious, the third, musical. The piece moves within these frames – inhabiting and then deconstructing one after the other, extracting meaning, and then destabilizing each, in a nomadic occupation of various culturally significant tropes. In a strategy of evocation and withdrawal, approach and then evasion, the piece brilliantly encompasses a broad range of potent and grounding cultural narratives, and orchestrates these into a narrative specific to the Internet.

It is worthwhile discussing the nature of these three kinds of sounds. According to Hansen and Rubin, 'the clicking of the relays (draws) attention to the location of the sound sources' and this engages a form of 'everyday listening, while the overall soundscape, composed of "pitch, harmony and rhythm" engages the listener in a form of "musical listening"' (Ibid.). 'Everyday listening' lacks the predictive power of musical listening, because, within western tonality, tones conform to patterns which are easily recognizable yet almost infinitely recombinant, thus 'musical meaning' is far less dependent on context, intention, and the reciprocal participation of speaker and listener that defines 'conversation'. The interplay between the everyday and the musical is essential to the dynamics of the piece. The everyday 'click' casts a hail of small, mechanical sounds over the installation space, and these function as a sonic surrogate for Internet chat, reanimating the conversational experience despite the fact that there are no actual voices. The mechanical quality of the sounds also harmonizes (literally and figuratively) with the synthetic voice in the production of the familiar, and highly symbolic, musical form of the chant. The chant enables vast movements across fictional spaces and cultural conceits, and bringing into play the god's eye, panoptical and omnipresent view of the cathedral, and the democratic, but also religious concept of the Mass, and it does this within the frame of music – the organizing system for transforming sound and noise into an aesthetic form.

As Rubin and Hansen noticed in their research for this project, the listeners' ability to understand the various streams of vocalized text was enhanced by 'assigning different pitches to voices from different speakers' and using 'monotone or chanting' voices to both separate voices within an installation space, and provide a 'more cohesive mix' (Ibid.). In the piece, multiple voices are orchestrated into a something akin to a liturgical chant, in the same way that the multiple voices of a choir are unified by pitch and separated by harmony.[19] While the often meaningless content of the LED screens might be organized via musical structures, at the same time its origins in the conversational mode of Internet chat rooms is retained: the piece still adheres to the common perception of the Internet as a democratic, secular forum, a part of everyday life, non-exclusive, non-hierarchical and essentially multiple. Just as the ocean of discourse that constitutes the Internet is rendered meaningful through software agents that search out text corresponding to pre-given words and phrases (such as 'I am'), the pitched voices and musical tones are 'de-musicalized' through the non-musical sounds of clicking on the one hand, and the computerized voice/s that speak the text, on the other. In this way the synthesized voice evades adopting any kind of subject position. Although the voice has a slightly British accent and is obviously male, it can only obliquely occupy the position of the priest – the singular voice that one would expect to hear surrounded by the multiple voices of the chant, the archetypical voice that intercepts and speaks for the multitude. In this way the synthetic voice has potentially damaging associations with theistic dogmatism.

However, there are other global, panoptic and authoritarian eyes and voices that that the piece must deter if it is to retain its conversational, democratic, cyber-spatial atmosphere. The piece has been described for instance, as 'speaking the world's thoughts' (Bouman 2003, 119).[20] What could 'speak' the thoughts of another? What could, to use Hansen and Rubin's terminology, 'sonify' the conversations that are rendered only in text, and in so doing help explain the 'patterns' and dynamics inherent in the multiple simultaneous conversations on the Internet? Of course the goal of pattern recognition immediately raises the spectre of Internet surveillance and tracking programs like Echelon, a point discussed by Margot Bouman in her article *The Machine Speaks the World Thoughts*. Bouman convincingly uses *Listening Post* to critique the Orwellian model of total surveillance. As she writes:

> The paradox that Listening Post brings out is that the more technologically advanced the computer search engine, or the greater the machine's success at retrieving global consciousness … the more unmanageable it becomes… Meaning is imposed from without, with little relationship to the raw data it filters (2003, 120).

---

19   'The layers of pitched voices take on the quality of a chant or liturgy as they blend with each other and with the reverberating tones. By the end of the [aural] scene, as many as forty voices can be heard and about 2/3 of the 110 displays show messages' (Rubin and Hansen 2002, 3).

20   Adam Gopnik describes the Listening Post as 'a machine that can tell you what the world is thinking – that actually listens to the world, reads its mind, and tells you exactly what's up in there' (cited in Bouman 2003, 119).

But there are other ways that *Listening Post* intersects with the spectre of surveillance. A 'listening post' is, after all, a station for intercepting electronic communications, a position from which to listen or gather information, and more ominously, a point near an enemy's lines for detecting movements by sound. If not a functional surveillance system in itself, the piece can be seen as a compositional device for transforming the Orwellian connotations of intercepting and tracking internet chatter – for many years now associated with terrorist activity – into the far more benign reading of a system simply 'speaking the worlds thoughts'. It is no coincidence that the 'chat' of chat rooms populating the Internet, is only a couple of letters away from the dark connotations of 'chatter'. Chatter has only recently been associated with electronic communications and threats against national security – the term has a much older association with rumour, babble and hearsay, and describes meaningless, unfounded and in-credible speech. The central ingredient in chatter is noise, both the noise of rumour, and the noise associated with interference in the transmission of speech and information generally. Noise has archaic associations with cacophony with its origins in excrement and evil, with the stench of nausea, and with the sound of the rabble, the mass that always threatens to disrupt the social order.

The transformation of chatter into chant through art and using sound is of the same nature as the transformation of vocal tone into useable, quantifiable, information. Both are processes for eliminating noise – the noise of fear (which is replaced by quasi-religious enchantment), and the noise of human-to-human conversation (which is replaced by a simple equivalence between affect and acoustics). Both are also mechanisms for producing affect, if not by calling forth the pseudo transcendentalism of a musical or religious sublime, then by mapping the acoustics of tone onto the contours of algorithms that have themselves, been charged with the impossible task of defining emotion. But bodies and voices engage with these technologies in ways that trouble the clear transmission of meaning and/or affect, evading either metaphysical or computational articulation, disappearing, like a mirage in the desert, as they are approached. To the extent that affect can be delivered in the form of physiological signals or acoustic markers, it can also be performed, worn on the sleeve in a caricature of the affective profiles used to sense, and potentially manipulate, the user's emotional state. If then, as this paper suggests, human emotion is inaccessible by any system of representation (linguistic, physiological, acoustic etc) at the same time it ceases to move, becoming a 'still' once it is 'captured'. Like a photograph or a sound bite, the 'motion' in emotion enters another mode of circulation, an economy of simulation where affect is, in a sense, always at the same time dis-affected.

Note

Parts of this chapter have appeared in Convergence: The Journal of Research into New Media Technologies (Winter-Spring 2005).

References

Cassell, J., Bickmore, T., Billinghurst, M., Campbell, L., Chang, K., Vilhjálmsson, H. and Yan, H., (1999), 'Embodiment in Conversational Interfaces: Rea'. Available online at: <http://citeseer.ist.psu.edu/cache/papers/cs/2011/http: zSzzSzgn.www.media.mit.eduzSzgroupszSzgnzSzpublicationszSzCHI99.pdf/ cassell99embodiment.pdf> Last accessed 8 June 2008.

Bickmore, T. and Cassell, J., (2005), 'Social Dialogue with Embodied Conversational Agents', in J. van Kuppevelt, L. Dybkjaer, and N. Bernsen (eds.), *Natural, Intelligent and Effective Interaction with Multimodal Dialogue Systems* (New York: Kluwer Academic). Available online at: <http://www.ccs.neu.edu/home/ bickmore/publications/bickmore.cassell.class.pdf> Last accessed 8 June 2008.

Bouman, M. (2003), 'The Machine Speaks the Worlds Thoughts' *Parachute* 112, 109–125.

Derrida, J. (1992), 'Ulysses Gramophone', in D. Attridge (ed.), *Acts of Literature* (New York: Routledge), 256–306.

Dolar, M. (2006), 'A Voice and Nothing More', in Slavoj Žižek (ed.), *Short Circuits* (Cambridge: MIT Press), 13–31.

Dreyfus, H. (2001), *On the Internet* (London: Routledge).

Eagle, N. and Pentland, A. (2003), 'The Relationship Barometer: Mobile Therapy for Couples', *IEEE Computer Magazine*, September. Available online at: <www. lifewear.org/resources/Eagle,_Pentland_-_The_Relationship_Barometer.pdf> Last accessed 8 June 2008.

Madan, A., Caneel, R. and Pentland, A. (2004), 'Voices of Attraction', AugCog Symposium of HCI, 2005, MIT Media Lab: Affective Computing Group. Available online at: <http://web.media.mit.edu/~anmol/TR-584.pdf> Last accessed 8 June 2008.

Mitchell, R. and Thurtle, P. (eds.) (2004), *Data Made Flesh: Embodying Information* (New York: Routledge).

Picard, R., Vyzas, E. and Healey, J. (2001), 'Toward Machine Emotional Intelligence: Analysis of Affective Physiological State', IEEE *Transactions Pattern Analysis and Machine Intelligence*, 23:10, 1175–1191. Available online at: <http://citeseer.ist.psu.edu/cache/papers/cs/22323/ftp:zSzzSzwhitechapel. media.mit.eduzSzpubzSztech-reportszSzTR-536.pdf/toward-machine- emotional-intelligence.pdf> Last accessed 8 June 2008.

Picard, R.W. and Klein, J. (2002), 'Computers That Recognize and Respond to User Emotion: Theoretical and Practical Implications', *Interacting With Computers* 14:2, 14.

Picard, R. (1997), *Affective Computing* (Cambridge, Massachusetts: MIT Press).

Richards, C. and Snelgrove, M. (2002), Patent draft for *Method and Apparatus for Finding Love*, US Patents Office (pending), USA.

Rubin, M. and Hanse, B. (2002), 'Listening Post: Giving Voice to Online Communication'. Proceedings of the 2002 International Conference on Auditory Display. Kyoto.

Chapter 14

Judith Merril Moving In and Out of This World: Urban Landscape Encounters of a Science Fiction Personality in the Sixties and Seventies

Dianne Newell and Jolene McCann

Once, twice, three times, [I found clubs that] felt as if they included all I knew of the universe. They had people in them with some of that tear-shimmering vision, and each of them let me feel comfortable and at-home inside for a while – a place a little smaller than the whole starry-skied cosmos to settle in – a place, say, to look out the window from. But the thing about a window is, it's in a wall.

(Merril and Pohl-Weary 2002, 21–22)[1]

Introduction

A prominent American science fiction (sf) star in the late 1960s when she left the US and moved to Toronto, Judith Merril's ability to locate and represent her emotional experiences and interactions with landscapes (cultural, emotional, historical, physical) was out of the ordinary. Sf writers tend to focus on the environment, to ask 'What If?' and to work in dialogue with others. But even within sf circles, the emotional content of her fiction and non-fiction work, fuelled by her physical mobility and her revolutionary desire to find windows onto the universe, stood out. Here we explore Merril as a case study of the ways in which connections between space, time and emotion have come together in the making of a human life. She relocated and moved around in the 1960s and 1970s, curious, energized and optimistic, always in pursuit of the 'excitement of ideas' and seeking new portals for discovery. We also look at her experience with recurring feelings of fear in times when she found herself helpless to act on her ideas. Fear is a powerful and essential emotion for our own safety and survival. It also absorbs or saps energy and can be debilitating when experienced for extended periods of time, as happened with Merril. Enabling our project are her unique reflections on her movements in space and time: the voluminous collection of personal papers,

---

1   From an unpublished note written by Merril c 1970.

which she maintained intact throughout all her moves, and her award-winning memoir, *Better to Have Loved: The Life of Judith Merril*.[2]

Judith Merril (1923–1997) was a central, socially radical powerhouse in the extraordinary world of modern sf, first in (and around) New York City in the 1940s and 1950s, briefly in London and Washington, DC in the mid-1960s and Japan in the early 1970s, and finally in Toronto, her permanent home since 1968. As a youth she was drawn to counter cultural movements of the day: the early Zionism of the 1920s and the Trotskyism of the 1930s and 1940s. Over her lifetime, she took on successive, multiple roles as a free-lance science fiction writer, editor, anthologist and critic, and after her move to Canada, documentarist, translator and memorist. A respected sf writer of women- and family-centred sf novels and short stories in the 1950s (see Cummins 1992, 2001; Stableford 1979; Merril 2005),[3] the vision she offered of the future of gender relations was not one of finality but one of uncertainty and possibility. Her 1958 space travel story, 'Wish upon a Star', for example, is told through the eyes of a teenage boy born on the *Survival*, the colonizing ship launched during the brief Matriarchy at the beginning of World Government on Terra (Earth). After generations in space, the women are for practical reasons still in charge of the voyage. But as the ship is finally about to land on an uninhabited but habitable planet, the boy ponders the possibility of a return to an 'Earth type' nuclear family, and the masculine independence that traditionally goes with it: '*House. Family. Inside-outside.* They were all words in books. Hills, sunsets, animals. *Wild* animals. Danger. But now he wasn't afraid; he *liked* the thought. Wild animals, he thought again, savouring it. Houses, inside and outside; inside the family; outside, the animals' (30). Merril's prestigious anthology series of the year's best sf ran from 1956 to 1968, and her influential role as a critic editing the *Magazine of Fantasy and Science Fiction* books column lasted from 1965 to 1969. In a watershed period between 1967 and 1974, she introduced British experimental sf writing to American sf circles with her edited collection of British 'New Wave' authors, *England Swings SF* (1968) and made the important decision to leave the US in 1968 in protest over the American policy on Vietnam, deliberately leaving the American sf community behind. At that point, her role as editor transitioned into catalyst, her influence on science fiction stretching into Canada and Japan. In Toronto, she moved around in the city's downtown core of students and immigrants, briefly living and working in the 'free university'

---

2   *Better to Have Loved* won a Hugo Award in 2003 in the category best 'Related Book'. The Hugo Award is the premier award for excellence in the sf field.

3   Merril's first and most popular sf novel, *Shadow on the Hearth* (1950), was adapted as 'Atomic Attack', screenplay by David Davidson, directed by Ralph Nelson, for Motorola TV Playhouse, ABC, 18 May 1954, and her sf short story, 'Dead Center', was selected for the prestigious mainstream literary series, Martha Foley (ed), 1956, *The Best American Short Stories, 1955*. In Canada, several of her sf writings from the 1950s were adapted and dramatized on CBC-Radio *Ideas* and as stage plays, including two productions by Theatre Passe Muraille, Toronto.

experiment at Rochdale College and founding the Spaced Out Library with her personal collection of sf literature. In working visits to Japan in 1970 and 1972, she learned about the Japanese language and culture and interviewed individuals for a series of free-lance radio broadcasts in Canada.[4] Most interesting for her personal emotional transformation was her experimentation with top Japanese translators, exploring novel, collaborative approaches to translation as movement between language and meaning. Returning to Toronto, she transitioned into a free-lance documentarist for radio, shifting her focus to the Canadian Broadcasting Corporation (CBC) studios and audio recording technologies, developing her 'ear' and exploring her voice. By the 1980s, she relocated her energies to the nascent Canadian sf community[5] and gave thought to her own life in the field. Her final years were spent researching and remembering her life's journey for a memoir that had been 'taking shape, slowly, through many transformations' (Merril 2002, 10).

Walking her readers through her emotional transformations and engagements with landscapes was an exercise that seems always to have connected Merril's changing physical locations to her emotional states and working and living spaces, even to her work habits. 'I was once a writer of science fiction', she writes, and 'that means my practice has always been to make the environment as important as the characters in a story' (Merril and Pohl-Weary 2002, 11). So, with Merril, there are multiple environments – the ones in which she lives and moves among (physical, emotional, cultural) and the ones that she creates in her writings.

Mobility

Urry in 'The Place of Emotions within Place' locates emotion in landscapes and in movement, and even speaks of the 'emotion of movement', because, he argues, places are 'massively' interconnected through movement even to the point that 'the language of landscape is ... a language of mobility, of abstract characteristics' (2005, 81). Conradson (2004, 104) points to the emotional gains to be had from

---

4   Merril, as scriptwriter, interviewer and narrator, produced for CBC Radio a ten-hour series, 'Japan: Future Probable,' *Ideas*, various dates 1971–1975; a five-half-hour series, 'Growing Up in Japan', *Radio Schools, Kaleidoscope*, 1973 and a one-hour program, 'Women of Japan', *Ideas*, 1972.

5   The Spaced Out Library that she founded with the Toronto Public Library in 1970 (renamed the Merril Collection of Science Fiction, Speculation and Fantasy in 1990 to honour Merril's donation) became the Canadian gathering place for sf fans, booksellers and writers. Merril was also a powerhouse in Toronto's celebrated literary festival, Harbourfront, organizing, for example, the 'Out of This World' Reading Series on Science Fiction (1980). She edited the first anthology of Canadian science fiction and fantasy, *Tesseracts* (1985); founded Hydra North in 1985; organized the First Annual Canadian Science Fiction Writers' Workshop in 1986; and was a founding member of both Science Fiction Canada and the Canadian Science Fiction Foundation.

mobility and 'environmental encounters': 'places do have the capacity to shape our feelings', both during and after our visits (2005, 107). Following Thrift on the notion of ecologies of place, Conradson (2005, 106) asks us to think of ecology as 'the broad spectrum of entities that comprise a place, and the interactions between them'. At the end of the day, mobility and relocation have emotionally significant outcomes; they foster reflexivity, according to Urry (2005, 81), and even are, Conradson (2005, 114) argues, 'significan[t] … for the reflexive management of one's emotions'. Feminist scholars would add, as Sidonie Smith has done, that mobility/travel also 'facilitates speedy escape … [p]hysical freedom translates into psychological freedom' (2001, 170).[6]

Breaking with Past and Place: Toronto in '68

When in the spring of 1967 Judith Merril returned home to the United States from her extended trip to London to collaborate with 'New Wave' sf writers, she walked into a scene of growing social and political unrest, especially over the US involvement in Vietnam. For two decades she had relished the activist, collaborative, rhizomatic nature of the sf genre, but now she lost faith in the ability of science fiction to respond to the revolutionary climate of the late 1960s. Feeling helpless to alter the social and political conditions in the US triggered an extended, severe bout of emotional paralysis. It was a cyclical phenomenon typically associated with her struggles as a writer, editor and activist to communicate alternative ideas. Merril described her condition in letters to writer-friends variously as 'waver[ing] in sort of immobilized anxiety for months',[7] 'a terrible depressive lethargy',[8] 'peculiar stiffness of body' and a feeling of being 'locked into a statue'.[9] The fate of her large collection of personal papers and sf books and magazines caused additional anxiety. She confessed to 'a neurotic fear of losing materials' (McCann 2006, 22), but also the need to escape from them: 'my accumulated books and files occupy so much space [at home] that I periodically have to flee',[10] reminding us that working at home has the potential to create a type of neurotic domestic landscape – you have to ignore it, organize it, or flee from it. Merril's subsequent quest for a radically different landscape – a new country and experimentation with non-written modes of communication – led to her relocation in November of 1968 to Toronto as a political and intellectual refugee (see Faber 1998, and also Kurlansky's work (2004) on the phenomenon of '1968' as an international turning

---

6   For a discussion of Merril as a type of 'travel writer', see McCann (2006).
7   Merril to Brian Aldiss, August 1968, vol. 1, file 15, Judith Merril Fonds, MG30-D326, Library and Archives of Canada (hereafter 'Merril Papers').
8   Merril to Dennis Lee, 25 September 1968, vol. 9, file 50, Merril Papers.
9   Merril to Walter Miller, 14 November [1974], vol. 39, file 21, Merril Papers.
10   Cited in Merril to Brian Aldiss, n.d. September 1968, vol. 1, file 13, Merril Papers.

point for radicalism). Merril tapped into a geographical metaphor to express her need to relocate: 'I needed to get far enough away from the centres of power to decide what do with my uncomfortable portion of it' (Merril and Pohl-Weary 2002, 162), recalling to mind Davidson, Bondi and Smith's observation (2005, 1) about the power of emotions – whether negative or positive – to transform our lives, 'creating new fissures or fixtures'.

In breaking with the past, Merril rejected her US citizenship and escaped from the positions of power and professional assignments she had accumulated there. She was, like her vision of an imaginary trip that introduced her anthology of British New Wave writing that year, 'heading out of sight' vis-à-vis the US and the American sf community.[11] Merril promised her readers that – in the language of the late 1960s – they were in for quite a 'trip': the book is 'an action photo, a record of process in change/a look through the Perspex porthole at the momentarily stilled bodies of the scout ship boosting/fast and heading out of sight into the multiplex of inner/outer space ...' (1968, 9–11). She moved to Toronto, 'and for that matter Canada' for the reasons she had committed most of her adult life in speculative fiction: 'I wanted to change the world' (Merril and Pohl-Weary 2002, 172).

Owram identifies Toronto, and in particular Yorkville Village and neighbouring Rochdale College and the University of Toronto's downtown campus, as the centre of a developing counter-culture and youth movement in Canada at that time (Owram 1997, 211, 294–5). Hagan recalls that Toronto's 'atmosphere was electric' and that the era was one of 'urban protest and uncertainty' (Hagan 2001, x, 69). Downtown Toronto was highly liveable in 1968, a place to which the celebrated American urbanist, writer and activist, Jane Jacobs, had also moved that year (and see Ley 1993). Merril quickly came to relate to Toronto as the place in the world she felt she had always been looking for; it had a culture, a 'feel' and a language beyond anything she had ever encountered.[12]

Toronto's Rochdale College was Merril's entry point, the revolutionary space she was looking for. This eighteen-story concrete high rise at the corner of Bloor and Huron streets, on the radical edge of the University of Toronto district, was the largest free university in North America at a time when these experiments in student-owned, student-run education linked with experimentation in co-operative living were 'in' (McCann 2006, 23; Merril and Pohl-Weary 2002, 173; and see Sharpe 1987). The year she spent living at Rochdale as the Resource Person on Writing and Publishing and the two more in which she continued to participate from outside stimulated her in the most fundamental way. Writing her memoir in the mid 1990s, she recalls the emotional rush: 'I cannot describe [my time at

---

11   John Hagan (2001, 29) categorizes the women who came to Canada as Vietnam war-resisters as women who accompanied draft-age men (and this would include the American urban activist Jane Jacobs, who came with her husband and draft-age sons); however, he singles out Judith Merril as an independent actor who 'resist[ed] the [US] in a very long-term way'.

12   Merril to Miller, 28 February 1975, 28 March 1975, vol. 39, file 19, Merril Papers.

Rochdale] any better than I might describe a twenty-five-years-ago loving act of sex'; 'it turned my life and psyche around' (Merril and Pohl-Weary, 2002, 174).

Being Elsewhere: Translating Japan in the Early 1970s

It was the people, ideas, language and culture Merril encountered after moving out of Rochdale, during a brief visit to Japan in 1970 for an international sf conference, that opened a 'storehouse' of experience and memory that seemed to transform her outlook completely,[13] so much so that when she returned home to Toronto she felt she had, 'not quite come home yet'.[14] Urry asks (2005, 77) in reference to the language of mobility: 'What are the different senses mobilized by being elsewhere?' In Merril's case, the sounds, gestures and structure of spoken and written Japanese language thrilled her, it seemed to offer a very open-ended mode of communicating and thinking about the future that helped her to express the ideas that were in her head (see Newell and Tallentire 2005). Writing excitedly to another writer, Merril explained that Japanese was a language 'with no way of making a categorical statement about the future (can only express intention, wish, probability, etc. No way to say "The sun will rise tomorrow"), also with no concept of "either/or" or "instead" – no words for the idea, apparently ... it all adds up to a relativistic culture: lacking both absolute and dualism in its basics'.[15] Not surprisingly, Merril gravitated to the Japanese sf *translators* at the conference rather than the international writers. Translators and written translation became the object of her deep curiosity and excitement and a focal point of her quest for new, experimental, collaborative modes of communication.

Desbiens and Ruddick (2006, 4) stress the existence of geographies of language, '[for] speaking one language over another is to call upon a broad set of people, places, and relationships; in short, to make particular geographies'. The translation of Japanese sf stories into English is said to be especially difficult because of the lack of a future indicative tense in the Japanese language, never mind the invented nature of the language and landscapes used in sf writing. That Merril would develop an approach to written Japanese translation as an experimental and a collaborative, multistage enterprise resonates with Desbiens and Ruddick's concern (4) that the challenges for translation go beyond 'mere' translation, and also address Desbiens' worry (2001, 1) that written translations can be culturally 'impoverished' because using conventional approaches translators 'often strive to eliminate every trace of native context'. The methods Merril insisted on were tedious and time-consuming, but rich in connection and desire to get to the heart of every meaning: 'The first step, I believe, is learning to travel in more than one direction, and think in more than one pattern, at the same time: learning to sustain

---

13   Merril to Tetsu Yano, 29 December 1970, vol. 20, file 19, Merril Papers.
14   Merril to Tetsu Yano, 9 October 1970, vol. 20, file 19, Merril Papers.
15   Merril to Katherine MacLean, 12 Oct 1970, v 10, Merril Papers.

awareness on many different levels of perception, and to communicate across all artificial barriers'.[16] Here, we see Merril building new geographies of translation/language/meaning.

Merril's involvement with translation of Japanese sf stories into English for a major book project with her as editor began in a typical Merril way, informally. She recalls that at the end of the conference she assisted one of the Japanese delegates to re-evaluate the work of a young English woman he had hired, as was the custom, to polish his very literal, translation of a Japanese sf story by Hoshi Shinichi. Merril thought the English fine, but knowing Hoshi's work, she could see why the story was a poor version of the original, for 'of course unless you can do it by evocation and suggestion, you're not translating Hoshi'.[17] So Merril started again from the original version of the story, searching collaboratively with Japanese translators for exact meaning and making it sound like something Hoshi Shinichi *would* write. Their process, which was slow and laborious, was also invigorating: 'And I would just say what else, you know, give me some other meanings for this, and then they would give me some, and I would eventually come up with an English phrase which they both would like and say "yes, that's right"…and at the end we were all tremendously excited, you know'.[18] This is how Merril, the communications connoisseur, got interested in the specifics of translation, and so planned her extended visit in 1972. She knew no Japanese but she picked up the syntax, knew a lot about sf writing and interacted with the Japanese translators in novel ways that got at the nucleus of the author's intent, rather than just the surface language: 'And it keeps going [back and forth] until they're satisfied and I am, so it's very slow, but we think we're actually getting translation. Not just information translation, but the best quality and style and cultural translation as well'.[19]

Stimulated on these trips by the language, culture and the dynamic circle of top Japanese translators of science fiction, Merril's emotional gain from her engagement with Tokyo was no less than a new sense of her life, what it was now and what it had been: she discovered that she was a translator in the broadest sense of the term. She was translating, she said, between the 'counter-culture' and the 'establishment', between Canadians and American political refugees in Canada and even between late sixties' visions of possible futures in North America, the UK and Japan. Even '[b]efore that, when I was writing s-f, I was in a sense trying to translate visions of possible futures for people trapped in concepts of the past – or trying to translate what I perceived as realities of the present (by means of images of the future, cast in literary forms of the past!) to people whose present-realit[ies] were different from mine'.[20]

---

16   Judith Merril, 'Essay on Translation', 1972 [for the *Japanese SF Magazine*], vol. 32, 'Science Fiction Correspondence, 1972–1980', uncatalogued, Merril Papers.
17   Merril Interview of Muto, 1972, 6, page 2, Merril Papers.
18   Merril Interview of Muto, 6, page 2.
19   Merril Interview of Muto 7 (13:05–14:15).
20   Judith Merril, 'Essay on Translation', 1972.

Merril's second visit to Japan was her last to that country. She had developed a deep personal identification with translation of Japanese stories into English as part of her lifework. The intense personal interaction that Merril needed, and always had throughout her career, offered her rich emotional rewards but created difficulties after she returned to Toronto. The method she devised for the translation project meant a strong, sustainable personal interaction was necessary. In the absence of such interactions, she fell into another severe bout of paralysis, aggravated by a 'translator's block'. The materiality of this paralysis is reflected in the state of the temporary living/working environment she moved into upon her return from Japan in 1972. The apartment was a severely disordered space that fostered, in this instance, accompanying emotional disarray, a dismal bachelor apartment in the basement at the back of a building at 42 Austin Terrace, with 'no light and air'. She wrote to a friend that she spent the entire winter exhausted and overwhelmed in this environment, sleeping and reading and replaying audiotapes she had made in Japan on a mattress on the floor, the dirty dishes and unanswered mail piling up, and watching the 'dust grow thicker on the piled-up possessions'.[21] Her occasional attempts over the years to make headway with the Japanese translation project usually started with a vow to get her Toronto living quarters in order.[22] She also surrounded herself with Japanese objects. For years she continued to sleep on a mattress on the floor, emulating her experience in Japan, and tried to eat half of her meals Japanese style and to follow a Japanese diet;[23] her local Japanese restaurant becoming 'a constant spending-lure'.[24] Despite all efforts, she found herself working alone on editing and completing the translations of the Japanese sf short stories in a pattern that she identified as 'disarray and binges', alternating 18-hour days with complete 'sloth'.[25] It was, she wrote, '[o]ne of those build-down things, where delays lead to silences creating blockages which create further delays and make communication IMPOSSIBLE – etc'.[26] She described the blockage as 'glacier-like', and explored that metaphor through intermittent correspondence with her Japanese translation team. But with the emotional and physical geographical proximity to her new circle of colleagues stretched and eventually broken, guilt encumbered the project for Merril, who after many years

21   Merril to Nikola Geiger, 25 October 1973, vol. 6, file 61, Merril Papers.

22   Merril to 'Asa-chan' (Hisashi Asakura), 9 November 1973, vol. 32, 'Science Fiction Correspondence, 1972–1980', uncatalogued; and see Merril to Miller, 3 July 1975, vol. 39, file 19, Merril Papers.

23   Merril to Miller, 17 December 1974, vol. 39, file 21, Merril Papers. The ideal diet for her was the ''traditional' Japanese [diet]: rice, seafoods, soya … lightly cooked fresh vegetables, fruits (raw), eggs … and only VERY occasional bits of meat'.

24   Merril to Miller, n.d. [1975], vol. 39, file 21, Merril Papers.

25   Merril to Miller, 19 October, 1974, vol. 39, file 22, Merril Papers.

26   Merril to David Bloggett, 1 October, 1975. vol. 32, 'Science Fiction Correspondence, 1972-1980', uncatalogued, Merril Papers.

was never able to afford to return to Japan or to complete work on more than a few individual stories.[27]

Eggs and Amoebas: The Maps

Conradson (2005, 105) contends that the self, in relation to various landscapes, is a 'distributed entity', observing that 'the faculties of memory and imagination permit the maintenance of connections across a range of times and spaces'. The feelings – of excitement, energy, and optimism – Merril had experienced in cities such as New York, Tokyo, and Toronto emerged, in part, through her mobility and direct engagement with the landscape, as well as her own special brand of limning and translating emotional geographies for old friends; for Merril, the latter was a way of re-seeing and re-visiting familiar places. On her 1972 visit to Tokyo, for example, in the space of what she called her 'Japan Journal', Merril included a hodgepodge of ecologies of place: maps of the underground, letters to friends, fragments of grocery lists and pamphlets from tourist sites.[28] She also sketched highly personalized maps of the neighbourhoods she inhabited, demonstrating Davidson and Milligan's (2004, 522) sense of emotions forming 'the fabric of... unique personal geographies'. Merril's map of the Kanda region demarcated the Tokyo and Kanda railway stations and her textual depiction labelled Kanda as 'the district of students, universities, [and] bookstores'. There are obvious links between this landscape and the Toronto university district where she lived, though in her written sketch, she connects the area only to the university district on the east side of New York City, a district with deep emotional links to her past life and to those of the handful of old American writer friends to whom she still wrote.

In the case of Toronto, Merril's sketch maps were designed to orient old friends to her most lived landscape: her neighbourhoods and living/work places in the downtown core, within walking distance from home to the Spaced Out Library – known locally by the very energetic-sounding acronym, 'SOL' – and, later, the CBC-radio studios. Simple maps accompanying her letters delineated Merril's precise, though ever shifting, position within the Toronto landscape. Objects and sites, including restaurants, markets, the CBC radio transmitter tower, ice rinks and parks, and most significantly, her sites of translation, namely SOL and CBC studios, dominate the more detailed of the sketches. She imagined her city for an old writer-friend who had never visited it, opening her map-like letter to him with the intriguing intention to 'spread out my territory as well as my person for your

---

27 Merril's second, extended trip to Japan had been financed by a publisher's advance for the translation project and by various Japanese sources that supported the book project, which created additional pressure for her to complete the translations.

28 Judith Merril, Japan Transcript, 12 March 1972, typescript draft by Judith Merril, vol. 30, file: 'Japan Journal, 1972', and vol. 8, file 25, 'International Science Fiction Symposium, 1970', Judith Merril Papers.

inspection'.[29] '[A]lmost all places are "toured"', suggests Urry (2005 82), and the emotions involved in the 'visual consumption of place' contribute, in some degree, to visitors' feelings and sensations about such places. Nowhere, however, is the fluidity of what Urry goes on to describe as 'the emotion of movement, of bodies, images, information, moving over, under and across the globe and reflexively monitoring places in terms of abstract characteristics' more applicable to Merril's map-making than in her amorphous characterization of her downtown Toronto 'territory' as looking 'something like an egg in an amoeba'. 'Depending on the day, the mood, weather and company, the amoeba [city] contracts or spreads itself out,' she writes in 1973. Meanwhile, '[t]he egg (or cell nucleus) changes shape and size only very slowly'. And she adds, it is small, '[only] about 2.5 miles long and 1.5 miles at its widest point' (Merril and Pohl-Weary 2002, 182–83) (Figure 14.1).

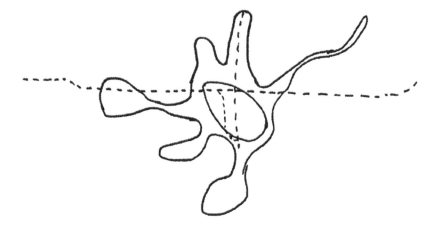

Figure 14.1   **Merril's sketch map of her Toronto territory as an 'egg in an amoeba', c. 1973, which appeared in Judith Merril and Emily Pohl-Weary, *Better to Have Loved: The Life of Judith Merril* (Toronto: Between the Lines, 2002)**

Source: Image courtesy of the Merril Estate.

If Conradson is correct in asserting that that the self emerges in relation to the people, events and 'inanimate things' that encompass a landscape (2005, 107), then Merril's translation of 'her' Toronto territory via the life-affirming egg metaphor is a perfect case in point. The people who lived in Merril's 'cell nucleus', were students and immigrants; they only moved out, she wrote, when 'they stop[ped] studying something' (Merril and Pohl-Weary 2002, 183). She sketched in words all

29   Merril to Miller, n.d., [1975].

that she related to, interacted with and depended upon in her work and her life there. Her accompanying essay description places the housing co-ops and communes, side-street and back-ally pockets and China Town within her amoeba-egg. The egg curves around the multi-cultural Kensington Market. She reads into that space memories of former urban haunts, picturing it as a hybrid of immigrant markets in London and New York (185). But it is SOL, her first significant communications/translation platform in Toronto after Rochdale College, and container for the paper version of her own life and work – in effect, an egg within her larger egg – that crowns her ovoid abstraction. SOL marked the 'tip' of her egg-shaped territory in so many ways, as she noted in 1973, that it was difficult for her to write about it 'at all' (185).

SOL was the piece of her old life that Merril had carried with her to Toronto. She had arrived at Rochdale in a rented trailer with 'eight thousand books and magazines', and 'fifteen file drawers of jumbled junk and value' (191, 173). She was adamant about the importance of bringing all her books and papers to Canada, but also aware of the emotional burden attached to doing so. To her Rochdale contact, the Canadian poet, Dennis Lee, she explained that 'attempting to sort [her books and files] before moving to Toronto is likely to become a Terrible Thing, taking up much more time and emotional energy than I want to give...'[30] The obvious tension between her 'neurotic fear' of losing her books and papers and her periodic need to escape from them signals the extent to which they had become pivotal objects in her self-landscape engagement. In talking about moving and managing her sf library and personal papers, Merril could have been talking about moving, unpacking and sorting out her life.

At Rochdale, Merril's disorganized accumulation of books and magazines and boxes of personal papers transformed into the first Spaced Out Library. When she moved out after a year, into a room in a downtown co-op house, she brought the library and her boxes of papers home with her. It proved an unnerving co-existence – 'my tiny room was completely filled with them ... I fell asleep at night waiting for the books to topple on me' (Merril and Pohl-Weary 2002, 126). Merril donated her library to the Toronto Public Library in the summer of 1970, which in exchange, and at her request for access to her all her books and personal papers, which she stored there, hired Merril as the resident paid consultant with an office for life. On the one hand, the arrangement stirred in her a mixture of joy, excitement, and sense of freedom.[31] It was 'too much of a dream come true', but it also meant separating from self, because 'too much of the paper-concretization of a quarter-century of my own life and work [is]on the shelves and in the file drawers there' (185). Ultimately, she welcomed the emotionally unburdening affect: 'I let Toronto Public Libraries build the walls elsewhere – where I could walk in – and

---

30   Merril to Lee, 25 September 1968.
31   Merril to Harvey Jacobs, 11 October 1970, vol. 8, file 30, Merril Papers.

out'. She was finally rid of 'the last (literal) protective barriers', she wrote to a fellow writer.[32]

While SOL would remain a constant in Merril's life in Toronto, her temporary shift to living and working around the CBC studios in the mid-1970s created a re-orientation in Merril's mental map and actual sketches of the downtown. The CBC studios in the mid-1970s functioned as a new place of excitement, experimentation and translation within 'her' Toronto. Merril wrote to American friends that 'the CBC at night and weekends and holidays is a haven'.[33] When she hung out there in the first half of the 1970s it was also an extraordinarily stimulating environment that operated much as a cooperative in which producers and writers pitched program ideas. She likened the place to 'a twenty-four-hour-a-day seminar' or 'a jam session' (Merril and Pohl-Weary 2002, 185).[34] The four-storey Brownstone CBC Radio building on Jarvis Street was her 'new ghetto'.[35] The reference here is to her days in the US as a writer of science fiction, a genre that was lovingly known to insiders as a 'literary ghetto'. So closely connected to the studios did she feel that she used the modifier 'we' to refer to CBC Radio; for instance, she boasted, '[w]e win a lot of prizes, feel like we say something, enjoy each other'.[36] When she moved into a new apartment building located at 433 Jarvis Street, across from the CBC studios, the move essentially replicated the close proximity between home and work from her Rochdale days and later, when she deliberately lived across or down the street from the various SOL locations. The Jarvis Street apartment also suited both her new, optimistic mood for equilibrium and her deepening affection for the heart of the city: 'I live in a large bright one-bedroom apartment, medium-rise, medium-new, medium-rent ... downtown'.[37]

Ecologies of Self, Place and Aging: The Memoir

Once Merril began thinking about her life in science fiction, we find a shift in temporal focus in her engagements. A life once so intensely immersed in speculative fiction, with about how things could be different in the future – with changing the world – turns to remembering, sifting and sorting the past. If, as Jones suggests (2005, 210), a way to represent emotional geographies is imaginatively, 'to think of the connections between our memory and our geographical imaginations', then Merril's final passion, her memoir project, should be viewed as an uncompromising

---

32   Merril to Miller, 14 November [1974].

33   Merril to Miller, 4 August 1975, vol. 39, file 22, Merril Papers.

34   She wasn't the only one to call the CBC studios 'home'. In the late sixties, Fetherling (212-13) surreptitiously went there in the off hours to write and he claims that another freelance writer actually lived there at night.

35   Merril to Miller, 4 October 1974, vol. 39 file 21, Merril Papers.

36   Merril to Miller, 4 October 1974.

37   Merril to Miller, 4 October 1974.

work of personal reflection not only on her lived experiences and movement, but on the emotional landscapes in which she sought to live them. Jones focuses specifically on what he terms the 'ecologies of self' that are concretely tied to place and to 'emotional attachments to place': 'Clearly remembering being-in-place, and perhaps remembering through place, through emotions of (remembered) place are powerful elements of emotional geographies of the self' (213). Merril's insistence on an affective model of life writing for a life in sf was crucial to examining the personal, intimate and even painful aspects of her life, thus fulfilling Buss's criteria for modern memoir, which 'works through the varied felt realities – the sensations and resulting feeling and thoughts of self as well as viewpoints, opinions, actions and reactions of significant others' seeking to 'revive the emotions and thoughts of the past – fears and desires, angers and delights' (2002, 15, and see Newell and Tallentire, 2007). When she began thinking about her memoir in the 1980s she moved her papers to Ottawa, donating them to the national archives, then consulting them time and time again as if her 'younger self' (Merril and Pohl-Weary, 250). Consulting her earlier self, Merril recalled events happening at specific points in time and space, mapping her wonderings and wanderings in ways that echo Jones's observation (2005, 206) that 'felt' moments are 'mapped into our bodies and minds to become a vast store of past geographies which shape who we are'.

Moving Out of This World

How does this account of one person's life resonate with emotional geographies more broadly? In the 1960s and 1970s Judith Merril's emotional engagements in time and space appear to explicate the links she made among the places she travelled over time, and between her locations within places and her treasured personal objects, emotional states and work habits, as well as her evolving role as a translator in 'everything' that she did. The books and personal papers that travelled with her to Rochdale College were key personal objects in her ecologies of self, at first cherished as representing the entirety of her life and work, then distanced, as protective barriers, walls, to be shed. Personal collections of publications and old correspondence files are like maps with which to navigate in relation to one's earlier self, and once the papers are left with a public archive, they become a record or extension of these personal geographies. Their public accessibility also creates future possibilities for writing and translation. The memoir itself is the last of a long list of just such 'translations' of emotional geographies of self, in different forums and by different means. Her emotional engagement with downtown Toronto, and especially Rochdale College (transformed her psyche), SOL (eliminated her protective barrier) and CBC studios (her new ghetto), gave 'shape' to her territory in Toronto, the egg in an amoeba. Her practice was to live in (or near) her workplaces – spaceship-like – fusing home-space and work-place. As a distributed entity, the radio documentary work for the CBC and 'half' Japanese lifestyle she adopted in Toronto kept her linked to former emotional engagements in Japan, and her 'Japan Journal' and maps of Tokyo neighbourhoods kept her connected to her old districts in London and New York. Perhaps mapping these places and objects for readers as she moved around allowed Merril to create the sort of 'permeable membrane between home and the foreign, domestic confinement and freedom on the road' (and even between the present and the future, because she makes her old correspondence available for future readers) that Lawrence (1994, 19) claims is a particular function of women's travel writings. At key moments Merril tended to withdraw, become emotionally overwhelmed and deeply immobilized with feelings of fear. Her life shows how breaking points can become productive fissures out of which, in Merrils's case, to launch searches for new windows on the universe in an almost an obsessive pursuit of the excitement of ideas.

Merril's life experience with urban landscapes can be thought of as cocooning herself in a succession of pockets (the 'clubs' to which she refers in the opening statement of this chapter) in which she could settle and feel 'comfortable and at-home'. Sometimes they were very small spaces – the confines of her room in a co-op house or, uncomfortably, the basement bachelor apartment, or, as at Rochdale College, shared living and working space within a single building. In other instances, the pockets were slightly larger, her combination of apartment and the place nearby where she worked (SOL and the CBC studios). Some pockets were larger still, special urban neighbourhoods – her imaginative Toronto 'egg' entity or the Kanda district of Tokyo. Only sporadically, when relocating these secure zones did she break out of this pattern. Through her writings, and even the language translation work and memoir project, Merril was able to imagine other kinds of emotional possibilities and connections, extending those geographies to imagined future places.

At the end, we see someone who usually moved across time and space with ease run out of places to go. Toronto, her long-time place of 'creative ferment' and hope – had failed her in the 1990s climate of political conservatism. She was feeling about Canada the way she had felt about the US when she left in 1968, only this time, she confessed, in an interview conducted six months before her death, 'there's no place left for me to go' (Merril and Pohl-Weary 2005, 264). Merril included the interview as the final chapter of her memoir, in the space of which she predicts a bleak future in which our civilization has reverted to a level only a step above the medieval dark ages. Yet, true to her roots in science fiction, she imagines a communications landscape filled with (pockets of) people and things that might add up to something: 'Just like in the Mad Max movies, pockets of technology will still function ... one guy has an airplane and knows how to fly it, someone else has a radio that works ... a laptop computer can work off the electricity in your skin ... some of the satellites will stay in orbit' (265). This joyless imaginative geography of the mid-1990s is a far cry from the dynamic emotional landscapes of districts of Toronto and Tokyo she sketched in the 1970s, and no match at all for the usual hopeful vision of people and technology she painted in her sf stories of the 1950s. Yet even in her 'death bed' vision of the future we see Merril engaged in translation, in the broadest sense. In Japan, she had discovered translation as movement between language and meaning, as geographies of language, and developed a translation method rich in connections and desire to get to the heart of every meaning, translating not just language, but 'heart'. To begin this process, one had to learn to travel in more than one direction and think in more than one pattern – at the same time. This seems a fair description of Judith Merril's life, moving in and out of this world.

Acknowledgments

Dianne Newell wishes to thank the Social Sciences and Humanities Research Council of Canada and The Peter Wall Institute for Advanced Studies at the University of British Columbia for funding this research.

References

Bondi, L., Davidson, J and Smith, M. (2005), 'Introduction: Geography's "Emotional Turn"', in Davidson et al. (eds.), *Emotional Geographies* (Aldershot, Ashgate), 1–16.

Buss, Helen M. (2002), *Repossessing the World: Reading Memoirs by Contemporary Women* (Waterloo: Wilfrid Laurier University Press).

Conradson, D. (2005), 'Freedom, Space and Perspective: Moving Encounters with Other Ecologies', in Davidson et al. (eds.), *Emotional Geographies* (Aldershot: Ashgate), 103–16.

Cummins, E. (1992), 'Short Fiction by Judith Merril', *Extrapolation* 33:4, 202–14.

— (2001), 'Bibliography of Works by Judith Merril', *Extrapolation* 42:3, 255–87.

Davidson, J., Bondi, L. and Smith, M. (2005) (eds.) *Emotional Geographies* (Aldershot: Ashgate).

Davidson, J. and Milligan, C. (2004), 'Editorial: Embodying Emotion Sensing Space: Introducing Emotional Geographies', *Social and Cultural Geography* 5:4, 522–32.

Desbiens, C. (2002), 'Guest Editorial. Speaking in Tongues, Making Geographies', *Environment and Planning D: Society and Space* 20, 1–3.

Desbiens, C. and Ruddick, S. (2006), 'Guest Editorial. Speaking of Geography: Language, Power, and the Spaces of Anglo-Saxon "Hegemony"', *Environment and Planning D: Society and Space* 24, 1–8.

Farber, D. (1998), *Chicago '68* (Chicago: University of Chicago Press).

Fetherling, D. (1994), *Travels By Night: A Memoir of the Sixties* (Toronto: Lester Publishing).

Hagan J. (2001), *Northern Passage: American Vietnam War Resisters in Canada* (Cambridge: Harvard University Press).

Jones, O. (2005), 'An Ecology of Emotion, Memory, Self and Landscape', in Davidson et al. (eds.), *Emotional Geographies* (Aldershot: Ashgate), 205–18.

Kurlansky, M. (2004), *1968: The Year that Rocked the World* (New York: Ballantine Books).

Lawrence, K. (1994), *Penelope Voyages: Women and Travel in the British Literary Tradition* (Ithaca, NY: Cornell University Press).

Ley, D (1993), 'Cooperative Housing as a Moral Landscape: Re-examining "the Postmodern City"', in J. Duncan and D. Ley (eds.), *Place/Culture/Representation* (London and New York: Routledge), 128–48.

McCann, J. (2006), '"The Love Token of a Token Immigrant": Judith Merril's Expatriate Narrative, 1968–1972' (MA thesis, University of British Columbia).

Merril, J. (1958), 'Wish Upon a Star', *Magazine of Fantasy & Science Fiction* 15:6, 87–100. Reprinted in J. Merril (1973), *Survival Ship and Other Stories* (Toronto: Kababka Publishing), 17–31.

— (1968), 'Introduction', in J. Merril (ed.), *England Swings SF: Stories of Speculative Fiction* (Garden City, NY: Doubleday).

— (2002), 'Transformations', in Merril and Pohl-Weary, *Better to Have Loved: The Life of Judith Merril* (Toronto: Between the Lines), 13–19.

— (2005), *Homecalling and Other Stories: The Complete Solo Fiction of Judith Merril*, Elisabeth Carey (ed.) (Farmington, MA: NEFSA).

Merril, J. and Pohl-Weary, E. (2002), *Better to Have Loved: The Life of Judith Merril* (Toronto: Between the Lines).

Newell, D. and Tallentire, J. (2005), 'Translating Science Fiction: Judith Merril in Japan', paper presented at the Academic Track, Worldcon 2005, Glasgow, available at the Science Fiction Foundation (UK) <http://www.sf-foundation.org/publications/academictrack/index/php>, accessed 1 December 2008.

— (2007), '"For the Extended Family and the Universe": Judith Merril and Science Fiction Autobiography', *Biography: An Interdisciplinary Quarterly* 30:1, 1–21.

Owram, D. (1997), *Born at the Right Time: A History of the Baby Boom Generation* (Toronto: University of Toronto Press).

Sharpe, D. (1987), *Rochdale: The Runaway College* (Toronto: Anansi Press).

Smith, S. (2001), *Moving Lives: Twentieth-Century Women's Travel Writing* (Minneapolis: University of Minnesota Press).

Stableford, B. (1979), 'The Short Fiction of Judith Merril', in F. McGill (ed.), *Survey of Science Fiction Literature*, vol. 4 (Englewood Cliffs: Salem Press), 2014–18.

Urry, J. (2005), 'The Place of Emotions within Place', in Davidson et al. (eds.), *Emotional Geographies* (Aldershot: Ashgate), 77–83.

Chapter 15

One Stone After the Other:
Geopoetical Considerations on Stony Ground

Alexandre Gillet

It reminded him how he had needed
A place to go to in his own direction
How he had recomposed the pines,
Shifted the rocks and picked his way
among clouds.
(Stevens 2006, 38)

Where the path ends
the changes begin
and the rocks appear
ideas of the earth.
(White 2003, 431)

Picking One's Way

The mere fact of finding oneself in an open space makes us bear in mind geographical matters of real significance. Being on one's own or not doesn't change fundamentally one's own position: we still have to find an orientation and choose a direction. Seized by our surroundings, here and there, sooner or later, we realize that even before we start moving again, we are being moved. We realize that even before place and space converse together through movement of the body and the mind, we are being situated in a wider horizon. Without doubt, an emotion takes place here. Of course, knowing that 'there hardly exists any non-emotional perception; somehow, man's experience is always pervaded by a certain tonality' (Anders 1950, 557). Simply that when we are ourselves on the move into an open space, walking or resting, this tonality is stronger than ever, and may bear us further than is customary.

At the heart of the 'erratic' lines composing this chapter – I'll come back shortly to say precisely what I mean by *erratic* – lies such a movement. But I must be clear, what is at stake here is not reaching a distant horizon or advancing through unknown landscapes. The aim here is 'simply' to open out our person, just as an emotion can do, and let it move into what will soon become not only an open space but an open world – an '*a*topia' (White 1993, 9) – where the human and the non-human are both mutually considered with attention. In fact, so important is the question of such an emotion that, along the way, it is possible that we may be brought to a certain standpoint. And from there imagine that the proper outlines of *human geography* would benefit from being lightly reconfigured. Either into what could be a more inclusive, a *'more than human' geography*. Or into a more directly open and less disciplinized *geography*. Without adjective (human, physical ...).[1]

This said, and before finally setting off, before 'undisciplinating' ourselves, I'd like to quote a few lines published in *Le Monde non chrétien* in April 1952. Those lines were dedicated to Eric Dardel's most well known book, *L'Homme et la Terre. Nature de la réalité géographique*:

> In our literature of thought, Mr Dardel occupies a unique place. Knowing as much about the field of human geography as modern philosophies, he never loses his bearing when he goes from geographical spaces to mythical spaces, and shows us round through the world as he sees, senses, touches and experiences it. Thus, he offers us the possibility of experiencing not a plenitude of knowledge, which, in fact, is not possible to man, but a plenitude of method. We feel at ease and secure with such a guide, a guide who reveals to us a hidden world at the exact moment where he makes us touch the world that our eyes see. It is an aesthetic and religious[2] impression which emerges from all these investigations in the external world (Meurice 1952, 242, my translation).

The idea that the geographer may be compared to a guide is not surprising in itself. That the geographer feels at ease also in philosophy is somehow another thing. That, in fact, the geographer, in the person of Eric Dardel, might not be so much a geographer as a thinker whose knowledge of geographical matters is great is again another thing... One could say much more about these few introductory lines dedicated to Dardel's work,[3] but you will have surely remarked how 'sensitive'

---

1 Among contemporary geographers, Claude Raffestin tells us in the most decisive way of the necessity of such a point of view. See in particular his text 'Théories du réel et géographicité' (1989).

2 Here, while no straightforward etymological reading should be made, the sense of *religare* (bringing together, binding) would seem the most appropriate to our context. Although, Jean Paulhan's opinion on the subject sounds like a very decisive warning. According to him, linguists do not know for sure whether 'religion' comes from the Latin *religare* or from *relegere* (to observe with great care, to respect, to venerate) or even from *relegare* (to take someone or something as responsible for). For more details see Paulhan 1953, 57–59.

3 In many respects the figure of Eric Dardel (1899–1967) is of interest to geographers. Above all because his 1952 book L'Homme et la Terre has, since its rediscovery, acquired a significant authority on the subject of geographicity, namely the nature of geographical reality, despite the fact, or maybe for this exact reason – truly it is difficult to say – that his definition of 'geographicity' stayed on a kind of threshold. André-Frédéric Hoyaux (2000, 194) also recognized this, when he said that Dardel's definition had a liminary character. Liminary but I would also say *liminal*. For when seen by geographers as 'a model of

and 'connected' these words are and that, as a matter of fact, is the exact point I'd like to bring to the fore. Eric Dardel's geography is very much concerned with an aesthetic and dynamic comprehension of the earth and its spaces. If we follow the main lines of his work, we enter into a work-field where thinking and sensing the reality in its concreteness (the 'real') is at the very core. Doing so, with Dardel as guide, and because as a person and as a geographer I do sometimes like to touch the world as much as the world touches me, I'll approach here what is nothing less than an *archaic* emotion – archaic in the sense of being related to the beginnings, to something beginning again – and one artefact among others that may trigger it: the cairn or, more generally speaking, the stone pile or mound.

A Moving Artefact

As it is commonly understood in French and English, what we mean by cairn are the heaps or mounds of *rough* stones that are most often found or seen erected in mountainous or desert areas along tiny and evanescent paths, at crossroads, on the edge of passes or on the top of some mountain peaks. Even if such a 'picture' seems quite clear, it is nonetheless interesting to have a quick look at the diverse meanings the word has in different cultural contexts.

Cairn, as a word, comes from the Gaelic *carn* meaning literally 'heap of stones' and stemming from *car*, the word for 'stone' or, more metaphorically, 'hard'.[4] Having said this, we can broaden the view a little. Cairn in French and English indicates not only the mound of stones, be it large or small, but more specifically and distinctively the tumulus, the boundary-mark, the way-mark, the beacon or reference point *(amers)*, the memorial, the reliquary, the cenotaph, *etcetera*. If I do not want to close this 'catalogue' too quickly, it is because today we should also, for instance, consider the new Land Art artefacts disseminated almost everywhere by artists and non-artists alike, but also publicity images more and more used out of their original context, to the point that sometimes the artefact turns purely, and simply, into an artifice. As when in a shop window, a cairn, made of plastic, serves as a display unit for menswear …

Having said before that this chapter had an erratic side to it – *erratic* in the sense of closely related to erring movements, knowing that one can err without getting lost or losing oneself, knowing that to err is to advance without being

---

action' (Raffestin 1989, 29), such geographicity plainly directs us towards the margins of geography. Gaining confidence that such margins, in being followed – as one follows a frontier path – rather than trespassed, would turn themselves far more interesting and richer in contacts than we may have first thought.

4   Interestingly enough, Gilbert Durand (1969, 143) reminds us that the word had an even more archaic use and would 'initially' come from the pre-Indo-European word for stone *kar* or *kal*.

forced to follow always the existing paths[5] – I must say that in my research I am much more interested in the 'simple' stone heaps as trail or way-markers, be they very small (only a few stones may suffice to make a cairn) and at first sight apparently not very attention-grabbing. As a matter of fact, if we follow a philological path into other languages, we'll find out quickly that in many of them the word 'cairn' is understood much more literally (and specifically) as a guide over rough and unsure terrains and areas. A few examples will be enough: the German *Wegzeiger* (one who/which shows the way) or *Wegweiser* (one who/which indicates the way) – both names being given, among others things, not only to the cairn but also to the guide who indicates, thus knows, the way; the Inuktitut *inunnguaq* meaning 'look as a human being' (as an Inuk, singular of Inuit), mostly referring to the anthropomorphic cairns of the Arctic or, even closer to us, the *inuksuk* meaning, according to Norman Hallendy (2000), 'act in the capacity of a human' – as a cairn the inuksuk being in fact the most fascinating trace of the coming and going of people; and, last but not least, the Greek *herma* related to the well-known figure of Hermes (Chittenden 1947), among other things, tricky god-guide to the travellers.

Even when I am more interested in trail-markers or rather 'trail-makers' – sometimes there are not even traces of the trail, as for instance on stony ground – there is something that links every cairn. Something which in no way gives up the idiosyncrasy of a particular artefact, that is, the cairn's architectonics: the dry-stone.

Two things must be said about dry-stone. But first of all, we may take a closer look at what the word actually tells us. In dry-stone building, neither mortar nor cement are needed to hold the stones together. Dry-stone building involves only stones and know-how; stones of any size and any sort, from nice looking boulders with a clean face to simple rubble. So *any* stone goes into such an artefact, be it a large or small dwelling, a wall or a cairn. When we take a closer look at such a technique – a technique that knows no waste, no loss – some very archaic articulation of time and space shows up. At the centre of any dry-stone artefact, in particular when we are considering the cairn, such an articulation is based on the ideas of 'perenniality' and 'participation'.

---

5   Erratic, be it a substantive or an adjective, is no easy term to grasp. Take the well known example of the erratics, those stones carried by advancing glaciers and deposited in different geological locales when they retreat. When you think of it, are they really *vagrant* stones deposited here and there, until the next glaciation put them on the move again? Or are they more simply 'out of place', as our eyes seem to confirm? In other terms, should a stone be named erratic because it stands out in its new surroundings, or because in some geological time (even recent) it has gone along with some glacier's advance? Fortunately, Paul Zumthor (1993, 206) give us some clues. In old French, to err did not mean to wander aimlessly. It meant simply to go or to continue on one's own way. Following this idea, what we name the 'erring ways' (of such a person but also of such an entity) may have more to do with actual movement than with error. Returning to what is of interest to us, we may, from now on, say that stones once carried by glaciers *were* erratics, whereas cairns simply *are*.

In brief, if we state something is 'perennial', we mean that it is something that *simply* lasts. We can't be sure that it has lasted long, just as we can't be sure it will last any longer. In this sense, perenniality must not be related too quickly to some sort of stability. Instead, perenniality is best seen as a kind of tension *(tonia)* which links the artefact and the person who 'encounters' it. A tension which calls for a kind of iteration: in this case, the adding of one's stone to the existing cairn by anonymous passers-by. In front of a perennial artefact, we feel compelled to participate most of the time. One of the reasons may be that the cairn shows us how fragile it actually is and how, in some way, it needs us. We may even imagine that this fragility is what actually, in return, best secures the cairn. The more fragile a cairn looks, the more one may be moved or even bound to participate in it, and the more it may last.

Hence direct participation is another very important aspect of the cairn's architectonics. Without participation, a cairn will not carry on its erratic nature: at the same time complete and incomplete, bounded and unbounded, finite and non-finite. From now on we could even say that, basically, the cairn is erratic because it is composed of stones in movement.

In this sense also, what matters is not whether the cairn is big or not, whether it is old or not. The cairn is no simple palimpsest showing layers upon layers of stones. On the contrary, stones set in will sooner or later go back to where they came from: the very ground where the cairn stands. To be picked up again.

So, what matters is the cairn's own rhythm. Whether it is apparently harmonious or not, made up sometimes out of moments of 'activity' and at other times from moments of 'calm'. Both moments being real events because something is going on. In this sense, the erratic nature of the cairn could easily be related with the idea of *eurythm*[6] – or 'good rhythm' – as it is revealed for example when the cairn keeps 'moving' itself (changing shape and height) and 'moving' the wanderer who encounters it.[7]

In my opinion, at the basis of this aesthetics is an emotion, or more precisely a very archaic e-motion *(emovere)*.[8] That emotion takes place, it is important to notice, *between* us and the cairn. This being best articulated (thought and felt)

6   Eric Dardel approaches the wider notion of eurhythmy in his text 'L'esthétique comme mode d'existence de l'homme archaïque' (1965).

7   There exist many ways for the cairn to expose a rhythm. Here of course we concentrate on the direct participation – by means of a stone laid – of the wanderer who encounters it. In this sense a single stone laid may have more than one impact on the artefact. If it can, in a first time, make it bigger, it can also modify its equilibrium. In other words, a single stone laid may, within a short time, bring more than one stone down. Having said this, I would agree that in fact it's quite impossible to verify any strict hypothesis, for the cairn's rhythm has a genesis which stems from a complex nexus of human and non-human factors. The latter including, among others, weather conditions, and passages of animals.

8   In a sense, what Edward S. Casey says (in his discussion of the power of place) about the *'e-motive* thrust' (Casey 1996, 23) of place, in other words 'its encouragement of motion' *(Ibid.)*, can be related to what I see at work in the architectonics of the cairn.

through the stone itself, the stone we first touch, grasp with the hand, take from the ground and add to the cairn.

I see such a grasping of a stone from the ground as fundamental in two ways. On the one hand, as we have seen earlier, because it is a direct participation in the cairn and allows its perenniality to go on or to be maintained. On the other hand, because of the simple fact that we touch some very rough material, maybe the most *antipodal* one we can think of: the stone.

Every geographer knows where the antipodes stand. Without denying such geographical evidence known at least since Pythagoras, we may turn to the poet and listen to his metaphorical words. For to hear him with care, is to be transported beyond what common sense takes for granted: 'The stone, being situated in the universe on the antipodes of humanity, speaks maybe the most persuasive language' (Caillois 1975, 13, my translation). What thus underlies such an opinion? We know that to state that something or someone is antipodal to us may, among other things, mean that we are opposed to it, one way or the other. But with Roger Caillois[9] we have to think more roundly, if I may say so. We have to accept that even though the stone is antipodal to human nature, it still is part of our world. A round world in which the antipodes are not far from becoming the most important point of all. The one which makes us think together, in a round and moving fashion, people, things and places. Bearing in mind that with any move made here, the antipodes move there.

Keeping that geographical digression in mind, we can now replace ourselves in the direct surroundings of the cairn. Here, the fact of touching the stone with the hand[10] brings a new thought to the fore: the idea that something may pass between thought and matter, mind and body. The French epistemologist Gaston Bachelard (1947, 25) would label this as an example of 'energetical dualism'. By this term he means a very active dualism, quite different from the Cartesian distinction between subject and object (*res cogitans* and *res extensa*). Gaston Bachelard (*Ibid.*, 24–25, my translation) is also keen on citing Novalis, the Romantic poet-thinker, to remind us, but without sharing his magical idealism, that in the act of touching, 'the substance touches us as much as we touch it, strongly or lightly'.

Let it be said, here, that we are not far from entering, abstractly and concretely, some geopoetical ground which may lead us as far as to point out the cairn no longer as an object – which a subject may eventually or possibly master – no

---

9   It is common knowledge to say that Roger Caillois (1913–1978) wrote extensively about 'simple' stones, stones in themselves. Much less known is his text on the inuksuit called 'Les hommes de pierres de l'Arctique canadien' (1976), where he turns his attention on the figurative cairns of rude stones. Suffice to say here that he is deeply moved by such anthropomorphic figures which nonetheless 'continue to belong to the ground' (Caillois 1976, 105, my translation) and whose 'solidity is not excessive' (*Ibid.*).

10   While Michel Serres (1985, 85) tells us the verb 'to touch' is acquainted with 'topology', he reminds us also that the verb 'to think' *(penser)* originates from the verb 'to weigh' *(peser)* (Serres 1987, 216).

longer as a thing[11] *(res)* in itself, standing in front of us quite unrelated to our comings and goings, but much more as a *moving figure* capable of connecting actively human and non-human realms.

These are, on the geopoetical path, the necessary coordinates for 'some-body-mind'[12] to enter a concrete geography.

A Concrete Geography

Through the encounter with the figure of the cairn, but also with the stone we touch and lift from the ground to weigh it in the palm of our hand[13] and set in the cairn, we may, from now on, consider that an accrued and renewed perception and recognition of the very place where we stand is happening. We may even go as far as saying that aesthetic lines are being drawn, and in doing so that the open space in front of us, before the cairn, gains a new substantiality, a new presence. This movement, in the words of Eric Dardel (1952, 8, my translation), could even be read as 'a geographical participation in concrete space'.

It is relevant here to say that for Ernst Cassirer, whose work Eric Dardel read thoroughly, and in opposition to the idea of geometry, aesthetic and mythical spaces are related in the sense that they are both 'perfectly concrete modes of spatiality' (Cassirer 1995, 112, my translation). Here, what I want to hint at is that working on such aesthetics may, in some way, bridge the dichotomy between mythical and geographical spaces which we may have first assumed when reading the 1952 *L'Homme et la Terre*'s review. In fact, if we look closely at the work of Eric Dardel, we'll see that beyond the two main concepts he develops, namely 'historicity' (in the Heideggerian sense of the word) and 'geographicity' *(géographicité)* – the latter term appearing in his work only in *L'Homme et la Terre* – the main concern of the geographer-scholar appears to be much more generally anchored in the work of his father-in-law, the missionary and ethnologist Maurice Leenhardt.[14] It

---

11   The following lines where Kenneth White, when speaking of Taoism, depicts the world in terms of flux and energy, must be read attentively: 'And even the notion of thing is false. A thing, this is a fixed, a rooted notion, whereas Taoism thinks in terms of flux, of energy. In a way, nothing is separated' (White 1982, n. p., my translation).

12   Kenneth White, quoted from memory.

13   'Matter and hand must be united in order to give the nexus itself of energetical dualism […]' (Bachelard 1947, 25, my translation).

14   There are more parallels between Maurice Leenhardt (1878–1954) and Eric Dardel than simply family ties. Both stayed out of the main vistas of their respective discipline of predilection. Even though Maurice Leenhardt came to be active in the institution – he held the chair of the Sciences of religion at the Ecole Pratique des Hautes Etudes after Marcel Mauss and before Claude Lévi-Strauss – he never became an ethnologist in the classic sense of the word. His work stays to this day quite marginal. Noteworthy enough, it was not as an anthropologist nor as an ethnologist that Maurice Leenhardt stayed in Neo-

is with that idea in mind, that we can now briefly explore Eric Dardel's view of the mythic.

First of all, and Eric Dardel makes this very clear, even if myth preceded 'history', it has not disappeared altogether. Secondly, history 'itself' has no place in myth, only 'event' has: myth is perennial, it 'transmits an experience of the perenniality of life' (Dardel 1984, 231). Finally and most importantly to us, Eric Dardel goes beyond Claude Lévi-Strauss' opinion on the subject. For him myth is not the 'radical inauthenticity' (in Charbonnier 1969, 64, my translation) that may be best found in fairy tales and thus easily acquainted with some kind of mystification. For him, the myth or more precisely the 'mythic' goes deeper, 'it is a way of living in the world, of orienting oneself in the midst of things, of seeking an answer in the quest of the self' (Dardel 1984, 226). He also adds that 'the mythic is not a prelogical, as opposed to a logical structure of the mind, but rather *another reading of the world*, a first coherence put upon things and an attitude that is complementary to logical behaviour' (*Ibid.*, 228, my italics).[15]

At this stage, it may be pointed out that this understanding of myth can be associated with another of Dardel's ideas, something which takes place more generally in his design to undermine, or possibly break down, conceptual limits which are too clearly assigned. The idea that fundamentally 'aesthetics lives at the heart of man's behaviour in the world, as his way of understanding the world which surrounds him and as his way of situating, comprehending and locating himself in the world' (Dardel 1965, 352, my translation). The term 'aesthetics' is best understood, in Dardel's work, as a *coherence* put upon things and beings. A coherence felt and thought or, so to say, not taken out of the experiencing of a specific and concrete context (rather than as an autonomous branch of knowledge or as a Kantian *Transcendentale Aesthetik*).

Rather than entering a general discussion of Dardel's approach to the mythic (which can be said to run all through his work, from *L'Histoire, une science du concret* (1946) to 'De la magie à l'histoire' (1967)), we will just keep in mind that for him myth is 'woven with emotionality' *(tissé d'émotivité)* (Dardel 1946, 91). This simple fact enables us not to lose ground in the apparent schism between geographical and mythical spaces. But on the other side, it needs to be recalled that for him the mythic also embodies the first 'rupture in [man's] being' (Dardel 1984, 229), between world and humanity, the first abstraction, the first 'irrealisation'. In this 'making the real unreal',[16] Eric Dardel will suggest, lies the 'source of all poetry and all culture' (*Ibid.*).

Caledonia between 1902 and 1926. It was as a missionary deeply interested in the lives of the autochthones.

15   The original goes as follows: 'Le mythique n'est pas une structure prélogique de l'esprit, opposée à une structure logique, mais *une lecture autre du monde*, une première cohérence mise dans les choses et une attitude complémentaire du comportement logique' (Dardel 1954, 52, my italics).

16   In the French original, Eric Dardel says 'irréalise[r] le réel' (Dardel 1954, 54).

We can see that such a concrete geography also kindles, in its turn, a place for abstraction, but I would say an abstraction still related to the actual presence of the things concerned. In other words, the concreteness we see in Eric Dardel's view of the mythic – and in his related opinion on aesthetics – truly calls for *another* reading, but also understanding, of the world that is not cut off from what is studied and experienced.

With this in mind, let us finally come back on some stony ground and again meet the cairn and its moving and open figure.

An Open World

In the actual encounter with the cairn – this precise cairn made of beautiful plates of limestone, and no other – I can imagine that different logics co-habit in our present geographies, among others the very important one that was, is and might be of primordial importance to us in a near future. This logic being non-dichotomic and at the same time post-humanistic. We might call it, along with Kenneth White,[17] *geopoetics*.

What exactly is geopoetics? Most simply what the word tells us: geo-poetics, a poetical thought and practice interested in the caring for the Earth and in the making of a world. More precisely, 'poetics', says Kenneth White (1996, 125), 'is concerned with the composition of a world'. Its 'aim is to open a world less centred on the human than the one we know of' (White 2005, 89–90, my translation). I had better say it straight. When, in the Fall of 1979, during a journey along St Lawrence River and then towards Labrador,[18] Kenneth White (2004, 160) started talking about geopoetics, it had for him 'something to do with geography, certainly – maybe a new type of geography'.

What the geopoetician was first concerned with at that time – and still is – was some 'new world-sensation' (Ibid., 151). For, even though geopoetics may today best be addressed as a transdisciplinary approach to sciences, philosophy and

17   Born in Glasgow in 1936, Kenneth White lives in Brittany (France). From 1983 to 1996 he held the Chair of 20th Century Poetics at the Sorbonne. His introduction to geopoetics, *Le Plateau de l'Albatros*, was published in Paris in 1994, following in fact three very important essays: *La Figure du dehors* (1982), *Une apocalypse tranquille* (1985) and *L'Esprit nomade* (1987). Since the early 1980s Kenneth White has written his essays directly in French. Most of them are not yet available in English.

18   This journey took place just after Kenneth White, in April 1979, defended a State doctorate thesis entitled *Le Nomadisme intellectuel* (Intellectual Nomadism). Gilles Deleuze, one of the jurors, clearly admired the thesis. More surprising was his subsequent reading of the same work, as it appeared a few months later in the book co-authored with Félix Guattari *Mille Plateaux. Capitalisme et schizophrénie* (1980), more precisely in the chapter 'Traité de nomadologie'. For more details about Deleuze and Guattari's interpretation of the 'nomadic thought' and the position held by Kenneth White on the subject, see White's recent *Dialogue avec Deleuze* (2007).

poetry, it intends to stays firmly and concretely grounded in the experience of our environment.

Returning to the experience that concerns us here – having a stone in one's hand and thinking of setting it in the cairn – we well remember how important the contact and the tension are, for it to be perennial. It is time maybe to ask ourselves what a cairn 'is', and where we stand when we meet it, when, before it, we experiment with its architectonics, when we are being moved.

For me, I must say that the direct contact with the stone, even if it is rough, cold, non-human material, gives me the possibility of breaking down to almost nothing any form of atonia *(atonie)*, or 'affective indifference', without – and I play with the words, and I make them, like the geopoetician, alliterate – taking away from me the nomadic idea of *neotonia*,[19] tension and curiosity towards the far off, the anew and the other, in short my *openness to the world*.

Emotion is never far off when we think of such an openness. In this sense, we could even think of a 'world's e-motion' (White 2002, 121, my translation), the latter offering us more than 'a simple titillation of some psychophysiological order, [here you are conscious] of an encounter between the human and the non-human' *(Ibid.).*[20]

If we think in more specific terms, our relation to cairns, but also to places and spaces, will change profoundly whether we walk a well marked path in nice weather or whether, on the contrary, we walk in and experience less known areas on almost 'uninscribed ground'. Our relation with our surroundings will result in a perceptive change but may also bring new thoughts to the fore. For in such a 'disencumbered space', there may be not only an empty space where we might lose ourselves (and loosen up the self), but also the 'locus of serious work' (White 2004, 154).

A Reconciliation, a Reconfiguration

To draw a more precise picture of that *landscape-mindscape* – which is how Kenneth White (1996, 43) sometimes comes to approach such a geography – I'm tempted to share some erratic considerations. Allow me to state that being simultaneously both erratic and more accurate is not being paradoxical:

On this tiny trail

with the idea in mind that

---

19   Jacques Ménétrier uses this word to mean 'openness to the world and to curiosity' (in White 1987, 47-48, my translation).

20   The full passage originally runs like this: 'Ce qui distingue l'é-motion du monde de cette émotion primaire, c'est à la fois une densité et une dimension. Prenez l'exemple que j'ai cité: celui d'un cri d'oiseau dans le néant salé. Vous avez là plus qu'une simple titillation d'ordre psychophysiologique, vous prenez conscience d'une rencontre entre l'humain et le non-humain' (White 2002, 121).

each stone I set in a cairn

one stone set on another

one way or the other

links me more and more closely to the world

makes me lose any confidence left that 'I' am

eventually making it difficult even to distinguish

to think apart

the stone from the cairn

the cairn from the figure

the figure from 'I'

'I' from the emotion

The experience alluded to through the poem may be related, in more compact terms, as a poetic sense of dissolution:

> […] it means a sense of relativity and topology. It means, globally, a heightened sensitivity towards the environment in which we try to live. In a society which is betting everything on quantitative information, it will be necessary to stress the importance of qualitative *en*formation and, further, the notion of *ex*formation: direct contact with the outside, the acquiring of a nonpanic, poetic sense of dispersion, disaggregation, dissolution (White 2004, 151).

In order not to finish with a rather disquieting quotation, I'd like to say that for the geopoetician, such a way of perception, such a way of thought,[21] may give us indications of some kind of far-and-close-ranged reconciliation, or, saying it in William Carlos Williams' own terms, a possible way 'to reconcile the people and the stones' (in White 1979, 12).

In such an open world, where emotion is not futile, where emotion transforms objects and things into moving figures, a figure like the cairn may teach us one thing more. Because its language is partly figurative, partly not, because its architectonics contains at the same time a logic of construction and a logic of dislocation, we are

---

21 '[…] a way of perception and a way of thought, a way of being oneself and deep in a sensation of the world' (White 2004, 61).

to weigh attentively any move we make. In other words, we must think and ask ourselves whether such a figure absolutely requires *our* participation.

In such an open world, emotion not only moves us, it transforms us too. Engaged into another reading of the world, we may soon perceive that if there is participation in progress in the cairn, it is a more inclusive *(englobante)* participation than we may have first thought. If we look closely, I mean on the terrain, a stony terrain where many cairns may live their way, at what wanderers actually do in presence of a cairn, we will soon figure out that most of them do not participate in it. Why so? Maybe because the cairn itself participates in ourselves: giving us indication of the way to take, making us pass and thus giving sense to what the world actually is.

Epilogue

Some time ago, in a second-hand bookshop in Edinburgh, I found the most unexpected book: its title, *Mount Analogue. An authentic narrative.* As an epilogue to this conversation about and around the cairn, as another mark of its erratic nature, I'd like to quote an extract from it, knowing that this novel, written between 1939 and 1943 (René Daumal[22] was 35 when he had to stop working on it) – mostly when he was staying either in the Pyrenees or in the Alps – remained unfinished, uncompleted, non-finite.

The poet, 'without being a sociologist or a psychologist', as his translator Roger Shattuck put it, still in early twenties, had already challenged and 'attacked Lévy-Bruhl's widely accepted thesis that the primitive mind is different in its operation from the civilized mind – a theory Lévy-Bruhl himself finally had to recant (not least because of the influence of Eric Dardel's own mentor, Maurice Leenhardt)'[23] (Shattuck 1959, 13).

With the cairn, *one stone after the other, one thought after the other*, it is not on primitive but on archaic ground that we are setting foot:

> When you strike off on your own, leave some trace of your passage which will guide you coming back: one stone set on another, some grass weighted by a stick. But if you come to an impasse or a dangerous spot, remember that the trail

---

22   René Daumal (1908–1944), apart from being a poet and the author of two 'novels', was the co-founder of the *Grand Jeu* group at the end of the 1920s. The *Grand Jeu* took explicitly, through the eponymous review, a singular position towards the Surrealist movement. Sharing some of its goals but at the same time keeping at a safe distance from it, also in order to challenge it. From the start, René Daumal wrote many essays whose innate radicality can still serve us as reference-points into any transdisciplinary movement.

23   This reorientation takes place mostly in his late *Carnets*, dated 1938–1939. Before that, Lucien Lévy-Bruhl (1857–1939) stayed clearly positivist, assigning to the 'primitive' mind mostly pre-logical traits.

you have left could lead people coming after you into trouble. So go back along your trail and obliterate any traces you have left. This applies to anyone who wishes to leave some mark of his passage in the world. Even without wanting to, you always leave a few traces. Be ready to answer to your fellow men for the trail you leave behind you (Daumal 1959, 106).

Considering what we have discussed earlier, those words are not only a poet's words. Those words are poetical outlines of a work and of a world.[24] I take them as an indication not to define the cairn too eagerly, too quickly. Maybe there is no more dangerous move for it, and for us, than the one which would come to enclose (and mutilate) it into one definition, one single meaning. Erratic is its nature, erratic must possibly become our way of thinking and approaching it.

Acknowledgements

I am grateful to Joyce Davidson, Liz Bondi and Mick Smith for their help and their on-going comments on earlier versions of this chapter.

References

Anders, G. (1950), 'Emotion and Reality', *Philosophy and Phenomenological Research* 10:4, 553–562.
Bachelard, G. (1947), *La Terre et les Rêveries de la volonté* (Paris: José Corti).
Caillois, R. (1975), *Pierres réfléchies* (Paris: Gallimard).
— (1976), 'Les hommes de pierres de l'Arctique canadien', *Diogène* 94, 92–108. Published simultaneously in English under the title 'The Stone Men of the Canadian Arctic', trans. R. Rowland, *Diogenes* 24:6, 78–93.
Casey, E. S. (1996), 'How to get from Space to Place in a Fairly Short Stretch of Time: Phenomenological Prologema', in S. Feld and K.H. Basso (eds.), *Senses of Place* (Santa Fee: School of American Research Press).
Cassirer, E. (1995), 'Espace mythique, espace esthétique, espace théorique', in J. Carro and J. Gaubert (trans.) *Ecrits sur l'art* (Paris: Le Cerf), 101–122.
Charbonnier, G. (1969), *Entretiens avec Lévi-Strauss* (Paris: René Julliard et Librairie Plon).
Chittenden, J. (1947), 'The Master of Animals', *Hesperia* 16:2, 89–114.
Dardel, E. (1946), *L'Histoire, science du concret* (Paris: P.U.F.).
— (1952), *L'Homme et la Terre. Nature de la réalité géographique* (Paris: P.U.F).

---

24   'The poet is the man who wants to make a world. I can't accept a poet on any lesser terms' (White 1963, 12).

— (1954), 'Signification du mythique. Le mythique d'après l'œuvre ethnologique de Maurice Leenhardt', *Diogène* 7, 50–71.

— (1965), 'L'esthétique comme mode d'existence de l'homme archaïque', *Revue d'histoire et de philosophie religieuses* 45:3–4, 352–357.

— (1967), 'De la magie à l'histoire', *Revue d'histoire et de philosophie religieuse* 47:1, 58–68.

— (1984), 'The Mythic: According to the Ethnological Work of Maurice Leenhardt', in A. Dundes (ed.) *Sacred Narrative. Readings in the Theory of Myth* (Berkeley, Los Angeles, London: University of California Press).

Daumal, R. (1959), *Mount Analogue. An Authentic Narrative*, trans. R. Shattuck (London: Vincent Stuart).

Deleuze, G. and Guattari, F. (1980), *Mille Plateaux. Capitalisme et schizophrénie* (Paris: Editions de Minuit).

Durand, G. (1969), *Les Structures anthropologiques de l'imaginaire* (Paris: Bordas).

Hallendy, N. (2001), *Inuksuit. Silent Messengers of the Artic* (Vancouver: Douglas & McIntyre).

Hoyaux, A.-F. (2000), 'Habiter la ville et la montagne: essai de géographie phénoménologique sur les relations des habitants au lieu, à l'espace et au territoire' [Doctoral thesis] (Grenoble: Université Joseph Fourier).

Meurice, E. (1952), 'Eric Dardel: L'Homme et la Terre. Nature de la réalité géographique (revue de livres)', *Le Monde non chrétien* 22, 242–244.

Paulhan, J. (1953), *La Preuve par l'étymologie* (Paris: Editions de Minuit).

Raffestin, C. (1989), 'Théories du réel et géographicité', *EspacesTemps* 40–41, 26–31.

Serres, M. (1985), *Les cinq Sens. Philosophie des corps mêlés – I* (Paris: Grasset).

— (1987), *Statues* (Paris: Bourin).

Shattuck, R. (1959), 'Introduction', in R. Daumal, *Mount Analogue. An Authentic Narrative*, trans. R. Shattuck (London: Vincent Stuart), 1–18.

Stevens, W. (2006), *A l'instant de quitter la pièce. Le Rocher et derniers poèmes. Adagia*, trans. C. Malroux (Paris: José Corti).

White, K. (1963), 'Poetry and the Poet (Extract from a Lecture)', *Soap Box* 6, 11–13.

— (1979), 'Le Nomadisme intellectuel. Tome 2. Poetry and the Tribe' [State doctoral thesis] (Paris: Université de Paris VII).

— (1982), 'La poésie, évidemment', *Fanal* 16, n. p.

— (1987), *L'Esprit nomade* (Paris: Grasset).

— (1993), 'Poetry, Anarchy, Geography', *Open World* 3, 8–10.

— (1994), *Le Plateau de l'albatros. Introduction à la géopoétique* (Paris: Grasset).

— (1996), *Coast to Coast* (Glasgow: Open World, Mythic Horse Press).

— (2002), *Le Champ du grand travail. Entretiens avec Claude Fintz* (Bruxelles: Didier Devillez Editeur).

— (2003), *Open World – Collected Poems 1960–2000* (Edinburgh: Polygon).

— (2004), *The Wanderer and his Charts* (Edinburgh: Polygon).

— (2005), *L'Ermitage des brumes: Occident, Orient et au-delà suivi de L'Anorak du goéland* (Paris: Dervy).

— (2007), *Dialogue avec Deleuze* (Paris: Isolato).

Zumthor, P. (1993), *La Mesure du monde* (Paris: Seuil).

Chapter 16

The Steppe

Alphonso Lingis

Let us stop admiring our heads for a moment, Georges Bataille said, and look our feet. Our big toe is the distinctively human part, the only new body part in the human ape; it stabilizes our upright posture.

Just look at our feet, Bruce Chatwin said; they are long and set parallel; they are made to move on ahead.

At the southern end of Patagonia, constantly battered by the icy winds of the Antarctic Ocean, at a place the Spanish called Punto Gualichó – Devil's Point – we visit a cave on whose walls are painted dashes and serpentine lines, circles and rows of dots. There were also some stencilled outlines of human hands. Some archaeologists have dated them at 9,000 years ago. They can only guess at what the designs might signify. It does seem likely that these broken and serpentine lines and these dots recall the vast distances the painters had walked to get here.

Twelve thousand or more years ago, during the last Ice Age, small bands of humans crossed what is now the Bering Strait into the North American continent – or, as now believed, ventured across and then down the coast in small boats. They pushed on, seeing hundreds of miles of glaciers finally give way to the terminal moraines and the green forests and prairies inland. They ventured down the Central American peninsula and lingered at the mouths of rivers that descended the coastal deserts from the Andes; they entered the vast jungles of the Amazon. They pushed on, across so many hundreds of miles of desert and jungle and desert again until here, before the stormy waters of the Antarctic Ocean, they could go no further.

Our feet are made to move on ahead. Onward into what? Into – space? If we speak with abstract terms like 'space', we are also going to use abstract terms like 'spirit' and 'freedom'. We will picture ourselves as existing in empty space, and free to move on ahead or in any direction or not move at all.

In fact, the space about and ahead of us is nowhere empty; it is filled with light, radiant, vibrant, tinted, or sombre, or it is filled with the darkness of the night. The light opens before us, marking out the rooms, the corridors; the radiance of the day summons us outdoors, marking the paths and the fields, outlining the contours of things. We do not look at the light as we look at things; it orders our movements: our eyes follow the light. If we wilfully turn our backs to its springtime summons and spend the day peering at the pale glow of our computer screen, we do so with a sense of non-compliance and defiance.

The outlying surfaces and contours of the earth summon forth our steps; we walk by following its directions and directives. The sliding waves of the sea direct us to divest ourselves of our primate upright posture and slip and slide and flip like dolphins.

Immanuel Kant took space to be something known, all of its properties and shapes codified by the first complete science, geometry. He said that we have from the first an intuition of this space, before our organism registers colours and shapes, and sounds and pressures. This intuition is a kind of sight. Vision, we currently say, is the distance sense. Kant argued that we have an *a priori* intuition of pure space, empty space, which precedes and makes possible our perception of extended things in space.

But when we awaken in our sleeping bag laid in a meadow we do not look at the luminous medium, the warmth about us, the stable ground beneath us. We do not gauge their horizontal and vertical axes or survey their extent. Instead, we feel the vibrant light and the warmth in which we find ourselves and feel the stability of the ground in our own body. Feeling still opaque and torpid, we close our eyes, reassured by the reality of the warmth and the underlying stability, feeling contentment, contented with the warm and stable medium in which we find ourselves. Or we feel the warmth and the radiance of light energizing us, the stability of the ground anchoring and supporting our steps: we stir, we stretch, we get up, we move.

Our movement is vibrant with a feeling of the ambient warmth or cold, a feeling of the radiant, thin or misty, or obscure and feeble light or of the fathomless darkness, a feeling of the damp or the aridity, a feeling of the stability of the ground, or of the sliding buoyancy of the water. We move about because we exist in a luminous and warm medium, supported by the solidity of the ground, like fish move because they exist in water and birds move because they exist in the open air and the winds.

As fish swimming about in the lake come upon a tadpole and swallows circling the air come upon a swarm of gnats, we come upon, in the water or in the light, those fish and those swallows and our eyes settle for a while on them, following them with pleasure. Things are not just scattered about in empty space; we find them on the ground, which supports them in their places and also supports our positions and our movements toward them. Or we find things floating in the air, or drifting in the light. While we spend eight hours a day wielding tools to frame out a house or asphalt the road, we also feel that we are just moving back and forth in the spring freshness or the summer heat.

There are things that are not objectives, or stepping-stones or implements to reach objectives, things that just summon us to go on, to go on. On my last day there in the highlands of New Guinea, a Papuan gave me his precious possession, an oval black stone strangely marked with thick white lines, and thin white lines intersecting in the zones the thick lines squared off. He gave me to understand that he carried it with him whenever he left his compound to go into the jungle, and was giving it now to me to guide me on my onward wanderings. He had seen that every day I wandered about without any discernible objective or goal, and understood that I had wandered over the planet to New Guinea without any task or project. Now whenever I happen to look at this stone, it makes me long to go on wandering.

Philosophers, generalizing and flattening the words of current language, use the term 'experience' to just designate any state of consciousness, but in the current language an 'experience' designates some encounter or event that arouses vehement emotions. Passions grasp at once the astonishing, enthralling, or terrifying cast of an event or situation that is suddenly before us and disconnect it from the outlying field of continuities and regularities that occupy everyday consciousness. They fix attention and set in motion a restlessness of the eyes that pass back and forth from the focal object to the background and jump from one feature to the next. Vehement passions also disconnect us from the others about us; our behaviours and utterances are no longer picked up from others, passed on to others. My rage or my terror or my grief are wholly mine and I am wholly in them; they produce that bounded core of intensity, vibrantly aware of itself, that is the ego.

Impassioned experiences, which erupt, endure, and come to an end, contain a vivid sense of time – of a trajectory of personal time cut off from the public or common time as from the time of nature. The time of wrath or shame is that of the immediate past still flanking the present. This immediate past is my own past, while the events of the social world and of nature continually recede into the common and anonymous past. This immediate past is disconnected from the expanse of the past phases of our life with which we are intermittently in touch through recall, with which we lose contact through forgetfulness. It is right there, an obsession pressing upon the present.

The imminent future we face in fear or hope is cut off from the field of possibilities and eventualities with which we are in touch through intermittent acts of calculation, planning, and reasoned decision. The imminent future is a different zone or medium of time, not simply the continuation of the world's presence, the universal and anonymous future of the continual oncoming of things and events.

Impassioned experiences break off from the continuity of time and are not pushed on by its resumption. At the end of a trip we will recount them right away; they are not exactly remembered: they are still hovering like storms over the present.

In the eighteenth century the very word 'passion' disappeared from scientific, psychological, and literary discourse. It was replaced by the words 'emotion' and 'sentiment'. The new discourse of emotions and sentiments conceive these as affective states that are properties of the ego; the ego is held to be able to feel them, observe them, and control them. Today passions are admitted only in popular novels the citizens read and television melodramas they watch in the immobility of their apartments. It is only vicariously that they know passions. Novels and television melodramas have made us think that passions are aroused by persons or by events or things. But in reality when we look back over the years, or when we look back over an extended trip, we see that our most intense passions were aroused by the epiphany of the light, or the most intense darkness, by the air, by the ground or by the sea.

I got very sick in Bali, consulted several Balinese and then Western doctors, and finally was diagnosed with hepatitis. There is really no medication for hepatitis: I would have to wait it out, weeks, maybe months. I moved to a small hut by the sea. The Balinese are not at all sea-going people; the Wallace Trench that separates Bali from Lombok is the deepest channel in the ocean, seven miles deep, and its current is very powerful, patrolled by great sharks. The Balinese live upland and terrace the mountain slopes, irrigating them from the crater lakes of volcanoes. But at the end of the day, they descend to the ocean shore and seat themselves on the dunes, hundreds of them, where they sit silent, silenced by the descending tropical sunset. There are always clouds about this very mountainous island, and the setting sun blazes the skies and the banks and layers of clouds with colours found nowhere on terrestrial things in a spectacular epiphany, each evening different and each minute changing. I too sat among them, rapt by this utterly gratuitous ethereal splendour that evaporated my misery and my sense of myself and that was profoundly healing – and transforming. Today, thirty years later, I can still remember many of the people and events I encountered those months in Bali, but their memory no longer retains the emotional charge. Yet how vivid still is the cosmic glory of those tropical twilights!

Trekking in Patagonia, across landscapes of grass so wiry that sheep have ground their teeth to the gums on them and have to be slaughtered at two years old, landscapes without a single tree because no tree could stand in the unceasing 60 km/h winds that race across this narrow dagger of land plunging into the Antarctic Ocean – but it was that wind that penetrated the trekker and filled him with its fury and its exhilaration. You could not think of Patagonia without passionately longing to go back, back to that wind.

From Patagonia you take a ship to Antarctica, a big ship because a smaller one could not cross these stormiest of seas under winds that here race around the planet uninterrupted by any land barrier. Unfortunately a big ship has a lot of service staff, daily lectures on oceanography, on famous Antarctic explorers, on whales and penguins, and in the evening long meals at big tables where hostesses circulate to keep everybody frenetic with talk. But after asking to have your meals taken in your cabin you get such a reputation as a misanthrope that soon nobody talks to you, even greets you any more. Then you gaze in silence at the glaciers imperceptibly flowing into the ocean, ice millions of years old, compacted under enormous weight so that the crystalline structure of the ice is changed, you gaze at fields of light frozen in prismatic colours of violet, blue, indigo that glow in your heart like immobilized and inhuman ecstasy.

Twice I crossed the Atlantic by ship, and for ten days saw only the ocean surface in constant undulation and knew all existence in undulation. Pleasure, voluptuousness, selflessness, communion are all blended together in giving oneself over to rhythm, which we know in dance, and in music. The oceanic experience is dance and music stripped of everything narrative or anecdotal and become fundamental and immense.

Practical initiatives bend our attention to paths, connections, and obstacles. Passions flood across paths, ignore the linkages of instrumental causality, are not stopped by obstacles and do not settle down on accessible objectives. That compulsion to obey the summons of the morning or the desert light, the tropical warmth or the Arctic cold, the surface of earth extending support ahead indefinitely or the surface of ocean dancing off into the horizons – is driven by a fundamental passion. A primal and imperious passion in the small bands of people who during the last Ice Age trekked across the Aleutians and did not stop until, on the fjords of Tierra del Feugo, they could go no further.

To go from the eastern United States to, say, Mongolia is to go as far as you can on our planet, to the opposite flank. The flight went from Baltimore to Chicago, then to Seoul where you had to overnight in a hotel, then onward to Ulaan Bataar. The thrill of the first men who mounted on horses, the thrill of moving across vast distances, is in passenger jetliners emptied out by the reduction of the sense of distance. Passenger jetliners are high-tech engineering contrivances to make sense perception illusory: motion of 980 km/h is felt as fixity in the stationary seat in the felt immobility of the airplane. If you can look outside during daylight, the occasional mists of clouds below enable you to gauge neither the 980 km/h movement of the plane nor the 1669 km/h spinning of the earth beneath. Already superrich tourists are travelling to outer space, to know 190 hours of movement at 39,000 km/h reduced to an experience of immobility. In a two week trip by camel into the featureless Sahara extending under blank skies you know rhythms – the rhythmic pace of the camel over the rhythmic undulations of the dunes, rhythms in your nervous circuitry and brain that induce the serenity of yogis and mystics. There are no rhythms in the cabin of the transcontinental jetliner and in your body the seat position in which you are held builds up muscle tensions that cannot be released. You do the opposite of meditation: your paste your mind with trivial and inconsequential distractions. You watch the in-flight movie or you read, not a thought-engaging book but a light novel. The 12,250 kilometres to Mongolia are eighteen hours of restless impotent urges to do something and the heavy lethargy that smothers those urges.

Finally you arrive in Ulaan Baatar.

Although the Chinese have annexed the southern half, and the Russians a big chunk in the north, today's independent Mongolia is still huge, three times the size of France. There are only two million Mongols here, half of whom now live in Ulaan Baatar; the rest are nomads. When they rent out a four-wheel vehicle it comes with the driver, though he speaks no English and you of course no Mongolian. But without a Mongol-speaking companion it would be impossible to use the vehicle: once outside the capital there are no paved or even graded roads and of course no road signs.

You asked the only Mongol you knew, the desk clerk at the hotel (what slight knowledge!) and he told you where to find somebody who had a four wheel drive – it turned out to be a Russian make, 29 years old – and the guy who had the car said the driver knew Mongolia from one end to the other and had friends everywhere. Of course he would say that. And of course if he knew the country and had friends everywhere, he could leave you anywhere with his cronies to cut your throat just to get your camera.

Trust is taking what is not known as though it were known. What is currently but confusedly called 'trust' in the world of governmental, media, and commercial institutions is based on, and increased by, knowledge. Like the operation of chemical combinations or physical collisions, knowledge of a sampling of their operations in the past is taken as knowledge of their operations in the future. A government agency or a news program or a manufacturer builds up or rebuilds 'trust' by broadcasting its performance in the past. This produces in us something like *confidence* based on knowledge of how the system or institution works. But there is strictly speaking *trust* only in the measure that we are aware that the future is not known. Before submitting to surgery, we find out all we can about the knowledge, competence, and experience of the surgeon, but in the end we realize that he is capable of misjudgement in diagnosis, neglect of available research, lapse of skill, inattentiveness, or malice. Trust makes contact with a singular individual, with the real individual, a power of initiative unto himself but also innately endowed with skills to hide what he does not know, to deceive, dominate, and trap others.

Two skydivers leap from an airplane flying 210 km/h at 12000 feet; the one without a parachute falling 150 km/h who must hold out his arms to slightly slow his freefall, his buddy jumping right after has his parachute and must dive toward him to hand it to him. Each has full knowledge of what he is doing and has maximum training, but it takes so very little for them to not connect and for the one to plummet to his death – some small nerve firing or blockage that skewers the convergence of the two men free-falling, some emotional impulse of jealousy or spite unrecognized in the one with the parachute but that is always there in the closest of friendships and lovers.

Trust is the strong surge of feeling that connects us with our fellows; it is the exhilarating leap that connects with another despite ignorance. Trust is impulsive and immediate. You do not decide to trust someone: you just trust him, or you do not.

There he is, now, you at once extended your hand to his, and at once entrusted yourself to this stranger who does not know a word of English. But he understands your hand and your smile and the buoyant steps with which you follow him to the vehicle. The leap of trust is an exhilaration. There is nothing more exhilarating that trusting a stranger with whom you have no religion, ethnic, or moral community or language in common with him.

Over the course of the summer you did a circuit of some 2,000 kilometres in the centre of the country and the Gobi desert. You drove across utterly flat steppe, where never was the least bush to be seen, and usually along the horizon on one side or another mountain ranges utterly devoid of trees or bushes, and, since it was summer, devoid of snow. Mountains that materialize time and patience rather than shape place and landscape. The winds are always on the move, across these steppes of central Asia where they find no barrier for thousands of miles. Overhead the legendary blue skies, most often without a trace of a cloud. When you get up in the morning and look out upon the luminous space without landmarks or obstacles, without promises or threats, you feel such an invitation to move on. You stop for lunch arbitrarily, one spot equivalent to any other. The driver follows seasonal horse or camel trails. You cross from one end of Mongolia to the other without seeing a single fence. What sights there are are also on the move, though also not going anywhere in particular: herds of horses, yaks, sheep and goats, and in the Gobi desert, two-humped Bactrian camels, that know no fences or lassos or brandings.

How unlike the vital, basically practical space philosopher Martin Heidegger described: here there are no things taking form as the hand reaches out to grasp and manipulate; there are no paths blazed toward objectives, there are not distances made determinate by practical initiatives. The fathomless depths of ground and warm air and luminous sky are just there when you open your eyes in the morning; it is a pure contemplation that perceives them. There are not purposive movements that establish a here-there axis; the distances yawn open and emanate an irresistible invitation to your body to move on, and on.

How you understand the nomadism of people here, moving with the herds as they graze across the steppe. How you understand the abrupt impulses of men in the springtime to mount their horses and go on and on to distant lands, eventually plundering sedentary settlements in China, Persia, the banks of the Danube, moving on. In the evidence of their immense, fundamental patience, you understand the Mongols you see are the most passionate of men. These furies, avidities, and exultant destructions of sedentary cupidities, disconnected from the immemorial passage of the seasons and the years, press down on them like an immediate and obsessive past.

And hovering still too over you. Moving on each day, you are joined by the herdsmen who as the last Ice Age was melting down drove their herds northward and pushed the peoples up north yet more northward and over the Bering Strait to filter down into the American hemisphere, by those warrior-nomads who thundered westward to push the Teutonic barbarians who overran and trashed the Roman empire, by the tumultuous horsemen who joined the hordes of Chinggis Khan racing over to Korea, across China, down to Java, over to Hungary, putting all these economies and cultures into contact and circulating their insights and inventions from one end to the other of the most extensive continuous empire the planet has ever seen, closing off the Silk Road and forcing the silk-coveting and spice-hungry Europeans to their boats to circumnavigate the Cape of Good Hope to overrun India and Indonesia and cross the Atlantic to plunder the Aztecs, Maya, and Inca. With these warrior-nomads, in 1347 the Black Death entered Europe, transmitted to humans by bites from fleas carried by marmots, squirrels, and rats, and decimated a third of its population. (This summer too twenty-four people had contracted the bubonic plague in the Altaï province while you were there.) It is here that everything originates; this so empty steppe in Central Asia is the vortex from which the most momentous events of human history were driven. In this now most unpopulated country on the planet, human remains have been found dating back 500,000 years.

As the days go by more fragments of long-forgotten pages of history books read in your schooling on the other side of the planet revivify; now everyday these ghosts of Mongols past accompany you. This steppe has become for you a world-historical site, the world-historical site from which there is more to understand than from any other. You cannot wander these steppes without hearing the thundering hoofs of Mongol nomad armies. Yet the great Khans and their warrior-nomads left no traces here of the world-historical events they launched. In Europe you have trekked Roman and in America Inca roads; here there are only the tracks of yak and camel herds that the winds will efface a day or so after they and you pass. Here, under the fathomless blue of the sky revered by the Mongols as their only deity, there are no sacred enclaves, no ruins of ancient citadels, no aqueducts, no bridges, no ruined fortresses, no monuments marking great victories or imperial tombs; the burial place of Chinggis Khan himself has never been found, despite American and Japanese pilots searching with infra-red scans and teams of geologists combing every square foot of the site recorded in the ancient texts. Day after day you move on, and on, where the trackless distances ahead are also distances in time. For we never just exist in the present. When you awaken in the morning you find yourself in yesterday's landscape, when you move you move in the landscape of hundreds and thousands of years ago.

The four-wheel lurches down tracks and sloshes through rivers. When it gets stuck, eventually nomads arrive on the two-humped Bactrian camels to pull it out of a river. At day's end, you are welcomed by nomads into a ger. ('Ger' is the word you use: 'yurt' is a Persian and Russian word and not used by the Mongols.) A ger is some 15 feet in diameter; the circular wall is made of a wooden lattice frame that can be collapsed together for transport, the roof made of 90 to 110 spokes like a Chinese umbrella; the whole covered of felt made of the white wool of sheep pounded into sheets. It can be taken down and put up in a half hour. The gers are as unmarked as the steppe, the same size, shape, and natural-wool colour from one end of the country to the other. But inside you find a little palace, the ground is covered with carpets, the walls covered with intricately embroidered hangings in bright reds, blues, purples, and yellows, designs traditional in this family and designs marking its history, and designs and colours materializing the ingenuity and taste of the weaver. Around the walls there are intricately carved chests for the family possessions. There is a Buddhist shrine veiled with the smoke of incense sticks. You enter into the traditions and memories of clan and family.

There is a low table immediately covered with salted milk tea, breads, cheeses, curds, and pastries for the guest. Soon the ger is crowded with relatives and friends arriving, each addressing you with a warm smile and ceremonious words of greeting. Chunks of dried dung are lit in the iron stove in the centre with a chimney pipe that extends up and out, pots of rice are cooked, lamb or goat is stewed. You can tell nothing of your wanderings and encounters, but they, regularly refreshed with glasses of fermented mare's milk, use the occasion to recount vividly to one another events of their day, tales told by passers-by they encountered, recalling tales from long ago. You do not have to know the language to recognize that, seeing a narrator hold the attention as his voice enchains cadences, rises to a thunderous or slows to a whispered climax, seeing the surprise, delight, or satisfaction of a shared memory. Someone begins playing a stringed instrument rather like a mandolin. They will still be drinking fermented mare's milk and telling tales when sleep overtakes you.

By day you come to see the practical space of the nomads; the herds typically of 200 horses and yaks – the horses like the yaks kept for the milk, mare's milk fermented, yak milk made into curds and cheese – flocks of sheep and goats, Bactrian camels. They are milked and then they go off to graze the steppe; in the evening the sheep and goats are brought to the ger camp where prowling wolves can be driven off. The horses selected for riding have to be fitted with horseshoes, which have to be refitted every few weeks in this rocky terrain. Sheep, goats, and camels are shorn; the wool is carded, woven, pounded into felt; blankets, garments are woven. The goats yield cashmere wool, which is the principal export of Mongolia. Men and boys train in the manly arts of archery, wrestling, and horse racing; there will competitions on the frequent festive occasions. The ger will be dismantled and the camp over as the herds move on to new areas, often over two or three hundred kilometres.

This practical layout is set in the social arena you discovered the evening before. A nomad is never alone, even when riding horseback or herding the flocks; the ger is an always temporary shelter, to be regularly dismantled, but it harbours clan and ancestral and family traditions, arts, and history; the nomad camp is the arena of circulation of people, where every event becomes a story that surprises and entertains, where places and events from long ago and far away are held in narratives retold and shared. The ever-shifting camp is a community of humans and also Bactrian camels, horses, sheep, and goats. In exchange for their milk the Mongols devote their labours to them, protecting them from lynx and wolves, rescuing those who have strayed or fallen into gullies, washing and binding their wounds, guiding them to summer pastures in the glacier-melting watered mountains and to winter pastures in the valleys.

In the far West, in the region of the highest Altaï Mountains, there is a community of Kazakhs. They too are nomads, they too live in felt gers, though of different proportions, they too are herdsmen, following their herds higher into the mountains as the snows melt. Their territory is extended by eagles. The eagles are taken as babies from the cliff nests and trained as falcons. It is female golden eagles that are kept, for they are the stronger hunters. They have fierce talons; the hunter perches her on his hand wearing a thick leather glove and sleeve. The eagle weighs some 20 pounds, after a few minutes you find you are propping your hand with your other hand to hold her. The eagle hunters are a knighthood; they wear black coats with silver belts and hats made of the tails of fifty mink. When winter comes and the animals are in new thick fur, the hunters mount their horses and climb the mountains and release the eagles who soar over the mountains and return with foxes, lynx, even wolves. Eagles live 30 or 40 years; after ten years the hunters release their eagles into nature so that they can find mates and reproduce.

Looking across the steppe, you see that this, the highest country (with the coldest capital) was visibly hoisted up by geological cataclysms: the great flat steppes visibly once gigantic lake beds drained as the vast south Asian continental plate ground its way under the Siberian continental plate wrinkling up all those mountains and bursting all these volcanoes. From time to time you come upon deep open gorges where fault lines came apart. You see the sculptures that the winds and glacial melts have carved in these mountains. (How blind that conceit of David Hume's epistemology, arguing that we don't see causality!) You see geological time and causality in what from a great distance looked like a thick line of white fog shimmering under a mountain ridge. But it materialized into a sand dune, which has been measured 800 meters high, 12 kilometres wide and a hundred kilometres long – all in one long singing wall under the black flanks of the mountain ridge. You see the sands drifting on to it, and feel it give way as you pull yourself two thirds up to the top but then slide all the way down. In the Gobi desert you come upon pieces of petrified wood everywhere that exhibit still the ancient tropical jungles of long ago. In small museums in the four corners of the country you see some of the spectacular dinosaur skeletons found on the sites and some dinosaur eggs; a nest of dinosaur eggs were first found in the Gobi desert, demonstrating that they, already like birds, brooded them. The Mongols did not date and explain these features of the landscape as our science teaches us to do, but they observed them with the terms designating causes and events of their account of primeval times. You see the geological future too in the present, in the rivers your vehicle had to drive into because the river would leave any bridge built aside when the glacial snows melt in the coming spring, and in the unending winds shifting the contours of mountain and steppe.

Grasslands, the books called the steppes of Central Asia, but wherever you stop, for lunch or only for a pee, you look down and see a nap of short plants all now flowering in riotous colours; there are not ten percent grasses among them. Botanists have identified 600 plant species in the Mongolian steppes, and are still counting. An incredible biological diversity in what first looked like vast monotony unpunctuated with trees or bushes. Why so many species? Wherever you stop, you find different groupings. There are in fact innumerable biological niches, formed by different soil compositions, different water sources. When you are on the move, you see the limitless distances of the steppe; when you stop your fascinated eyes are held within small geological niches, small ecosystems whose flowers glitter like scattered crumbs of jewels, where you are enveloped by a nature as abundant and diverse as that of the tropical rain forests. You take out a magnifying glass and between the smallest plants pass into a wonderland of minute lichens and microplants. *Wolffia globosa* and *augusta*, but .6 mm long, produce exquisite flowers and are very prolific; one plant could produce 1 nonillion (1 and 30 zeros) in four months, a volume of flowering plants equal to the size of Earth. The minute ecosystems are crawling with exquisite translucent mites with semispherical eyes honeycombed with lenses like minuscule diamonds directing their sparks of passion, of fear and avidity and astonishment, whole tundras you were about to crush with your foot. The lizard *Sphaerodactylus ariasae* is but 16 mm long. Back in your house you keep a fish tank, and how much pleasure it brought into your house, marvelling over the courtship dances of the guppies and the sensitive filaments of tiny coral animals. How you have come to care for that fish tank and care for its inhabitants. One day you had taken that magnifying glass in your home and looked at the cushion of the sofa and seen scurrying between its fibbers a biological universe as exquisite. The radius across which you move each day, whose vistas and textures and perfumes keep you alive with sensuous pleasure, is in fact one stratum of fractal layouts and ecosystems just under and also just beyond what the naked eye can see. In them there are no objectives you can want to reach and acquire, no things to detach and refashion with tools and stamp with the spirit, make yours and annex to your identity, will, and status. For you are in Mongolia, in the winds and under the featureless blue sky, 12,000 kilometres from self-respect and self-importance, summoned to go on and on and on.

Index